Writing Windows VxDs and Device Drivers

Karen Hazzah

CMP Books

CMP Books
an imprint of CMP Media, Inc.
600 Harrison Street
San Francisco, CA 94107
www.cmpbooks.com • books@cmp.com

Distributed in the U.S. and Canada by:
Publishers Group West
1700 Fourth St.
Berkeley, CA 94710
ISBN: 0-87930-438-3

LS #4 11-00

R&D Developer Series

Preface

This book is primarily for developers who need to write a non-standard device driver, either as a VxD or as a DLL. (A non-standard device is anything except a display, keyboard, mouse, serial port, or printer.) This second edition expands the coverage of VxDs, with particular attention to the issues raised by new Windows 95 features, like Plug and Play.

While not intended for a beginning programmer, it is my intent that this book will be accessible and useful to a wide range of readers. If you have written a device driver or device interface code for DOS or some other operating system, you should be comfortable with the material in this book. To get the most from this book, you should have a strong working knowledge of C. You should also be able to read 80x86 assembly, although this edition uses far less assembly than the first edition. A strong grasp of how segments are used by DOS compilers and assemblers will be helpful. You do not need to be a Windows application programmer. In fact, you'll find the code in this book bears a much stronger resemblance to conventional DOS code than to the typical Windows application.

A Step-by-Step Approach

Windows can be an overwhelmingly complex environment. My goal in this book is to help you understand which parts of that environment are really critical to each different type of driver. Each chapter introduces a new driver, and each chapter introduces only as much new material as you need to understand the new example. I've tried to keep each example driver as simple as possible so that the critical features are nearly self-evident. Most of the example code is written in C and embedded assembly using VC++ 4.0. Where necessary, code is written in assembly using Microsoft Assembly (MASM) v6.1. The code disk includes a library of wrapper functions that allow VxDs to be coded almost entirely in C.

Which Version of Windows?

This book covers both Windows 95 and Windows 3.x (Enhanced Mode). The focus is on Windows 95, but almost all of the material also applies to Windows 3.x. In most chapters the differences between the two versions are minimal and Windows 3.x considerations are simply highlighted in a separate section at the end of the chapter. In a few chapters the differences are larger. In these chapters I fully describe both versions, each in a separate section.

About the Book

This book is partitioned into two major sections. Part I (Chapters 2 through 12) covers the Windows execution environment and VxDs. Part II (Chapters 13 through 19) covers DLL-based drivers. Within each part, the chapters are ordered so that each builds on the prior chapters. Once you have read Chapter 1 and decided whether you need to build a VxD or a DLL, you can decide how to read the rest of the book. Nearly everyone should read Chapters 2 and 3. These chapters describe those portions of the Windows architecture that are important to device driver writers. The topics covered in these chapters are important to both VxD and DLL developers. Those readers who are rusty on selectors, descriptors, page tables, and the other architectural details of the 80x86 family of processors will want to read and refer to Appendix A as they read Chapters 2 and 3. Throughout the book, I assume you are comfortably familiar with the architectural information in Appendix A. Finally, if it bothers you to have certain implementation details hidden, you may want to read portions of Appendix B as you study the first example drivers. This appendix is the primary reference for the assembly language "wrappers" used throughout the text.

Companion Files on the FTP Site

NOTE: The full source code, wrapper library, and so forth for this book are now available on the publisher's ftp site at `ftp://ftp.cmpbooks.com/pub/hazzah.zip`; logon as "anonymous" and download the file. Any references to the "companion code disk" in the book now refer to the files available on the ftp site.

Table of Contents

## Chapter 1

Introduction

What is a Driver?

In its broadest definition, a "driver" is a set of functions that manipulates a hardware device. One way of categorizing drivers is by how these functions are packaged. In the DOS world, a "driver" can be a module that is linked into the application .EXE, or a "driver" can be another piece of software which is completely separate from the application (a DOS device driver or a TSR). In the world of Windows, a "driver" can be a module that is *dynamically* linked into the application .EXE (called a DLL), or it can be completely separate from the application (called a VxD).

Privileged and Non-privileged Packages

Another way of categorizing drivers is privilege. Some operating systems, such as UNIX and Windows NT, prohibit applications from manipulating hardware directly. In these environments, only privileged pieces of code known as "device drivers" are allowed to interface to hardware. Applications that need to control hardware must use the services provided by these drivers.

Windows too supports a privileged driver package. In Windows, these device drivers are called VxDs. However, Windows does not *require* hardware support to be contained in a VxD. In Windows, a surprising amount of hardware support is contained in DLLs, not VxDs. In Windows, DLLs that interface to hardware are often called "drivers".

Driver Interfaces

Yet another way of categorizing a driver is by the interface it presents to the application and the OS kernel. All Windows NT drivers use the same exact interface to the NT kernel. The kernel in turn provides a standard interface which applications can use to call any driver (open, read, etc.). The privileged driver package in Windows, the VxD, is different. Although all Windows VxDs use the same kernel interface, there is no standard interface to a VxD from the application level. Instead, each VxD defines its own application interface.

Some Windows drivers packaged as DLLs interface to the kernel and are required to export a specific interface to the kernel. Such drivers are sometimes called "system drivers". However, note that the interface used by the system keyboard driver looks very different than the interface used by the system display driver. Other driver DLLs have no required interface to the kernel at all, and the driver developer has a free hand in designing whatever kernel interface and application interface he wants.

What Kind of a Driver Do I Need to Write?

Clearly there are many different kinds of "drivers" under Windows. Exactly which type of driver you need to write depends on several interrelated factors:

- the version of Windows (3.x, 95),
- the class of hardware device (keyboard, network card, custom A/D board),
- the kind of hardware interface (I/O ports, interrupts), and
- the performance requirements (throughput, interrupt latency).

Collectively these four factors will determine whether you write your driver as a DLL or as a VxD.

What Class of Device?

The first factor that will narrow down the decision is the class of device you're supporting. Windows dictates a specific driver type for many device classes, so if you're supporting one of these, there is no decision to make. Windows dictates both the packaging of the driver (DLL or VxD) and its interface. Table 1.1 shows the device classes that Windows directly supports and the type of driver required.

As Table 1.1 shows, for most classes of device, both Windows 3.x and Windows 95 require exactly the same type of driver(s). The two exceptions are network adapters and block devices, neither of which was supported directly by Windows 3.x (DOS drivers were used instead), but both of which now require a VxD under Windows 95.

Both a DLL and a VxD are required to support most device classes, with the bulk of the work done in the DLL. You should also note that Driver DLLs are always16-bit components — even under Windows 95, where native applications and DLLs are 32-bit instead of 16-bit.

The multimedia drivers were first introduced in Windows 3.1, where they were implemented as DLLs that conformed to a new message-based interface. A driver DLL that conformed to this interface was called an "installable driver", and exported a `DriverProc` (similar to the `WindowProc` of a Windows application) and responded to messages such as `DRV_LOAD`, `DRV_OPEN`, `DRV_INSTALL`, and `DRV_CONFIGURE`. This interface provided the user with a standard mechanism for installing multimedia drivers through the Control Panel. The new interface also provided the operating system with a standard way of loading, enabling, and disabling multimedia devices.

Table 1.1 Devices that require a particular type of driver.				
Device Class	**Windows 3.x**		**Windows 95**	
	16-bit DLL	**VxD**	**16-bit DLL**	**VxD**
Display	`DISPLAY.DRV`	`VDD.VXD`	`DISPLAY.DRV`	`VDD.VXD`
Printer	`PRINTER.DRV`		`PRINTER.DRV`	
Keyboard	`KEYBOARD.DRV`	`VKD.VXD`	`KEYBOARD.DRV`	`VKD.VXD`
Mouse	`MOUSE.DRV`	`VMD.VXD`	`MOUSE.DRV`	`VMD.VXD`
Serial/Parallel Port	`COMM.DRV`	`VCD.VXD`		**VCOMM** port driver
Multimedia	installable driver DLL		installable driver DLL	
Network	not a Windows driver, but a DOS device driver or TSR (e.g. NDIS 2.0 or ODI)			NDIS 3.0 MAC driver
Block Device (Hard Disk, CD-ROM)	not a Windows driver, but a DOS device driver			layered block device driver

During the reign of Windows 3.1, the installable driver DLL soon caught on as a driver interface for types of devices other than multimedia. However, Microsoft is now pushing VxDs as the recommended driver type. Interestingly, multimedia drivers under Windows 95 remain as 16-bit installable drivers. Luckily, developers of multimedia drivers don't have to worry about thunking issues as other 16-bit driver developers do, because Windows itself contains the required thunking layer (just as it contains thunks for lots of other Windows pieces that remain 16-bit, such as USER and GDI). See Chapter 18 for a discussion of thunking.

What Kind of Hardware Interface?

If you are *not* writing a driver for one of the device classes in the table above, then Windows does not dictate either the driver package (DLL or VxD) or the interface. Since for either package you're going to design your own interface, the choice is between DLL and VxD. The next factor to consider when choosing a package is the hardware interface to your device:

• Is the device I/O-mapped or memory-mapped?

• Does the device generate interrupts?

• Does the device use DMA?

It is very easy to talk to an I/O-mapped device from a DLL, both under Windows 3.x and Windows 95. If your device is I/O-mapped and doesn't generate interrupts or DMA, the best choice for you may well be a DLL.

On the other hand, talking to a memory-mapped device, handling hardware interrupts, and performing DMA all are *possible* from a DLL, but only *easy* under Windows 3.x. Under Windows 95, only 16-bit DLLs are capable of these three operations. Native Windows 95 applications are, of course, 32-bit, not 16-bit, so if you use a 16-bit driver DLL under Windows 95 you also need to develop a separate "thunk layer" DLL. This thunk layer converts between the 16-bit world of your driver DLL and the 32-bit world of native Windows 95 applications that use your driver.

Because of the extra work required to develop the thunk DLL, if you're supporting Windows 95, there are only two reasons to consider using a driver DLL instead of a VxD. One, if you're supporting a very simple I/O-mapped device that doesn't use interrupts. In this case, you can write a simple 32-bit DLL that accesses the device. Two, if you've already written a 16-bit DLL driver for the device. In this case, add a thunk layer and you'll have Windows 95 support.

You should also consider how fully you wish to support the capabilities of the newer buses. Windows 95 includes built-in support for Plug and Play devices — which includes PCI, PCMCIA, and VL-Bus. To get full support, the driver for a Plug and Play device must be a VxD and interact with the Plug and Play Configuration Manager (also implemented as a VxD). See Chapters 10 and 11 for a full discussion of Plug and Play and the Configuration Manager.

If you choose to write a driver DLL instead of a VxD for your Plug and Play device, you'll have to use bus-specific BIOS methods to obtain your device's configuration information. And since most of these BIOS calls require using a software interrupt, and software interrupts aren't supported from 32-bit code (see Chapter 13 for an explanation of why this is so), your DLL must be 16-bit with a thunk layer. Thunk layers are discussed in Chapter 18.

What are the Performance Requirements?

Actual hardware access time, for both IO-mapped and memory-mapped devices, is roughly the same from either a driver DLL or a VxD. However, interrupt response time, also known as interrupt latency, is much faster (orders of magnitude) for a VxD. So if your device generates a lot of interrupts and/or doesn't have much buffering, you'll probably want to write a VxD.

Summary

With the information in this chapter, you should be able to reach a preliminary decision about what type of driver you need to develop. If a DLL will meet your requirements, then you can probably skip Chapters 4 through 12, for now, and focus on the DLL information in the second part. If you plan to develop a VxD, you will want to focus on the information in Part I.

In either case, you should probably browse through Appendix A sometime before you have finished reading Chapter 3. Throughout the book, I will assume you are comfortably familiar with the architectural information in that appendix.

In either case, whether you plan to develop a VxD or a DLL, the next two chapters lay an important foundation. Chapter 2 explains the basics of Virtual Machines. Chapter 3 explains how Windows exploits the 80x86 architecture to implement its Virtual Machines.

Part 1

Windows Execution Environment and VxDs

Chapter 2

The Virtual World of Windows

Windows 95 runs three different types of applications: DOS applications, Win16 applications, and Win32 applications. To overcome the potential incompatibilities among these types of applications, Windows executes them on virtual machines in virtual environments. When developing applications, Windows programmers can usually ignore the distinction between the virtual environment and the real environment; to most applications, the virtual environment *is* the real environment.

Writing a VxD, however, is a different matter, because a VxD runs in a supervisor context — meaning it runs outside of any of the virtual machines. In fact, a VxD becomes a part of the software which *implements* the virtual machine. Thus, the VxD writer needs a more complete understanding of how the virtual environment differs from the physical environment and how Windows creates the illusion of the virtual machine. A full understanding of the virtual machine is especially important to programmers who are developing VxDs that need to manipulate resources in an application's virtual environment, as many are.

This chapter explains the salient aspects of the Windows architecture, including how virtual machines are implemented, the major characteristics of the virtual environments, and the characteristics of the supervisor environment.

What is a Virtual Machine?

A virtual machine is a system-created illusion; virtual resources are emulations of hardware (and sometimes software) resources. To qualify as a virtual resource, the emulation must be so complete that the typical program can be written just as if the hardware were real, not emulated. For example, virtual memory systems use disk space, system software, special processor capabilities, and relatively small amounts of physical memory to emulate systems with enormous quantities of physical memory. The emulation is so convincing that programs running in a virtual environment can be written just as if the entire virtual address space were actually populated with physical memory. Such a memory system is said to have been "virtualized".

When a system virtualizes all, or nearly all, program-accessible resources, it creates a "virtual machine", or VM. Program-accessible resources include processor registers, memory, and peripheral devices (display, keyboard, etc.). The real reason behind the use of virtual machines under Windows is to support existing DOS applications. A DOS application assumes it is the only application running and often accesses hardware directly, uses all of available system memory, and uses all of the processor time. Since under Windows the DOS application is not the only one running, Windows creates a virtual machine for the application to run in: access to hardware is trapped and may be redirected, disk space may replace physical memory, and the VM is "put to sleep" while other VMs get processor time.

The definition of Virtual Machine is: A task with its own execution environment, which includes its own

- address space,
- I/O port space,
- interrupt operations, and
- processor registers.

Virtualizing this much of a machine while still executing the bulk of the code directly requires specialized processor support. The 80386 (and upwardly-compatible descendants) includes sophisticated processor support for address translation, demand paging, I/O trapping, instruction trapping, and interrupt trapping.

The main supervisor process, called the Virtual Machine Manager (VMM), uses these hardware capabilities to create not just one virtual machine, but several independent virtual machines, each with its own virtual execution environment. All Windows applications (both Win32 and Win16) run a single VM, called the System VM, whereas each DOS application runs in its own independent VM. Each of these virtual environments can differ substantially from the underlying physical machine.

Multitasking Model

Windows 3.x and Windows 95 use slightly different multitasking models. In Windows 3.x, the VMM preemptively multitasks among VMs. The VMM scheduler picks a VM and executes it for an assigned time slice, and when the time slice is up, the scheduler executes the next VM. This execution switch is transparent to the application — after all, some of the time-shared applications are DOS applications, which certainly aren't written to support multitasking.

Although VMs are unaware of this preemptive timeslicing, the Windows 3.x VMM itself is unaware that multiple Windows applications might be running in the System VM. To the VMM, all Windows applications are part of the same task. A higher layer "kernel" in the KERNEL DLL takes care of non-preemptive multitasking among the Windows applications in the System VM.

Because the Windows 3.x VMM scheduler deals only with VMs, the benefits of preemptive multitasking are realized only by users running DOS programs inside Windows. Badly behaved Windows programs can and do prevent other Windows applications from running, because the Kernel layer scheduler uses *non-preemptive* multitasking.

Windows 95 changes all that, bringing the benefits of preemptive multitasking to Win32 applications also. In Windows 95, the tasking unit is something new called a thread. Each DOS VM has a single thread. Within the System VM, all Win16 processes share a single thread, while each Win32 process has its own thread. In addition, each Win32 process may itself be multithreaded. In a multithreaded Win32 process, the main thread creates additional threads during execution.

In Windows 3.x the VMM switches execution among VMs, and when the System VM is run, a higher layer chooses which Windows application runs within the System VM. In contrast, the Windows 95 VMM switches execution among threads, not VMs, and it's the lowest layer, the VMM, that chooses which thread to run in the System VM. Since DOS VMs are always limited to a single thread, sometimes I'll simplify and say that the Windows 95 VMM "runs a DOS VM" — while technically speaking, it's running the single thread within that DOS VM.

Virtual Memory through Demand Paging

Because Windows supports multitasking, it's easy to imagine situations where the total amount of memory used by all running programs is greater than the actual memory present in the system. An operating system that limits a user to running just a couple of programs because he only has a small amount of physical memory might be useful, but not nearly as useful as one that somehow lets him run lots of programs. This problem is hardly unique to Windows, and the solution — demand paged virtual memory — isn't unique either: mainframe operating systems have had it for years.

The term *virtual memory* refers to a system that makes more memory available to applications than physically exists. "Demand paged" refers to a specific type of virtual memory. In a "paged" system, the operating system and processor divide the address space into blocks of uniform size, called pages. Windows uses a page size of 4Kb, since that's what the processor supports. "Demand" means that the virtual memory used by a program is associated with actual physical memory "on demand". Only when the program reads, writes, or executes a location on a page in virtual memory do the processor and operating system intervene to associate a page of physical memory with the virtual page.

The operating system and the processor work together to implement demand paging. When a program is loaded, Windows first allocates pages in virtual memory to hold the program, its data, and its resources. However, these are pages in virtual memory only, not in physical memory. The pages are marked as "not present" in physical memory. When the program actually attempts to execute or read from a not-present page, the attempted memory access triggers a processor exception called a page fault. (An exception is a condition that causes an immediate transfer of control to an exception handler, which is almost always part of the operating system.) The Windows page fault handler then allocates physical memory for that page and restarts the instruction that caused the page fault. The restarted instruction doesn't cause a fault because the page is now present. This fault handling is completely transparent to the application, which doesn't realize that all of the memory it's using is not present in physical memory at the same time.

The other half of demand paging is swapping pages to and from disk storage. Even though Windows delays allocating physical memory until it's actually used, at some point all physical memory will have been used. When the page fault handler finds that it can't allocate a page because physical memory is exhausted, it frees up a physical page by writing that page out to disk. The page fault handler then loads the needed page into the newly vacated physical page. Later, when the swapped-out page is accessed and causes a fault (it's definitely not present; it's on disk), the page fault handler first allocates a page (swapping out yet another page if necessary) and then checks to see whether this new page was previously written to disk. If it was, it copies the page contents from disk to physical memory. When the instruction is restarted, the swapped-out page is once again present in physical memory, with exactly the same contents as before.

Processor Modes

In order to create and maintain virtual machines, the VMM exploits special characteristics of the 80386 family of processors. These processors can operate in any of three modes: protected, real, and V86. Windows 95 utilizes two of the modes: protected mode and V86 mode.

The processor mode determines several important execution characteristics, including

- how much memory the processor can address,
- how the processor translates the logical addresses manipulated by software into physical addresses placed on the bus, and
- how the processor protects access to memory and I/O ports and prevents execution of certain instructions.

Windows 95 requires an 80386 processor, or one of its upwardly compatible descendants: 80486, Pentium, Pentium Pro. From now on when I use the term "processor", I mean one of these processors. I'll also use the terms "32-bit protected mode" and "16-bit protected mode" to refer to the processor when it is in protected mode and executing either 32-bit or 16-bit code, respectively. Although technically these two aren't "modes" in the same sense that V86 and protected are (i.e. this behavior isn't controlled by bits in the flags register), the size or "bitness" of the executing code has such an effect on the processor's behavior that 32-bit protected mode can essentially be considered a different mode than 16-bit protected mode.

Protected Mode

The biggest difference between 32-bit and 16-bit protected mode is the amount of addressable memory. In 16-bit protected mode, total addressable memory is only 16Mb. In 32-bit protected mode, the processor can address 4Gb, which is 2^{32}. Although 4Gb is such a large number that systems have nowhere near that much physical memory, such a large address space is still useful when the operating system provides virtual memory.

Although this difference in total address space is certainly important, what's more important is the difference in segment size — the maximum amount of memory addressable at once. Appendix A explains segments and other features of the Intel 80x86 architecture. In 16-bit protected mode, segments are limited to 64Kb (2^{16}), and developers working on large programs must be aware of segments. In 32-bit protected mode, segments can be 4Gb in size — so large that most operating systems that utilize 32-bit protected mode, including Windows 95, make segmentation invisible to the programmer by creating a single segment that addresses all 4Gb. Applications then never need to change segments.

As used by Windows 95, both 32-bit and 16-bit protected mode use the same method to translate the logical addresses used by software into the physical addresses placed on the bus. The translation process has two steps. A logical address consisting of a selector and offset is translated first to an intermediate form, called a linear address, by looking up the selector in a descriptor table which contains the segment's base linear address. Then the linear address is translated into a physical address by a second step called paging. I'll explain this two-step translation process in much more detail later; for now, just remember that the first step uses a selector lookup to find the linear address, which is different than the first step used by V86 mode.

The term "protected mode" came about because it was the first 80x86 processor mode to provide mechanisms to control access to memory and to I/O ports, mechanisms which an operating system could use to protect itself from applications. These mechanism are all based on the concept of privilege level. Executing code always has a privilege level, which Intel jargon calls a "ring", where Ring 0 is the innermost and most privileged ring, Ring 3 the outermost and least privileged.

A code segment's privilege level is determined by the operating system, and this privilege level controls which areas of memory and which I/O ports the code can access, as well as what instructions it can execute. Ring 0 code — referred to as supervisor code earlier — can access any memory location or I/O location and can execute any instruction. If an application running at an outer ring attempts an action that its privilege level doesn't allow, the processor raises an exception.

V86 Mode

Whereas protected mode was invented to support bigger programs and more robust operating systems, V86 mode exists to emulate real mode, the only mode supported by the original PC and the only mode supported by DOS applications even today. This emulation allows operating systems like Windows to better multitask DOS applications. V86 mode has a 1Mb address limit like real mode. The V86 mode address translation, however, is a cross between real and protected mode. V86 mode takes the logical-to-linear translation method from real mode: the segment is simply shifted left by 4 bits. (Contrast this to the selector lookup used in protected mode.) V86 mode takes the linear-to-physical method from protected mode: paging. The paging is completely transparent to DOS applications.

To keep multitasked DOS applications from crashing the system, V86 mode supports some of the same protection mechanisms as protected mode. Any program running in V86 mode will cause an exception (transferring control to the operating system) if it attempts to execute certain "privileged" instructions, access certain I/O ports, or access forbidden areas of memory. Table 2.1 summarizes the 80386+ physical execution environments.

Windows Execution Environments

The Windows 95 architecture supports four fundamentally different types of processes: supervisor processes, Win32 applications, Win16 applications, and DOS applications. Windows 95 runs each of these in a different execution environment. An execution environment can be described by processor mode, privilege level, and "bitness", which is a fancy term for 16-bit or 32-bit. Table 2.2 summarizes the Windows execution environments.

Table 2.1 Physical execution environments associated with various 80386+ processor modes.

	32-bit Protected	16-bit Protected	V86
Total Address Space	4Gb (2^{32})	16Mb (2^{24})	1Mb (2^{20})
Segment Size	4Gb	64Kb	64Kb
Address Translation	logical to linear: selector lookup linear to physical: page tables	logical to linear: selector lookup linear to physical: page tables	logical to linear: segment << 4 linear to physical: page tables
Privilege Level	0 through 3	0 through 3	3
Protection Mechanisms	yes	yes	yes

Table 2.2 Windows execution environments associated with various process types.

Process Type	Processor Mode	Privilege	Bitness	Memory Model	VM
Supervisor	protected	Ring 0	32-bit	flat	outside all
Win32	protected	Ring 3	32-bit	flat	System VM
Win16	protected	Ring 3	16-bit	segmented	System VM
DOS	V86	Ring 3	16-bit	segmented	individual VM

The supervisor processes run in protected mode with Ring 0 privilege (the highest access privilege), so they are able to see and manipulate the actual hardware environment. That is, the supervisor processes execute on the actual machine, not on a virtual machine; or to put it another way, supervisor processes run outside of any VM. Of all the components that make up Windows 95, only the VMM and VxDs execute in the supervisor environment. All other components run in a VM.

The supervisor environment is 32-bit, so these processes can address 4Gb of virtual memory. Supervisor processes use only two selectors, both of which address 4Gb. These two selectors differ only in their attributes: one is marked executable and loaded into CS; and the other is marked non-executable and loaded into DS, ES, and SS. (These selector attributes are stored in the same descriptor table that stores the segment's base linear address.) This type of memory model, where segments are loaded once and never again, is called flat model, and makes segmentation essentially invisible to the programmer.

While supervisor processes run outside of any VM (on the real machine), Win32 processes run at Ring 3 (the lowest access privilege) in a VM. Furthermore, all Win32 processes run in the same VM, called the System VM. Win32 processes are 32-bit protected mode and use a flat memory model, like supervisor processes, seeing a 4Gb address space and for all practical purposes ignoring selectors and segments.

Win16 processes run in the same SystemVM as Win32 processes. Win16 processes run in protected mode with Ring 3 privileges but don't get the luxury of a flat memory model. Because they run in 16-bit protected mode, Win16 processes are still stuck with a 16Mb address space and must deal with selectors and 64Kb segments.

Each DOS process gets its own VM. A DOS process doesn't run in protected mode like all the other types of processes. Instead, it runs in V86 mode, the 80386 mode built specially for emulating an 8086. V86 mode means a segmented memory model with 8086-type translation plus the addition of paging. V86 mode also implies Ring 3 privilege, so access to hardware resources and interrupts is hidden and virtualized.

Why does each DOS process get its own VM, while all Win32 and all Win16 applications share the System VM? Because DOS processes are in general unaware that they are sharing the system with any other process, and so usually "take over" the machine. DOS processes do things like modify the interrupt vector table and write directly to the screen. Windows runs each DOS program in a separate virtual machine so that each one modifies only its own virtual interrupt vector table, and writes only to its own virtual screen.

Windows applications, on the other hand (both Win32 and Win16), *are* aware that other processes are running. They write only to their own windows, not directly to the screen, and use a DOS call to modify the interrupt vector table instead of modifying it directly. Windows applications don't need to be protected so much from each other as they do from the DOS applications that aren't aware of them. So Windows can safely run all Windows applications in the same virtual machine.

Summary

Windows can run Win32, Win16, and DOS applications and can multitask among them. It does this by running the applications not on the real machine, but in virtual machines. The Virtual Machine Manager, a supervisor process that runs on the real machine, provides each of the different types of applications with a different virtual environment. The next chapter will take a closer look at each of the four resources in a Virtual Machine — I/O space, interrupt operations, processor registers, and address space — and show how Windows utilizes specialized processor features to virtualize each.

## Chapter 3

How Windows Implements the Virtual Environments

The previous chapter introduced the concept of a virtual machine and the four components that make up a virtual machine: I/O space, interrupt operations, processor registers, and address space. It also described the virtual environments seen by each of the four different types of processes that run under Windows: Win32, Win16, DOS, and supervisor (VMM and VxDs). This chapter will take a closer look at how the VMM virtualizes each of the components in the VM, for each different type of process. (This chapter assumes you are familiar with the basic features of the Intel 80x86 architecture. See Appendix A for a review of the important aspects of the architecture.)

Trapping I/O Port Access

Both protected mode and V86 mode incorporate several features that an operating system can use to trap IN and OUT instructions and thus prevent an application from directly accessing an I/O-mapped device. Memory-mapped devices are accessed via any instruction that uses a memory reference, while I/O-mapped devices are accessed only via IN and OUT instructions. (For a more detailed discussion of I/O-mapped and memory-mapped devices, see Chapter 6.) Windows 95 uses a combination of two processor features, I/O Privilege Level (IOPL) and the I/O Permission Map (IOPM), to control VM access to I/O addresses.

In protected mode, every code segment has an associated Descriptor Privilege Level stored in the descriptor table. Each code segment also has a separate attribute for I/O Privilege Level, also stored in the descriptor table. When an IN or OUT instruction is executed in protected mode, the processor compares the segment's IOPL to the privilege level of the currently executing code segment (called CPL for current privilege level). If CPL < IOPL, the segment has enough privilege, and the processor executes the instruction. If CPL >= IOPL, the processor uses the IOPM as a second level of protection. The IOPM is a bit-mapped list of ports: a 1 bit means "access denied", and a 0 bit means "access granted", So if CPL >= IOPL and the IOPM bit for the specific port is clear, the instruction is executed. But if the IOPM bit for that port is set, the processor generates an exception.

As used by Windows 95, the IOPM is really the dominant privilege mechanism for *all* VMs. In DOS VMs, the IOPM determines the I/O privilege of the application because the VMM runs DOS applications in V86 mode where the processor ignores the IOPL and looks only at the IOPM when processing IN and OUT instructions. In Win16 and Win32 VMs, the IOPM determines the I/O privilege of the application because the VMM runs all Win16 and Win32 processes with CPL > IOPL. Thus, even though Win16 and Win32 applications run in protected mode where the processor tests the IOPL, the test always results in a further check "through" the IOPM.

By manipulating the IOPM, Windows 95 can trap accesses to specific ports while allowing uninhibited access to other ports. Windows 95 uses this ability to virtualize the physical device located at the trapped port address. By routing device accesses through virtual device drivers (VxDs), Windows 95 can maintain separate state information for each of the VMs that might use the device.

The VMM is responsible for maintaining the IOPM. VxDs call a VMM service to request that the VMM trap a particular port. When making this request, the VxD specifies a callback function, called a "port trap handler". The VMM responds to such a request by setting the port's bit in the IOPM. When a VM accesses that port and thus causes a fault, the VMM fault handler calls the VxD's registered port trap handler. This port trap handler can do anything in response to the I/O access: the VxD may ignore the instruction, may execute the instruction, or may substitute a value instead (e.g. OUT 3F8h, 01h might become OUT 3F8h, 81h).

Windows 95 and its standard component VxDs trap almost all standard PC I/O devices but never trap non-standard I/O addresses. Table 3.1 lists the port locations trapped. A third-party VxD may trap other ports as well.

Table 3.1   I/O ports trapped by standard VxDs.

Windows 3.1

Port Address	VxD	Description
00–0F/C0–DF	VDMAD	DMA controller
20/21/A0/A1	VPICD	programmable interrupt controller
40/43	VTD	timer
60/64	VKD	keyboard
3F8–3FE/3E8–3EE/2F8–2FE	VCD	com port (COM1/2/3)
1F0/3F6	WDCTR1	hard disk controller (if Western Digital compatible)
3B4/3B5/3BA/3C0–3CF/3D0–3DF	VDD	VGA display

Windows 95

Port Address	VxD	Description
3F0/3F1/3F2/3F4/3F5/3F7	VFBACKUP	floppy controller
1F0–1F7	ESDI_506	hard disk controller
378/379/37A	VPD	printer LPT1
2F8–2Fe/3F8–3Fe	SERIAL	serial port COM1 and COM2
61	VSD	sound
3B4/3B5/3Ba/3D0–3DF/3C0–3CF	VDD	VGA display
1CE/1CF/2E8/x6EC–EF AEC–EF/xEEC–EF	ATI	miniport display PCI-specific VGA
00–0F/C0–DF/81/82/83/87/89/8A/83/87/89/8A	VDMAD	DMA controller
60/64	VKD	keyboard
40/43	VTD	timer
20/21/A0/A1	VPICD	programmable interrupt controller

Trapping Access to Memory-mapped Devices

While most standard peripherals are I/O-mapped, some are memory-mapped. Windows 95 relies primarily upon the page fault mechanism to virtualize access to memory-mapped devices. To trap references to one of these devices, the VxD virtualizing the device will mark the page corresponding to the device's physical address as "not present", and register its own page fault handler with VMM. When a process running in a VM tries to access that page, the access will cause a page fault. Instead of performing its default response and attempting to swap a page, the VMM fault handler will now call the registered page fault handler in the VxD that is virtualizing the device. The VxD handler can then decide what action is consistent with the requirements of the virtual environment.

The Virtual Display Device (VDD) uses this mechanism to virtualize the video frame buffer. When a DOS program writes to the video buffer at logical address B000:0000, the output doesn't appear on the screen because the VDD marks that particular page "not present". Instead, accesses to the video frame buffer are trapped by the VxD's page fault handler and redirected to another location in physical memory. This redirection causes writes to the video buffer to appear in a window instead of on the full screen. The VxD in Chapter 8 uses this same mechanism to arbitrate access to another memory-mapped device, a monochrome adapter.

Trapping Interrupts and Exceptions

In addition to trapping memory and I/O references, Windows 95 traps certain "privileged" instructions. "Privileged" instructions are those that could be used to bypass the processor's protection features or that could interfere with the integrity of the virtual machine. Privileged instructions include: those that affect the processor interrupt flag (CLI, STI, POPF, IRET); software interrupts (INT n); and those that load descriptor tables (LLDT, LDGT, LIDT). For the most part, Windows 95 traps these instructions to protect the integrity of the VM. In the instance of the INT instructions, Windows 95 exploits the trap to transparently intercept DOS and BIOS calls.

Processes running in a VM execute with Ring 3 (least privileged) permissions. Code executing at Ring 3 causes an exception when executing one of these "privileged" instructions. When this exception is raised, the processor switches to Ring 0 and then transfers control to an appropriate handler.

More precisely, each segment has an associated Descriptor Privilege Level (DPL). This segment privilege level determines the privilege level of most instructions (e.g. LLDT, LGDT). However, a few instructions (those which affect the processor's interrupt flag) derive their privilege level from the IOPL, not the DPL. When a Ring 3 process executes STI or CLI, for example, the processor will raise an exception only if CPL > IOPL.

One of the more significant differences between the System VM environment and the DOS VM environment relates to these IOPL-based privileges. While the 80386 architecture supports trapping of CLI and STI in both protected and V86 modes, Windows 95 does not trap the STI and CLI instructions in V86 mode. The VMM purposely sets CPL = IOPL for DOS applications, so that CLI and STI do *not* cause an exception. Apparently the designers decided the overhead of trapping all STIs and CLIs was a bigger performance penalty than they were prepared to pay. The behavior of CLI and STI in Windows applications (both Win16 and Win32) is different. By definition, IOPL=0 when running in protected mode, so CLI and STI do cause an exception. The VMM's exception handler then disables or enables the virtual interrupt flag for the system VM, but the processor interrupt flag itself is not affected.

Processor Registers

Virtualizing the third resource, processor registers, is trivial when compared to the mechanisms required to virtualize I/O port space and interrupt operations. The VMM maintains a virtual register data structure for each VM, and each time the VMM switches from executing one VM (say, VM1) to executing another VM (say, VM2), it first saves the state of VM1's registers in VM1's virtual register structure then updates the actual processor registers from VM2's virtual register structure before executing VM2.

A Closer Look at Linear Addresses and Paging

The previous chapter introduced the different processor modes and the address translation used in each. Before explaining how Windows virtualizes the address space, this chapter will examine, more closely, the two-step address translation mechanism used in both protected and V86 modes.

As viewed by software, an address has two parts, a selector and an offset. (Or in V86 mode, a segment and offset.) This form of address is known as a *logical address*. When software references this address, the processor translates the logical address into an intermediate form called a linear address, and then to a *physical address* which is actually placed on the bus and decoded by memory or a device.

In V86 mode, this first level translation, logical to linear, is very simple. The segment is shifted left by 4 bits and the offset is added in to form a linear address. In protected mode there is no arithmetic relationship between the logical address manipulated by the software and the corresponding linear address. Instead, the processor uses the selector portion of the logical address to index an entry in the Descriptor Table. Each entry in this table is a descriptor, a data structure that holds the base address of a segment. The processor translates the logical address to a linear address by using the selector to index the appropriate descriptor, extracting the base address from the descriptor, and adding that base address to the offset portion of the logical address. The resulting sum is a *linear address*. This process is depicted in Figure 3.1.

The next level of translation, from linear address to physical address, involves another set of data structures: the page directory and the page tables, sometimes collectively called "the page tables". Together, these structures map every 4Kb page of linear address space onto some 4Kb page of physical memory. (With virtual memory, though, this page of "physical memory" can exist either in RAM or on the hard disk.) Windows makes extensive use of the page tables to remap physical memory to meet the varying needs of each type of process, as well as to implement virtual memory. Once again, there is no arithmetic relationship between linear memory and physical memory.

The "page tables" are a hierarchical arrangement of a root page directory, multiple page tables and multiple page table entries, as illustrated in Figure 3.2. Each Page Table Entry (PTE) maps a 4Kb page of linear memory to a physical address. A group of 1024 PTEs forms a page table, which maps 4Kb*1024 = 4Mb of linear memory. A group of 1024 page tables forms a page directory, which maps 4Mb*1024 = 4Gb, all of linear memory.

Thanks to the hierarchical encoding of the data structures, the linear to physical translation can be implemented quite efficiently in hardware. To the processor, a linear address isn't merely a number between 0 and 4Gb — it's actually three bitfields: a page directory index, a page table index, and a page offset. Adding together the address of the root page directory table (stored in the CR3 register) and the page directory index bits, the processor finds a page directory entry. Inside this entry is the address of a page table. Adding together the address of this page table and the page table index bits, the processor finds a page table entry. Inside this PTE is a physical address. Adding together this physical address and the final bitfield, the page offset, the processor forms a final 32-bit physical address.

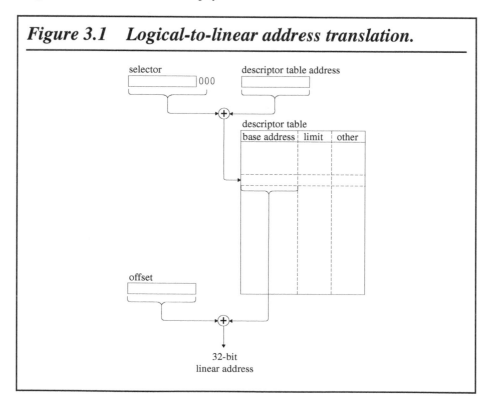

Figure 3.1   Logical-to-linear address translation.

Figure 3.2 **Illustrates how bitfields from the linear address are combined with Page Table Entries (PTEs) to construct a physical reference.**

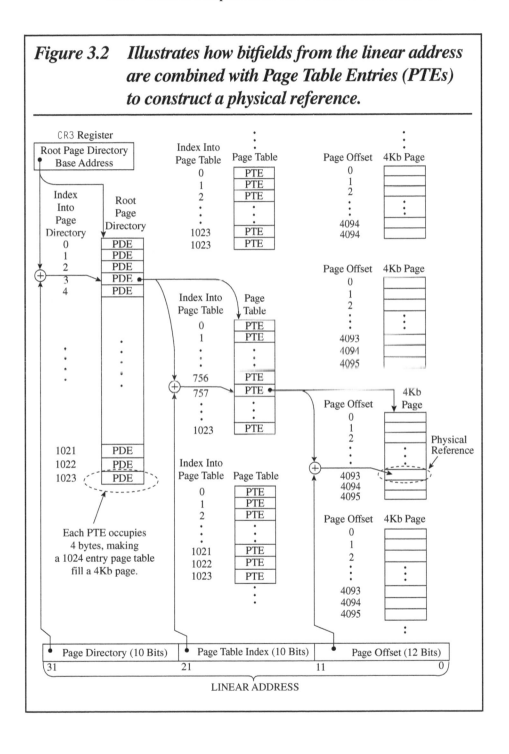

Competing Address Requirements of Win32, Win16, and DOS Applications

Windows 95 multitasks Win32, Win16, and DOS applications. Each of these three types of processes expects to see an address space with different characteristics. By address space, here I mean *linear* address space, not actual physical address space. When running under Windows, applications are not even aware of physical addresses — the generation of physical addresses by the processor happens "beneath" them.

Win32 Address Requirements

Every Win32 application has a 4Gb address space, which is completely separate from the address space of all other Win32 applications (Figure 3.3). By "completely separate", I mean it is literally impossible for one Win32 application to access the memory of another Win32 application. However, each Win32 application shares some of its vast 4Gb address space with other system components, like system DLLs and VMM/VxD code. Since all Win32 applications will be using these components, it makes sense to share these common components, instead of having a separate copy of each of these in physical memory. All Win32 applications can access the shared system components, but they can't access each other.

Win16 Address Requirements

Win16 applications have very different address space requirements than Win32 applications. Win16 applications expect a smaller address space (about 2Gb), and they expect to share this smaller address space not only with system components but also with all other Win16 applications as well (Figure 3.4). This shared address space is the main reason Win16 applications are less robust than Win32 applications. A Win16 application can obtain a selector — by accident or by design — to a segment belonging to another Win16 application and use that selector to write into the other application's data segment. Many Win16 applications rely on this shared address space, so in order to be backwardly compatible, Windows 95 must run Win16 applications in a shared address space.

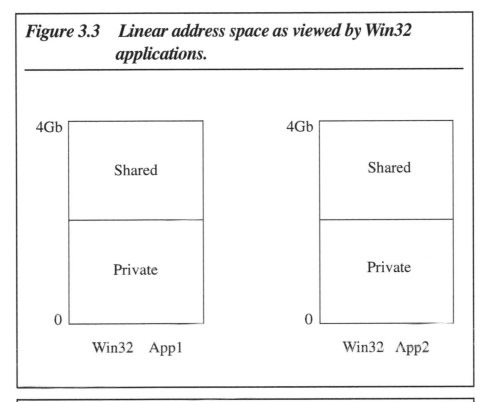

Figure 3.3   Linear address space as viewed by Win32 applications.

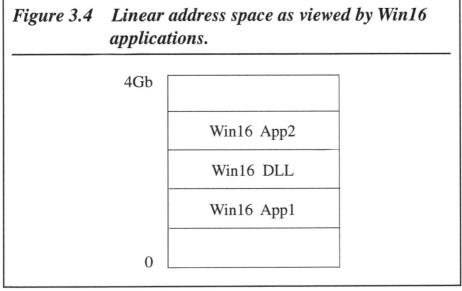

Figure 3.4   Linear address space as viewed by Win16 applications.

DOS Address Requirements

Windows 95 runs DOS applications in V86 mode. In this mode, the processor can only generate linear addresses in the 0–1Mb region. When a DOS application runs under Windows 95, it sees certain system components in its address space: TSR or device drivers loaded before Windows 95 began, the interrupt vector table and BIOS data areas in low memory, and "DOS" itself — COMMAND.COM. When Windows 95 runs multiple DOS applications, all of the DOS applications will see exactly the same set of system components (Figure 3.5). These DOS system components are shared among the multiple DOS applications, meaning they appear in the address space of each DOS application (somewhere below 1Mb), but only one copy of each is in physical memory.

Satisfying Address Requirements of Win16 and DOS Applications: How Does Windows 3.x Do It?

Windows 3.x doesn't run Win32 applications but it still needs to handle Win16 and DOS applications. These applications have exactly the same requirements under Windows 3.x as under Windows 95: Win16 applications run in a shared address space, DOS applications in linear 0–1Mb.

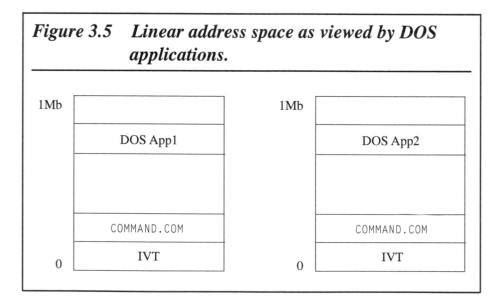

Figure 3.5   Linear address space as viewed by DOS applications.

Under Windows 3.x, all types of processes — Win16, DOS, and supervisor — share the same 4Gb linear address space. In fact, they really share less than 4Gb, because Windows 3.x uses only a little over a half of the 4Gb address space. Windows 3.x uses a small portion of the lower half (below 2Gb), and a larger portion of the upper half (above 2Gb). (If these numbers sound unusually large, remember, they are linear addresses, not physical addresses.)

The Windows 3.x VMM loads processes into linear address space in 4Mb chunks. The vast majority of all processes live in the upper half of the linear address space (2Gb and above). Supervisor processes — VMM itself plus VxDs —are loaded in the 4Mb starting at 2Gb. The VMM loads VMs immediately above these supervisor processes (Figure 3.6).

Figure 3.6 Linear address space under Windows 3.x.

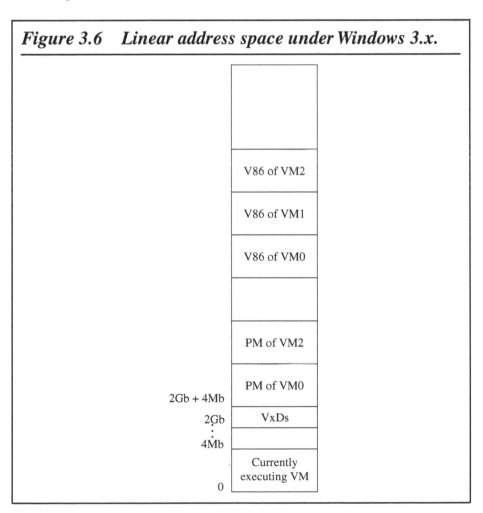

If a VM switches processor modes, Windows 95 will load both a protected mode component and a V86 mode component, each taking up at least 4Mb of address space. (Note that in Figure 3.6, VM0 has both a "PM" component and a "V86" component.) Although the System VM usually runs in protected mode, and DOS VMs usually run in V86 mode, VMs can and do flip modes. For example, all VMs, including the System VM, start in V86 mode. Once started, any VM can later switch to protected mode. In the System VM, the Ring 3 KERNEL module always switches into protected mode very early in the Windows initialization process. When a DOS-extended application runs under Windows, it too starts life in a VM in V86 mode, then the DOS-extender switches into protected mode.

Protected mode VMs, both the System VM and any DOS-extended VMs, switch back to V86 mode to access real mode DOS and BIOS services. Together, these DOS and BIOS services and TSRs make up the V86 mode component of the System VM, while the Windows applications, DLLs and system modules (KERNEL, USER, etc.) make up the protected mode component of the System VM. A DOS VM that runs a normal DOS application has only a V86 mode component. On the other hand, a DOS VM running a DOS-extended application has a V86 mode component containing DOS, BIOS, etc., and a protected mode component containing the DOS-extended pieces that run in protected mode.

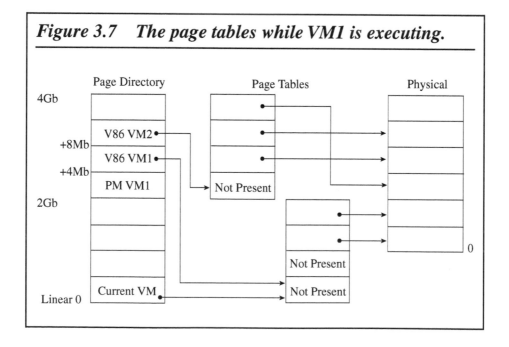

Figure 3.7   The page tables while VM1 is executing.

Figure 3.8   The page tables while VM2 is executing.

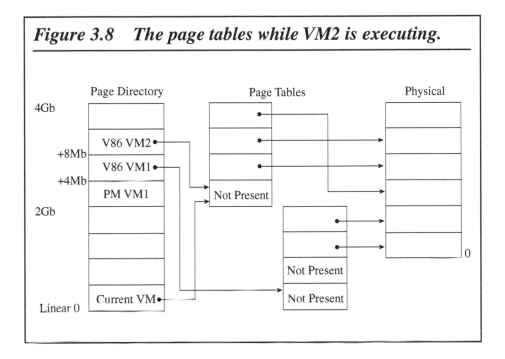

Copying Without Copying

When we say the VMM "copies" the V86 component down in linear address space, it sounds like a vast amount of memory is being copied. In reality, only a single pointer (32 bits) is copied, thanks to the hierarchy of the page table structures. Figure 3.7 shows an example load configuration. The VMM uses a single page directory, which maps the entire 4Gb of linear address space. Each of the 1024 entries in the page directory maps 4Mb (4Gb/1024 = 4Mb). Look first at the two entries labeled V86 VM1 (at location 2Gb + 4Mb) and V86 VM2 (at location 2Gb + 8Mb). Each of these two page directory entries points to a different page table, and the two page tables contain different PTEs.

Now look at the lowest (0–4Mb) entry in the page directory (labeled Current VM). Note that this entry points to one of the two page tables I just described. In Figure 3.7, VM1 is the currently executing VM, so the 0–4Mb (Current VM) entry points to the page table for VM1. To switch to VM2, the VMM merely updates the first entry of the page directory table, causing it point to VM2's page table instead of VM1's page table. Changing a single 32-bit entry in the page directory table accomplishes a "move" of 4Mb in linear memory.

After the switch (Figure 3.8), VM2 is visible at two different locations in linear memory, one below 1Mb (0–4Mb) and one above 2Gb (2Gb + 8Mb). The VMM can now begin executing the V86 component located below 1Mb and still retain access to the copy above 2Gb.

In V86 mode the processor can only generate linear addresses below 1Mb. Because of this restriction, the V86 component of the currently executing VM must live below 1Mb. More precisely, the currently executing V86 component must *occupy linear address space* below 1Mb. The active V86 component may be located in any part of physical memory — as long as the page tables properly map that physical image into the correct region of linear address space.

Thus Windows must remap the lower 4Mb of linear address space each time it runs a different VM. Only one active V86 component may occupy the linear space below 1Mb at any one time. Windows keeps a copy of *all* VM components (active and inactive) above 2Gb, but once a VM becomes active, Windows must "move" its V86 component to the lower portion of linear address space. Windows exploits the page-mapping hardware to effect this "move" without performing a copy. (See the sidebar "Copying Without Copying" on page 31.)

Thanks to the magic of the page mapping hardware, a single physical copy of a VM component can be visible at two different positions in linear address space at the same time. Windows uses this page table trick to make it more convenient for Ring 0 code to manipulate the V86 component. Windows constructs the page tables so that each V86 component appears at two locations in linear memory: once below 1Mb and once above 2Gb. These "aliased" page table entries allow Ring 0 code to manipulate a V86 component without testing to see if the component is part of the currently executing VM.

To summarize: Windows 3.x loads both a V86 and a PM component for each VM. These components always reside above the 2Gb boundary in linear address space, and the active V86 component is also mapped into the region below 1Mb. To switch VMs, Windows simply switches page tables (see the sidebar). Because Win16 processes run in the same VM, switching from one Win16 process to another does not involve any change in the page tables. In fact, the Windows 3.x VMM doesn't know anything about the multiple Win16 programs running in the System VM.

Satisfying Address Requirements of Win32, Win16, and DOS Applications: How Does Windows 95 Do It?

Although Windows 3.x uses only a small portion of the 4Gb linear address space, Windows 95 uses all of it. Windows 95 divides this 4Gb into several different regions, called arenas (Figure 3.9):

- private arena,
- shared arena,
- system arena, and
- DOS arena.

The private arena, from 4Mb–2Gb (almost half the entire 4Gb) is used for Win32 application code, data, and resources. This arena is private because it's mapped to a different location in physical memory for each Win32 application. So when Win32 App1 accesses linear address 4Mb, it accesses one set of physical locations, but when Win32 App2 accesses the *same* linear address 4Mb, it accesses a different set of physical locations. Windows 95 achieves this magic by switching the 511 page directory entries that map linear 4Mb–2Gb. When executing Win32 App1, these page directory entries point to one set of page tables (Figure 3.10). When Windows 95 switches to execute Win32 App2, they point to another set of page tables (Figure 3.11).

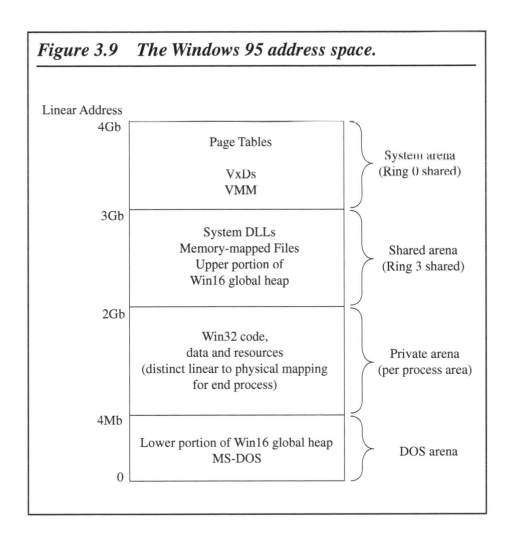

Figure 3.9   The Windows 95 address space.

By changing the private arena PDEs, Windows 95 protects Win32 applications from each other. The page table entries used for Win32 App1 simply don't contain the physical addresses used by App2, and the page table entries used for Win32 App2 don't contain the physical addresses used by App1. App1 and App2 are each literally unable to touch the other's resources.

The shared arena, located at 2Gb–3Gb, contains all Ring 3 code and data that must be shared. This arena hosts both Win32 system DLLs (because all Win32 applications need to share them) and all Win16 processes (because Win16 processes depend on a shared address space). Windows 95 implements the shared arena by more clever use of the page directory: Windows 95 never switches the 256-page directory entries that map linear 2Gb–3Gb. No matter what process is running, linear 2Gb–3Gb always maps to the same location in physical memory.

The system arena is at the top of address space, from 3Gb–4Gb. Windows 95 uses the system arena exclusively for supervisor (Ring 0) components: the VMM and VxDs. This arena is shared also, in exactly the same way as the shared arena, by never switching the page directory entries that map 3Gb–4Gb.

Figure 3.10 **Before the switch — when Win32 App1 is executing, the page directory's 4Mb slot points to a page table whose PTEs point to pages 2, 3, and 4 in physical memory.**

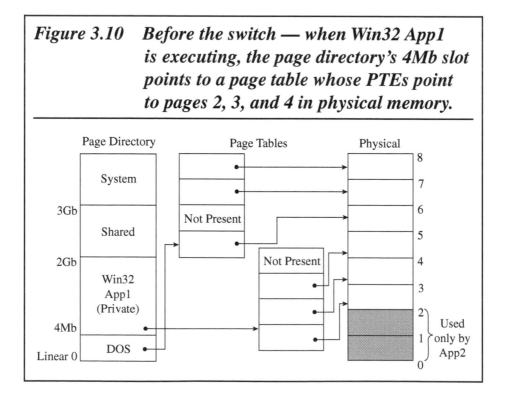

Many operating systems prevent user-mode components from accessing system pages directly by setting the Supervisor bit in the PTEs for system pages, which causes a page fault to occur if a system page is accessed from user-mode. Windows 95 does not use Supervisor bits at all, which makes it easy to pass data between a VxD and an application — the VxD can just give the application a pointer, which is directly usable by the application. (I'll explain this technique in detail in a later chapter.)

The DOS arena, at linear 0–4Mb, is devoted to DOS applications and a small portion of the Win16 heap. As stated earlier, DOS applications must reside here because they run in V86 mode and thus generate linear addresses below 1Mb. A small portion of the Win16 heap must also be below 1Mb, for use by Win16 applications and system DLLs allocating memory for communication with DOS, TSRs, etc.

Figure 3.11 *After the switch — when Win32 App2 begins executing, the page directory's 4Mb slot points to a different page table, whose PTEs point to pages 0 and 1. The page directory entries for the shared regions above 2Gb remain the same.*

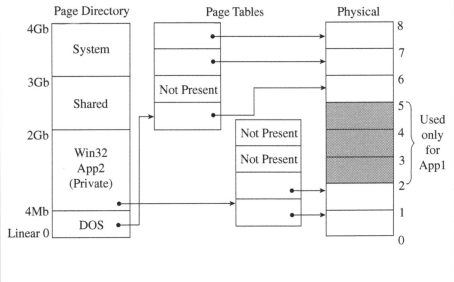

Windows 95 manages the page directory entries for the DOS arena in the same way that Windows 3.x did. With every VM switch, the V86 component of the currently executing VM is copied in linear space, from above 2Gb to below 1Mb, by simply changing the first entry in the root page directory.

Windows 95 makes more extensive use of page directory switching than Windows 3.x. Each time a different Win32 process is executed, the Windows 95 VMM switches the page directory entries for the private arena, leaving the page directory entries for the shared and system arenas alone. And each time a different VM is executed, the Windows 95 VMM switches the single page directory entry for the first 4Mb.

Summary

This chapter has explained how the VMM creates appropriate virtual environments for Win32, Win16, and DOS applications. The VMM utilizes several privilege-related processor features to virtualize access to IO-mapped and memory-mapped devices, as well as to control execution of privileged instructions. The VMM also utilizes the processor's paging features to provide each type of application with the linear address space that it expects. The remaining chapters in this section will focus on VxDs, the supervisor processes that assist the VM in creating and maintaining the virtual world of Windows.

## Chapter 4

Introduction to VxDs

Although VxD is an abbreviation for Virtual Device Driver, a VxD can be much more than a device driver that virtualizes a particular device. Some VxDs do virtualize a device. However, some VxDs act as a device driver, but don't virtualize the device. Some VxDs don't interact with any device; they exist merely to provide a service to other VxDs or to applications.

VxDs may be loaded along with the VMM (statically loaded) or on demand (dynamically loaded). In both cases, though, the VxD cooperates closely with, and shares execution context with the Virtual Machine Manager (VMM). This special relationship with the operating system gives a VxD powers that are unavailable to Windows and DOS applications. VxDs have unrestricted access to all hardware devices, can freely examine operating system data structures (such as descriptor and page tables), and can access any memory location. A VxD can also trap software interrupts, trap I/O port and memory region accesses, and even intercept hardware interrupts.

Although Windows or DOS applications may be able to do some "low-level" tasks (such as trap software interrupts), an application is always limited. For example, a Windows application can trap software interrupts issued by another Windows application — but not interrupts issued by a DOS application. A VxD would see all interrupts, regardless of source.

To support this level of integration with the VMM kernel, both statically loaded and dynamically loaded VxDs

- conform to a standard structure,
- register their services with the VMM, and
- service at least parts of a special message protocol.

This chapter explains how VxDs are loaded and how each type of VxD conforms to these fundamental requirements of a VxD. The following chapters show how VxDs can be used to implement different device-related capabilities.

VxD Loading

Windows 95 supports both statically loaded and dynamically loaded VxDs. Statically loaded VxDs are loaded when Windows initializes and remain loaded for the lifetime of Windows. If a VxD is used only by a particular application or exists only to provide services to certain applications, the memory it occupies is wasted when the VxD isn't actually in use. Static loading is particularly annoying for VxD developers, who must exit and restart Windows before they can test a change to a VxD.

Windows 95 supports two methods for static loading. The first, also supported by Windows 3.x, is to name the VxD in a device= statement in SYSTEM.INI. The second, new for Windows 95, is to add a Static VxD named value (e.g. Static VxD = pathname) to the registry, under the subkey \HKLM\System\CurrentControlSet\Services\VxD.

Dynamically loadable VxDs aren't loaded automatically when Windows initializes but are instead loaded and unloaded under the control of either an application or another VxD. For example, Plug and Play VxDs (discussed in detail in Chapter 10) must be dynamically loadable because Windows 95 supports runtime removal and reconfiguration of hardware. The VxDs that support this kind of hardware must be able to be loaded and unloaded as necessary.

Dynamically loadable VxDs are also useful as drivers for devices that are used only by a particular application. When the application needs to use the device, it loads the VxD. When the application is finished with the device, it unloads the VxD.

Statically and dynamically loaded VxDs respond to slightly different sets of VMM messages. Some messages are seen only by static VxDs, some are seen only by dynamic VxDs, but most are seen by both. In fact, it is easy to write a VxD that supports both methods of loading, simply by responding to both sets of messages.

Basic Structure of a VxD

Although VxDs use the 32-bit flat memory model, VxD code and data are still organized into segments. (In fact, a base plus offset addressing model is a necessary architectural component if a machine is to efficiently load and execute relocatable modules.) VxDs use these types of segments:

- real mode initialization,
- protected mode initialization,
- pageable,
- locked (non-pageable),
- static, and
- debug only.

For each of these segment types, there is a code segment and a data segment, so a VxD could have a total of 12 segments. The real mode code and data segments are both 16-bit (segmented model), and all other segments are 32-bit (flat model).

The real mode initialization segment contains code that is executed early in the Windows initialization sequence, before the VMM switches into protected mode. This early initialization phase gives each statically loaded VxD an opportunity to examine the pre-Windows real mode environment, and then decide whether the VxD should continue loading. By returning with an exit code in AX, the VxD can tell VMM to continue loading the protected mode portion of the VxD, to abort loading of this VxD, or even to abort loading Windows.

Most VxDs don't need a real mode initialization routine, but the PAGEFILE VxD, included as part of VMM.VXD, illustrates a possible use of one. PAGEFILE uses several DOS (INT 21h) calls to find out if the SMARTDRV DOS device driver is loaded. If not, PAGEFILE returns from its real mode initialization routine with Carry set, so that VMM never calls PAGEFILE's protected mode code.

After the real mode section of each statically loaded VxD has been executed, VMM switches into protected mode and gives each statically loaded VxD an opportunity to execute the code in its protected mode initialization segment. The protected mode initialization code can also return with an error code to tell VMM that the VxD has failed to initialize. If a VxD reports an initialization failure, the VMM marks the VxD inactive, and never calls it again.

Both real mode and protected mode initialization segments are discarded after initialization is complete. These segments are loaded before the first VxD is initialized and not discarded until all VxDs have finished initialization.

Most of a VxD resides in one of the other segments. In a statically loaded VxD, these other segments exist until Windows terminates. In a dynamically loaded VxD, they remain present until the VxD is unloaded. As their names suggest, a pageable segment may be paged to disk by the Virtual Memory Manager, while a locked segment will never be paged out. Most VxD code and data should be in a pageable segment, to allow the Virtual Memory Manager to swap out VxD pages and free up physical memory. Only the following items should — and must — go in a locked segment:

- The Device Control Procedure (the VxD's main entry point).
- Hardware interrupt handlers and all data accessed by them.
- Services that may be called by another VxD's hardware interrupt handler (referred to as asynchronous services).

Static segments are used only by dynamically loadable VxDs, which are discussed later in this chapter. The static code and data segments of a dynamically loadable VxD will not be unloaded when the rest of the VxD is dynamically unloaded but will remain in memory.

The VMM loads debug-only segments only when the system is running under a system debugger like WDEB386 or SoftIce/Windows. By partitioning debugging code into a debug-only segment, developers can always build the same executable, including the debug code without any run-time code overhead. The VMM will load the debug code when a system debugger is present, but omit it during normal load cycles (i.e. when no system debugger is present).

The Device Descriptor Block

The Device Descriptor Block, or DDB, is the VMM's "handle" to the VxD. The DDB includes information that identifies the VxD and a pointer to the VxD's main entry point. The DDB may optionally include pointers to other entry points, used by either applications or other VxDs. Table 4.1 shows the fields of the DDB structure that are initialized by the VxD. The VMM finds the VxD's DDB, and thus the main entry point, as soon as it loads the VxD by looking for the first exported symbol in the module.

Even when written in C, a VxD has no `main` procedure. Instead, the Device Control Procedure field in the DDB contains the address of the main entry point into a VxD. After real mode initialization, all calls from the VMM come to a VxD through this entry point. The VMM uses this entry point to notify a VxD of state changes in VMs and in Windows itself, and VxDs do their job by reacting to these events. (I'll discuss these events in detail a bit later.)

The DDB Device ID field is used by the VMM to identify the VxD. In particular, the VMM relies upon unique IDs to correctly resolve exported PM and V86 API entry points. Here are the rules for choosing a Device ID.

- If your VxD is a direct replacement for an existing VxD, use the ID of the existing VxD from the VMM header file.

- If your VxD is not a direct replacement, and it exports any entry points to DOS or Win16 applications or to other VxDs, you must apply to Microsoft for a unique ID.

- If your VxD doesn't replace a standard VxD and doesn't export any entry points to DOS or Win16 applications, you can use the `UNDEFINED_DEVICE_ID` constant defined in the VMM header file.

Table 4.1   The fields of the DDB structure.

Field	Description
Name	8-byte VxD name
Major Version	of VxD, not related to Windows version
Minor Version	of VxD, not related to Windows version
Device Control Procedure	address* of Device Control Procedure
Device ID	same as ID of VxD being replaced, or unique value assigned by Microsoft
Initialization Order	usually `Undefined_Init_Order`. To force intialization before/after a specific VxD, assign an `Init_Order` in `VMM.INC` and add/subtract 1.
Service Table	address* of Service Table
V86 API Procedure	address* of V86 API Procedure
PM API Procedure	address of PM API Procedure
	*32-bit offset

If a VxD provides an API for Win16 or DOS applications, its DDB contains the address of the API entry point. The DDB contains one field for each type of API: the PM API field is the 16-bit protected mode entry point used by Win16 applications, and the V86 API field is the entry point used by DOS applications. Because there is only one API entry point for each of these types of application, VxDs typically use a function code in a register to determine the specific function needed by the caller (much like a software interrupt under DOS).

A VxD can also export an entry point for use by other VxDs. VxD documentation usually refers to this as a "Service", not an API. Services are different from APIs in that the DDB contains a field for a service table, not a single service entry point. A service table is basically a list of function codes and function addresses.

One other field in the DDB is sometimes used by a VxD, though the VxD does not initialize this field. The Reference_Data field allows the real mode initialization piece of a VxD to communicate with the rest of the (protected mode) VxD. When the real mode initialization code returns, the VMM copies the value in EDX to the Reference_Data field of the VxD's DDB. If the real mode code needs to communicate more than four bytes, it should allocate a block of memory with LDSRV_Copy_Extended_Memory and return the address of the block in EDX. The protected mode portion of the VxD can then use Reference_Data as a pointer to the allocated block.

Supporting Data Structures

The DDB is the only data structure actually required of a VxD by the VMM. However, VxDs typically service more than one physical device (e.g. multiple serial ports) and interact with more than one Virtual Machine. Most VxDs will need to create their own supporting data structures to store per-device and per-VM configuration and state information.

VxDs typically use one or more device context structures to store device-specific information like I/O base address, IRQ, etc. These device context structures can be allocated statically in the VxD's data segment (locked if used by an interrupt handler) or dynamically through VMM services.

In general, if the number of devices is always fixed, allocate the device structures statically, and if the number varies, allocate the structures dynamically. For example, all PCs have two DMA controllers, so Virtual DMA Driver (see Chapter 6) declares static device structures in its data segment, but the number of serial ports on a PC varies, so the serial port driver dynamically allocates a device structure as each serial port is discovered.

If you dynamically allocate your device structure at runtime, use the VMM service
_HeapAllocate, which is very similar to malloc. However, if your device structure
includes a large buffer (4Kb or larger), you'll want to include only a pointer to the
buffer in the device structure itself, and then allocate the large buffer separately using
_PageAllocate. The rule is to use _HeapAllocate for small allocations and
_PageAllocate for large allocations, where small and large are relative to 4Kb.

***Figure 4.1   Illustrates how Control Block Data (CBD)
can be used to save per-VM state
information for each multiple device.***

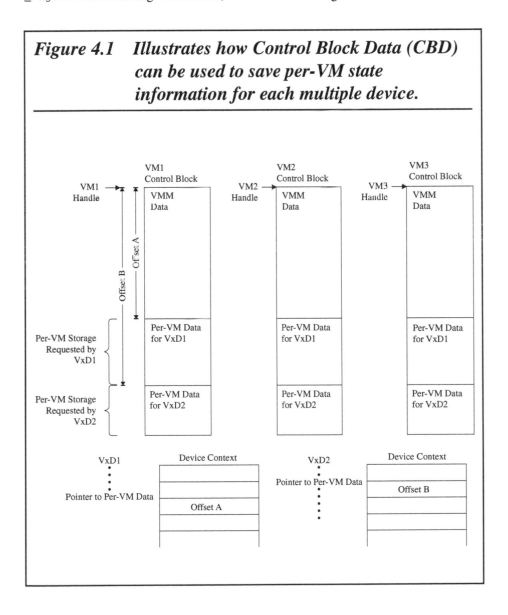

While managing per-device information is a familiar concept for device drivers, managing per-VM or per-device/per-VM information is less common. Fortunately, a VxD can ask the VMM to manage per-VM storage on behalf of the VxD. The VMM itself allocates and uses a Control Block for each VM. A VxD can use a VMM service to reserve its own per-VM data area within the VM Control Block.

To reserve this Control Block space, the VxD calls the VMM service _Allocate_Device_CB_Area during initialization, requesting a certain size block. The VMM will return the allocated block's offset within the entire Control Block. Once the VxD has requested this space, the VMM will reserve it at this same offset in every VM Control Block. Because the VxD will always have access to the current VM's handle, and the VM handle is actually the starting address of the VM Control Block, the VxD will always be able to get to this control block data. (I'll explain how the VxD gets the current VM handle in the next section.) Figure 4.1 shows how Control Block Data (CBD) can be used to save per-VM state information.

Just as VxDs have a need for per-VM data, some VxDs also have a need for per-thread data. The reason is that Windows 95 schedules threads, not VMs, and the System VM may have more than one thread. The mechanism for per-thread storage resembles that used for per-VM storage. A VxD allocates per-thread storage during VxD initialization by calling the service _AllocateThreadDataSlot. This service returns the offset of the thread data slot, relative to a data structure called the Thread Control Block or THCB. The VMM provides the THCB of the currently executing thread when it calls a VxD's Device Control Procedure with thread-related messages. A VxD can also get the THCB of the currently executing thread by calling the VMM service Get_Cur_Thread_Handle.

Unlike _Allocate_Device_CB_Area, which can reserve various size data areas, _AllocateThreadDataSlot always allocates 4 bytes of per-thread storage. If your VxD's per-thread data won't fit in 4 bytes, use these 4 bytes to store a pointer to a larger structure. Your VxD should allocate the larger structure when the thread is created (Figure 4.2).

To examine or modify the state of a VM, a VxD examines or modifies the fields in another important data structure, the Client Register Structure. This structure contains the VM's current registers and flags. Typically a VxD is interested in the VM state if it provides an API for use by PM or V86 mode applications. Such a VxD gets its input and provides its output through these client registers. The VMM sets EBP to point to

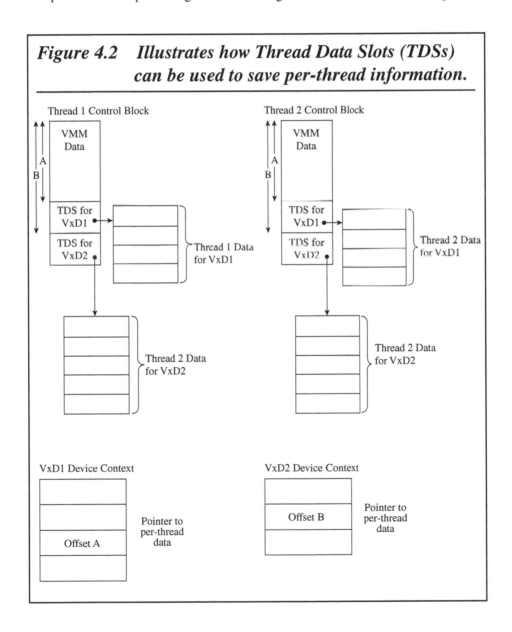

Figure 4.2   Illustrates how Thread Data Slots (TDSs) can be used to save per-thread information.

the Client Register Structure before calling the VxD API entry point, so most access to the Client Register Structure is done through EBP. A VxD can also find the Client Register Structure through the CB_Client_Pointer address found in the VM's Control Block. Figure 4.3 shows these relationships.

Figure 4.3   *Illustrates the relationship between the current VM handle, the VM control block, and the Client Register Structure.*

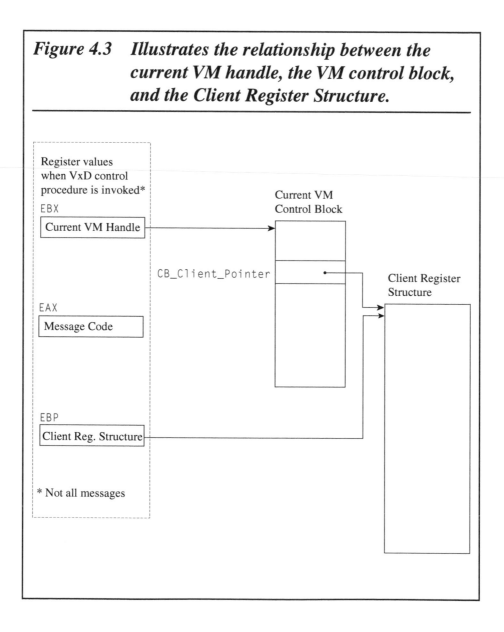

Event Notification

Once real mode initialization is complete, the VMM will notify the VxD about relevant events through a special message interface. To send a message to the VxD, the VMM obtains the address of the VxD's Device Control Procedure from the VxD's DDB and calls this procedure with a message code in EAX and the handle to the current VM in EBX. The control procedure then branches to message-specific code. The VMM uses this interface to notify the VxD of initialization activities, cleanup activities, and VM state changes.

Although the VxD message interface is conceptually similar to the WinProc message interface, the implementation is completely unrelated and incompatible.

The roughly two dozen messages can be divided into eight major categories. The messages and their categories are shown in Table 4.2. The messages in the initialization and termination categories are always sent in the order listed. A more detailed list of the messages and their register parameters and return codes can be found in the Windows 95 DDK documentation.

Table 4.2   The event notification messages that VMM sends to VxDs.

Message Category	Message	Description
System Initialization	Sys_Critical_Init	Interrupts disabled and remain so; minimal processing.
	Device_Init	System VM already loaded; VxDs do most initialization here.
	Init_Complete	Any processing needed after all VxDs do Device_Init.
System Termination	System_Exit	System VM destroyed, but still in memory.
	Sys_Critical_Exit	System VM no longer in memory; interrupts disabled.

Many VxDs process only a handful of these messages. The example VxDs beginning in the next chapter will illustrate the processing of the most commonly handled messages. Most of these messages mark important events in the life of either the VxD or a VM. The following section explains how the messages relate to the normal life cycle of a VxD and the VMs it services.

Message Category	Message	Description
VM Initialization	Create_VM	VxDs initialize per-VM data.
	VM_Critical_Init	Interrupts disabled.
	VM_Init	VM fully created; VxD can now call code in VM.
	Sys_VM_Init	Equivalent to VM_Init, but VM is System VM.
VM Termination	Query_Destroy	Abnormal VM termination: return Carry flag set if VM should not be destroyed.
	VM_Terminate	Normal VM termination; VM still exists so VxD can call code in VM.
	Sys_VM_Terminate	Equivalent to VM_Terminate, but VM is System VM.
	VM_Not_Executeable	Sent for both normal and abnormal termination; VM still in memory, but not executable.
	Destroy_VM	VM no longer in memory.
VM State Change	VM_Suspend	VM suspended by another VxD; VxD should give up any resources associated with the VM.
	VM_Resume	VM resumed from a suspend.
	Set_Device_Focus	VM has keyboard/mouse focus.
	Begin_PM_App	VM has started a protected mode application.
	End_PM_App	VM has ended a protected mode application.

Table 4.2 (continued) ***The event notification messages that VMM sends to VxDs.***

Statically Loaded VxD Initialization and Termination Messages

A statically loaded VxD is loaded when Windows initializes and is unloaded when Windows terminates. During Windows initialization, a statically loaded VxD will receive three messages, one marking each phase of Windows initialization. In response to any of the three messages, a VxD may indicate failure by returning with the Carry flag set. On such failure, Windows will unload the VxD, and the VxD will receive no further messages.

The first phase of Windows initialization is marked by the Sys_Critical_Init message. At this time, interrupts are disabled, so if your device requires uninterruptible initialization, do it here. If a VxD exports services to other VxDs, it should perform any initialization needed to carry out these services in the handler for Sys_Critical_Init, because other VxDs may call the exported services immediately after the exporting VxD processes this message. If a VxD virtualizes a memory-mapped adapter that can be used by DOS applications, then it should reserve pages in V86 address space here. (For example, the virtual display adapter reserves pages for the video frame buffer, usually at A0000h–C0000h, in each VM's address space.)

Table 4.2 (continued)	*The event notification messages that VMM sends to VxDs.*	
Message Category	**Message**	**Description**
Thread Initialization	Create_Thread	New thread is being created; allocate and initialize THCB data.
	Thread_Init	New thread has been created and is currently executing.
Thread Termination	Terminate_Thread	Thread is about to be terminated; release any thread-specific resources.
	Thread_Not_Executeable	Thread is being terminated and will not be executed again.
	Destroy_Thread	Thread has been destroyed.
Miscellaneous	Reboot_Processor	Handled only by Virtual Keyboard Driver.
	Debug_Query	Generated on behalf of debugger; VxDs display status.

All VxDs should defer any other actions until the next phase. Note that services such as Simulate_Int or Exec_Int, which execute code in a VM, are not available at this time because no VMs have been created yet. (I'll explain the role of Simulate_Int and Exec_Int in more detail in Chapter 12.)

The next message, Device_Init, notifies a VxD of the second initialization phase, which occurs after VMM has created the System VM. Most of a VxD's setup is performed during this phase. At this time, a VxD should allocate device context and Control Block memory, hook I/O ports, and hook interrupts.

Init_Complete marks the last phase of system initialization. Usually only VxDs that allocate pages in V86 address space need to respond to this message.

Windows also shuts down in three phases. When the system terminates normally (i.e. not in a crash), the System VM is terminated first, resulting in a Sys_VM_Terminate message. The System VM has not been destroyed yet, so Simulate_Int and Exec_Int services are still available if the VxD needs to execute code in the System VM. The next message in the shutdown sequence is System_Exit, which occurs during both normal and abnormal terminations. At this time, interrupts are enabled but the System VM has already been destroyed, so Simulate_Int and Exec_Int are no longer available. Most VxDs do their shutdown processing during System_Exit, shutting down their device. The last message is Sys_Critical_Exit, sent with interrupts disabled. Most VxDs don't process this message.

Dynamically Loaded VxD Initialization and Termination Messages

A dynamically loadable VxD doesn't see the system initialization messages (Sys_Critical_Init, Device_Init, and Init_Complete) because it hasn't been loaded yet when these messages are sent. However, the VMM provides an analogous message to a dynamic VxD during its loading procedure, Sys_Dynamic_Device_Init, and another message when the VxD is unloaded, Sys_Dynamic_Device_Exit.

A dynamic VxD processes the Sys_Dynamic_Device_Init message much as a static VxD would process the system initialization messages — by performing basic device initialization, hooking I/O ports, installing hardware interrupt handlers, etc. Note that certain VMM services are available only during system initialization and therefore may not be used by dynamic VxDs (see the Windows 95 DDK for a list of these services). A dynamic VxD may indicate that it failed to load by returning from the Sys_Dynamic_Device_Init message with the Carry flag set.

Although static VxDs receive several system termination messages, static VxDs are often careless about releasing resources during termination, since Windows itself is terminating. A dynamic VxD must, on the other hand, be very careful to free any resources it has allocated. This includes unhooking I/O ports, uninstalling hardware interrupt handlers, and unhooking services. In addition, a dynamic VxD must cancel all outstanding timeouts and events during `Sys_Dynamic_Device_Exit`, otherwise the VMM will end up calling code that is no longer loaded and the system will probably crash.

Static code and data segments can be used to solve some of the problems a dynamic VxD may encounter in releasing resources. For example, sometimes the VMM doesn't provide a "deallocate" service for a particular resource, and sometimes the deallocate may fail. In these cases, the code using this resource should be in the static code segment and shouldn't take any action unless the rest of the VxD is loaded. The VxD should also reuse the already allocated resource the next time the VxD is loaded, instead of allocating the resource again.

VM State Change Messages

Another set of messages tracks the life of VMs. Creation of a new VM also occurs in three phases, each with its own message: `Create_VM`, `VM_Critical_Init`, and `VM_Init`. For each of these messages, the VM handle is in `EBX`.

When the VxD receives the first message, `Create_VM`, it should initialize any data associated with the VM. `VM_Critical_Init` marks the next phase. An error response (returning with `Carry` flag set) to the `VM_Critical_Init` message will cause a VM termination sequence, starting with `VM_Not_Executeable`. (There is no VM termination sequence if `VM_Create` is failed.) The final phase of creation is `VM_Init`. At this time, the VM has already been created, and `Simulate_Int` and `Exec_Int` are available for calling software interrupts in the newly created VM.

A VM's destruction also takes place in three stages, again with the VM handle in `EBX`. A VM that exits gracefully results in a `VM_Terminate` message, which indicates the VM is "about to die". (An abnormal termination will first generate a `Query_Destroy`, see the following paragraph.) The VxD should take any action requiring `Simulate_Int` or `Exec_Int` here, while the VM is still present. The next phase, `VM_Not_Executeable`, occurs both during a graceful exit and an abnormal exit. The `EDX` register contains flag values that indicate the actual cause of termination.

These flag values are listed in Table 4.3. Because the VM has already been terminated, `Simulate_Int` and `Exec_Int` are not available. The last phase is marked by `Destroy_VM`. If a VxD doesn't care about the specific reason for VM termination and it doesn't need to use `Simulate_Int` or `Exec_Int`, it can choose to respond to only this final message.

Before the SHELL VxD shuts down a VM abnormally (typically in response to a user request), it will send a `Query_Destroy` message. A VxD can respond to this message with the `Carry` flag set to indicate the SHELL should not destroy the VM. In this case, the VxD should also inform the user of the problem, using the SHELL message services (covered in Chapter 8, in the "Checking Ownership" section).

In addition to VM startup and shutdown events, VxDs are also notified about scheduling events that change the currently running VM. `VM_Suspend` and `VM_Resume` messages are sent to VxDs as the VMM scheduler suspends and resumes execution of a VM.

Although the DDK documentation says to free any resources associated with the suspended VM on receipt of a `VM_Suspend`, only a few of the VxDs whose source is provided in the DDK respond to the `VM_Suspend` and `VM_Destroy` messages. The Virtual Display Driver (VDD) responds to `VM_Suspend` by unlocking the pages of video memory and to `VM_Resume` by locking the pages again. The Virtual Comm Driver (VCD) responds to `VM_Suspend` by clearing any pending serial port interrupt if the port is owned by the VM being suspended.

Table 4.3 Flag values contained in the *EDX* register that indicate the cause of termination.

Flag	Description
`VNE_Crashed`	VM crashed.
`VNE_Nuked`	VM destroyed while still active.
`VNE_CreateFail`	A VxD failed `Create_VM`.
`VNE_CrInitFail`	A VxD failed `VM_Critical_Init`.
`VNE_InitFail`	A VxD failed `VM_Init`.
`VNE_Closed`	VM closed properly then destroyed.

Thread Messages

Another set of messages tracks the life of threads, the unit of tasking used by the Windows 95 VMM scheduler. These messages are Create_Thread, Thread_Init, Terminate_Thread, Thread_Not_Executeable, and Destroy_Thread. However, these messages are not sent for the initial thread of a VM, only for subsequently created threads in a VM. As discussed in an earlier chapter, DOS VMs have exactly one thread each, so even though creation of a DOS VM *does* result in creation of a new thread, the VMM does *not* send a Create_Thread message. (It does however, send a Create_VM message.)

Threads are created and destroyed in stages, similar to VMs. The first message, Create_Thread, is sent early in the thread creation process. EDI contains the handle (THCB) of the thread being created (which is not the currently executing thread). A VxD can return with Carry set and the VMM will not create the thread. A VxD typically allocates and initializes any thread-specific data here. The extra allocation step is necessary if the 4 bytes of per-thread data in the THCB (allocated during VxD initialization) isn't enough. In this case, a per-thread structure is allocated during Create_Thread, and the per-thread data in the THCB is used to store a pointer to this newly allocated structure.

Once the thread has been fully created, the VMM sends out the Thread_Init message. EDI once again contains the handle of the newly created thread, but now the new thread is also the currently executing thread. A VxD should delay any initialization that requires the new thread to be the currently executing thread until it receives this message.

Thread destruction also involves multiple messages: Terminate_Thread, Thread_Not_Executeable, and Destroy_Thread. When the first message, Terminate_Thread, is sent, the thread is "about to be terminated", but is still capable of being executed. VxDs typically respond to this message by freeing any resources associated with the thread. The next message, Thread_Not_Executeable, is sent when the thread will no longer be executed. The last message, Destroy_Thread, occurs after the thread has actually been destroyed and gives VxDs a last chance to free thread-specific resources.

Windows 3.x Differences

Windows 3.x used only three types of segments: real mode initialization, protected mode initialization, and locked (non-pageable). The Windows 3.x VMM never swaps out any VxD code or data.

Windows 3.x doesn't support dynamic VxD loading, only static loading. Static loading is specified via a device= statement in the [386Enh] statement in SYSTEM.INI, just as it is under Windows 95.

Windows 3.x doesn't support threads. This means there is no need for per-thread data, no Allocate_Thread_Data_Slot, and no thread-specific messages.

Summary

Despite the hundreds of functions supported by the VMM and other VxDs, for many VxD applications you really don't need to know much more than what I've covered in this chapter. Unless you are doing something very special (like writing a replacement for the VMM), you'll probably never need more than a dozen of the functions in that API.

In the following chapters I'll show you how to build several practical VxDs. Even though these VxDs span a wide variety of applications, collectively they use only a few functions from the VMM/VxD API. As you easily can tell just by scanning some of the listings in the chapters ahead, VxDs don't have to be overwhelmingly complicated to be useful.

## Chapter 5

A Skeleton VxD

This chapter will introduce a "skeleton" VxD, one that won't have much functionality but will provide the basic framework for future VxDs. This skeleton VxD will simply monitor the creation and destruction of VMs and threads and will print out VM and thread information during these events. This output is sent both to the debugger and to a file, techniques that will be used in later VxDs to provide trace information for debugging.

This chapter will introduce you to two different approaches to developing VxDs in C: one using tools from the Windows 95 DDK and the other using the VToolsD product from Vireo Software. VToolsD gives you a big head start, automatically generating a makefile and a prototype C file. VToolsD also requires no assembly language modules. In contrast, the DDK-only process requires one assembly language file. This chapter will cover both methods but will focus more on the DDK-only process, since it is more complicated.

Tools for Building VxDs

In the days of Windows 3.x, VxDs were almost always written in assembly, simply because VxDs are 32-bit flat model programs and there were few 32-bit C compilers available. Now that 32-bit compilers are the norm, it's possible to write VxDs in C. However, your standard 32-bit compiler and linker won't be enough.

You'll also need the include (.h) files for VMM and other VxD services, as well as a special library for interfacing to the VMM and other VxDs. The routines in the library contain glue code that transforms the register-based interface used by VMM and other VxD services into a C-callable interface. The include files and the VMM library are available from two different sources: the Windows 95 DDK (Device Driver Kit), which is available as part of the Microsoft Developer Network Subscription, and the VToolsD toolkit.

Both the Windows 95 DDK and VToolsD come with the tools you need to write VxDs in C — just add a 32-bit compiler and linker. VToolsD explicitly supports both Borland and Microsoft compilers, while the Windows 95 DDK supports only Microsoft, although it can be coerced to work with Borland. VToolsD includes several other features which the Windows 95 DDK does not. One is QuickVxD, a VxD "wizard" that quickly generates a skeleton VxD, including C source, header file, and makefile. VToolsD also includes a C run-time library for VxDs. This alternate library is useful because a VxD can't just use the C run-time included with a 32-bit compiler; the standard compiler-provided libraries make assumptions about the run-time environment that don't hold true for VxDs.

Although the DDK technically provides all you need to write VxDs in C, VToolsD makes it much easier. The VMM "glue" library provided by both VToolsD and the DDK solves only half of the problem, allowing your VxD written in C to call VMM and other VxD services, which use register-based parameters. However, only VToolsD addresses the problem of register-based parameters in the other direction. The messages sent to your VxD's Device Control Procedure, as well as many callbacks (port trap, interrupt, fault handler, etc.), all call into your VxD with parameters in registers. When using the DDK, you must either write small stub functions in assembly or embed assembly statements directly in your C code in order to extract these register parameters. VToolsD, on the other hand, provides a "C framework" that passes these parameters on the stack and allows you to write message handlers and callbacks all in C.

Even if you don't use the DDK development tools, you may still find it very valuable. The DDK also contains the source code for about a dozen of the VxDs that ship with Windows 95. These VxDs range from the virtual display driver to the virtual DMA driver to the virtual NetBios driver. If you're planning to write a VxD to support new hardware that is similar to an existing device, you'll certainly want to invest in the DDK and modify the VxD for the existing device. Even if you're creating a brand new VxD, taking a look at other VxDs is a great way to learn, and the DDK is the only source I know of for non-trivial, real world VxDs.

You'll also need a debugger to get your VxD working, and the application-level debugger shipped with your compiler simply won't do. Only two products can debug VxDs: the WDEB386 debugger included with the DDK or SoftIce/Windows by NuMega Technologies. Whether to use WDEB386 or SoftIce is largely a matter of taste, money, and development preferences. Although both are powerful enough to debug VxDs, SoftIce has more user-friendly features: WDEB386 requires a terminal, SoftIce does not; SoftIce/Windows can debug C at the source level, WDEB386 shows you only assembly.

"DDK" Version Source Files

The "DDK" version of the SKELETON VxD consists of two source files:

- SKELCTRL.ASM, which contains the Device Descriptor Block (DDB) and Device Control Procedure found in every VxD;
- VXDCALL.C, provided free of charge by Vireo (makers of VToolsD), which contains a patch necessary to fix a bug in the Microsoft VC++ 4.1 compiler; and
- SKELETON.C, which contains the message handler functions called by the Device Control Procedure.

Although it's not absolutely necessary to place the DDB and Device Control Procedure in an assembly language file (VToolsD doesn't), I prefer to do so. These very small pieces are easily coded in assembly, and putting them in a C file would involve writing complicated pre-processor macros and embedded assembly.

As explained in the last chapter, when a C module calls a VMM or VxD service, an assembly language function is required to take parameters from the stack and place them in appropriate registers as expected by the specific service. The VXDWRAPS.CLB library in the DDK provides wrappers for some commonly used VMM and VxD services, but SKELETON.VXD uses several services that aren't contained in this library. The wrapper functions for these services are in the WRAPPERS.CLB library, provided in the \wrappers directory on the code disk.

This chapter will focus on how SKELETON.C (Listing 5.1, page 69) uses the functions in the wrapper library, not on the wrapper functions themselves. Refer to Appendix B for a complete description of WRAPPERS.CLB, instructions on how to add new VMM/VxD services to the module, and how to place these functions in a library.

If you're using Microsoft VC++ 4.1 to build your VxD, you'll need to link one more file, VXDCALL.C, into your VxD. Without this module, a bug in the 4.1 compiler makes it worthless for building VxDs. In a nutshell, the compiler generates incorrect code when enums are used in embedded assembly statements: the VMMcall macro in VMM.H uses enums. VxDs generated with this incorrect code causes the run-time error message, "Unsupported service xx in VxD xx".

The VXDCALL.C module provided free of charge by Vireo (makers of VToolsD) back-patches the incorrect code at run time. Compile the code once and simply link in the OBJ file to any VxDs built with VC++ 4.1. Note that you must also include the accompanying header file, VXDCALL.H, in all your VxD C source files.

Although Vireo provides VXDCALL.C on their web page (www.vireo.com), you don't need VToolsD to use VXDCALL.C. You need VXDCALL.C if you're using VC++ 4.1, regardless of whether your toolkit is the DDK or VToolsD.

The file SKELCTRL.ASM (Listing 5.2, page 71) provides the building blocks for SKELETON.VXD, and for the VxDs in later chapters. SKELCTRL.ASM can be easily adapted for use in other VxDs by changing DDB fields (for example, the VxD name) and adding/deleting messages from the Device Control Procedure as desired. The other file, SKELETON.C, contains the message handler functions, which implement specific VxD functionality, and will vary greatly from one VxD to the next.

Although the specific functionality of the C source file will vary for each of the VxDs in this and later chapters, each version of the C source file includes the same basic set of header files. The header files, and a description of each, are found in Table 5.1.

The makefile, SKELETON.MAK (Listing 5.3, page 72) is used to build SKELETON.VXD. The makefile compiles, assembles, and links all components needed to build SKELETON.VXD. After building SKELETON.VXD, the makefile runs the MAPSYM utility, which converts the linker map file into a symbol file usable by either the WDEB386 or SoftIce/Win debugger.

The compiler and assembler options (flags) are defined by the macros CVXDFLAGS and AFLAGS at the top of the makefile. Tables 5.2 and 5.3 explain the purpose of each of these flags.

Table 5.1   Header files for SKELETON.C.

Header File	Description	Directory
BASEDEF.H	constants and types used by other header files	inc32 of Win95 DDK
DEBUG.H	macros for enabling/disabling debug code	inc32 of Win95 DDK
VMM.H	constants and types for VMM services	inc32 of Win95 DDK
VXDWRAPS.H	function prototypes for VMM/VxD services provided in DDK (VXDWRAPS.CLB)	inc32 of Win95 DDK
WRAPPERS.H	function prototypes for VMM/VxD services provided by WRAPPERS.CLB	wrappers
VXDCALL.H	function prototype for Vireo VMMcall/VxD-call patch	wrappers
INTRINSI.H	function prototype for intrinsic string functions	wrappers

Table 5.2   Compiler options and flags for VxDs.

Option or Flag	Purpose
c	compile only (no link)
Gs	disable stack overrun checking
Zdp, Zd	name PDB file that stores debug and symbol information
Z1	suppress default C run-time library name in OBJ; prevents accidental link with unsupported C run-time
DIS_32	specifies 32-bit code, not 16-bit; used by some VxD header files
DDEBUG	enables debug macros and functions in some VxD header files
DDEBLEVEL=1	sets debug level to normal in DEBUG.H (choices are retail, normal, or max)
DWANTVXDWRAPS	disable some inline functions in VxD header files, forcing ones in wrapper library to be used instead

Table 5.3   Assembler flags for VxDs.

Option or Flag	Purpose
c	assemble only (no link)
coff	output file in COFF format; MS linker now uses COFF, not OMF
Cx	preserve case in publics and externs
W2	set warning level to 2
Zd	include line number debug information in OBJ
DIS_32	specifies 32-bit code, not 16-bit (used by some VxD include files)
DDEBUG	enables debug macros and functions in some VxD include files
DDEBLEVEL=1	sets debug level to normal in DEBUG.INC (choices are retail, normal, or max)
DMASM6	specifies assembler is MASM 6.x (used by some VxD include files)
DBLD_COFF	specifies COFF format (used by some VxD include files)

The DDB and Device Control Procedure:
SKELCTRL.ASM

The short assembly language module SKELCTRL.ASM (Listing 5.2, page 71) contains the DDB and a Device Control Procedure:

```
    .386p

    include vmm.inc
    include debug.inc

DECLARE_VIRTUAL_DEVICE      SKELETON, 1, 0, ControlProc, \
                            UNDEFINED_DEVICE_ID, \
                            UNDEFINED_INIT_ORDER

VxD_LOCKED_CODE_SEG

BeginProc ControlProc
    Control_Dispatch SYS_VM_INIT, _OnSysVmInit, cCall, <ebx>
    Control_Dispatch SYS_VM_TERMINATE, _OnSysVmTerminate, cCall, <ebx>
    Control_Dispatch CREATE_VM, _OnCreateVm, cCall, <ebx>
    Control_Dispatch DESTROY_VM, _OnDestroyVm, cCall, <ebx>
    Control_Dispatch CREATE_THREAD, _OnCreateThread, cCall, <edi>
    Control_Dispatch DESTROY_THREAD, _OnDestroyThread, cCall, <edi>

    clc
    ret

EndProc ControlProc

VxD_LOCKED_CODE_ENDS

    END
```

At the top of the file, the DDB is declared with the macro DECLARE_VIRTUAL_-DEVICE. This macro's parameters correspond one for one to the DDB fields described in the section "The Device Descriptor Block" in Chapter 4. SKELCTRL.ASM uses only the first six macro parameters, because the VxD doesn't export either a V86 or a PM API. Because SKELETON doesn't export an API or any services, it doesn't need a VxD ID, so SKELCTRL.ASM uses UNDEFINED_DEVICE_ID for the Device_Num macro parameter (Device_Num is the same as Device ID). SKELETON doesn't have any requirements for a particular initialization order, so it uses UNDEFINED_INIT_ORDER for the Init_Order macro parameter.

The last half of SKELCTRL.ASM defines the VxD's Device Control Procedure (ControlProc). A VxD's Device Control Procedure must be placed in the locked segment, so ControlProc is surrounded by the macros VXD_LOCKED_CODE_SEG and VXD_LOCKED_CODE_ENDS. ControlProc uses a series of Control_Dispatch macros to generate code for a basic switch statement. For example, the line

```
Control_Dispatch SYS_VM_INIT, _OnSysVmInit, cCall, <ebx>
```

translates to code that compares the message code in EAX with SYS_VM_INIT, and if equal, calls the function OnSysVmInit in the C module, passing the VM handle in EBX as a parameter.

That's enough information about SKELCTRL.ASM to allow you to make minor modifications to support other messages in your VxD. Appendix B contains further details, including more information on the macros Control_Dispatch and cCall. Appendix B also contains information about the wrapper library, WRAPPERS.CLB, which you'll need if you add other VMM/VxD service wrappers to the library. In the rest of the chapter, I'll concentrate on the real functionality of SKELETON.VXD, contained in SKELETON.C.

SKELETON.C (Listing 5.1, page 69) contains the message handlers for the SKELETON.VXD. The SKELETON VxD processes six messages relating to creation and destruction of VMs and threads: Sys_VM_Init, Sys_VM_Terminate, Create_VM, Destroy_VM, Create_Thread, and Destroy_Thread. Each time a VM is created, all VxDs are sent one of two messages: Sys_VM_Init for System VM or Create_VM for non-System VMs. VM creation also results in the creation of an initial thread, but no message is sent for this thread. Subsequent (non-initial) threads created in a VM do result in a message, Create_Thread. As discussed earlier in Chapter 2, each non-System VM is limited to a single thread, which means all Create_Thread messages are associated with the System VM.

SKELETON demonstrates this behavior by printing out both VM handle and thread handle values for the six messages. The VM message handlers (OnSysVmInit, OnCreateVm, OnDestroyVm, and OnSysVmTerminate) use the VMM service Get_Initial_Thread_Handle to obtain the thread handle of the initial thread created along with the VM. (This service is not supported by the DDK library VXDWRAPS.CLB, so its wrapper is in WRAPPERS.CLB). The thread message handlers Create_Thread and Destroy_Thread extract the VM associated with the thread from the thread handle — which is really a pointer to the thread's control block. One of the fields in the thread control block is the handle of the VM associated with the thread.

Each message handler function prints these VM and thread handle messages to the debugger and to a file. The functions use the DPRINTF macro to generate debugger output. This macro mimics the useful VToolsD function dprintf. The macro combines a call to two VMM services: _Sprintf, which formats a string; and Out_Debug_String which outputs the formatted string to the debugger. Both of the services are included in the DDK library VXDWRAPS.CLB.

The DPRINTF macro expands only if the symbol DEBUG is defined at compile time. Typically this symbol is defined via a compiler switch rather than a #define in a source file. For example, with Microsoft's compiler you would use -DDEBUG=1. If DEBUG is not defined, the DPRINTF macro expands to nothing.

To send the messages to a file, the message handlers use the IFSMgr_Ring0_-FileIO service. The IFSMgr is the Installable File System Manager VxD, the top level manager of all the VxDs that together form a file system. Most IFSMgr services are used by other VxDs that are part of the file system, but the IFSMgr_Ring0_FileIO service is useful to any VxD: it lets a VxD perform file I/O at Ring 0. The "Ring 0" part is significant because before the IFSMgr arrived with Windows for Workgroups 3.11, a VxD could only perform file I/O by switching to Ring 3, and each individual I/O operation (open, close, etc.) involved a sequence of several VMM services. Under Windows 95, it takes only a single call to IFSMgr to do each file I/O operation.

The IFSMgr_Ring0_FileIO service will not work correctly if used before the Sys_Init_Complete message.

Although the actual IFSMgr service uses a single entry point for all I/O operations (open, close, etc.) with a function code to distinguish them, it's more convenient to have a separate function call for each operation. When creating the wrapper functions in WRAPPERS.CLB, I took a cue from VToolsD and provided a different wrapper function for each: IFSMgr_Ring0_OpenCreateFile, IFSMgr_Ring0_WriteFile, etc.

During System VM creation, OnSysVmInit opens the file VXDSKEL.LOG with a call to IFSMgr_Ring0_OpenCreateFile. The IFSMgr_Ring0_OpenCreateFile interface mimics the INT 21h File Open interface, with parameters for filename, open mode (read, write, and share flags), creation attributes (normal, hidden, etc.), and action (fail if file doesn't exist, etc.). In fact, the mode, attributes, and action parameters use exactly the same values as the INT 21h File Open.

The IFSMgr adds two additional parameters to the Open call that aren't part of the INT 21h interface. One is a context boolean: if set, the file is opened in the context of the current thread and thus can only be accessed when that thread is current. The other parameter contains a flag bit which if set means "don't cache reads and writes to this file".

OnSysVmInit uses "create and truncate" for the action parameter, so that the log file is created if it doesn't exist or opened and truncated if it already exists. OnSysVmInit allows file caching (since I/O to the log file isn't critical) and uses FALSE for the context boolean, so that the VxD can do file I/O at any time without worrying about which thread is current. This allows the VxD to open the file during Sys_VM_Init when the initial thread of the System VM is current and then to write to the file with the same handle during another VM or thread message when another thread is current.

OnSysVmInit keeps the file open and stores the file handle in the global variable fh so that other SKELETON.VXD message handlers can also write to the file. The file is closed by the OnSysVmTerminate message handler when Windows shuts down.

All the message handlers, including OnSysVmInit, write to this already-open file using IFSMgr_Ring0_WriteFile. This function uses the parameters you'd expect for a write: a handle, a buffer, and a count. But where most file I/O functions update file position automatically with each read and write, IFSMgr_Ring0_WriteFile requires an explicit file position parameter, which means the caller must keep track of file position. SKELETON does this by initializing the global file_pos variable to zero and incrementing file_pos by the number of bytes written with each call to IFSMgr_Ring0_WriteFile.

IFSMgr_Ring0_WriteFile performs no formatting, it simply writes a raw buffer. So before calling IFSMgr_Ring0_WriteFile, each message handler first formats the buffer using the VMM _Sprintf service provided in the DDK library VXDWRAPS.CLB.

VToolsD Version

To generate the VToolsD version of SKELETON.VXD, I used the QuickVxD "wizard" included with VToolsD to quickly generate a prototype VxD. Using QuickVxD is simple. You fill in several DDB fields (name, ID, init order, etc.), select which messages your VxD will handle, specify whether or not your VxD supports a V86 or PM API, and which (if any) services your VxD provides to other VxDs.

I used the name SKELETON and left both the ID and init order with the default value, which was UNDEFINED. I selected six messages: Sys_VM_Init, Sys_VM_Terminate, Create_VM, Destroy_VM, Create_Thread, and Terminate_Thread. Then I clicked on "Generate" and QuickVxd generated a single C source file, a header file, a makefile, and a definition file (Listings 5.5–5.8, pages 73–77).

QuickVxD uses the name you specify for your VxD as the base filename. Because I chose SKELETON, the source file was named SKELETON.C. This file contained the DDB, the Device Control Procedure, and message handler stub functions for the six messages I selected. The message handler functions created by QuickVxD all follow the same naming convention: OnX, where X is the name of the message. For example, a message handler for Init_Complete would be called OnInitComplete (notice the underscore is removed). Parameters for the message handlers are message specific, but usually include either a VM handle or a thread handle, and sometimes an additional parameter.

To complete the VToolsD version of SKELETON.VXD, I added a few global variables and some additional code to each of the stub message handlers. The resulting SKELETON.C is shown in the following code. Sections that I added are delimited by comments. (Text continues on page 67.)

```
#define   DEVICE_MAIN
#include  "skeleton.h"
#undef    DEVICE_MAIN

//-----------begin section added to prototype
DWORD filepos = 0;
HANDLE fh;
//-----------end section added to prototype

Declare_Virtual_Device(SKELETON)

DefineControlHandler(SYS_VM_INIT, OnSysVmInit);
DefineControlHandler(SYS_VM_TERMINATE, OnSysVmTerminate);
DefineControlHandler(CREATE_VM, OnCreateVm);
DefineControlHandler(DESTROY_VM, OnDestroyVm);
DefineControlHandler(CREATE_THREAD, OnCreateThread);
DefineControlHandler(DESTROY_THREAD, OnDestroyThread);

BOOL __cdecl ControlDispatcher(
   DWORD dwControlMessage,
   DWORD EBX,
   DWORD EDX,
   DWORD ESI,
   DWORD EDI,
   DWORD ECX)
{
   START_CONTROL_DISPATCH

   ON_SYS_VM_INIT(OnSysVmInit);
   ON_SYSTEM_EXIT(OnSysVmTerminate);
   ON_CREATE_VM(OnCreateVm);
   ON_DESTROY_VM(OnDestroyVm);
   ON_CREATE_THREAD(OnCreateThread);
   ON_DESTROY_THREAD(OnDestroyThread);

   END_CONTROL_DISPATCH

   return TRUE;
}
```

```
BOOL OnSysVmInit(VMHANDLE hVM)
{
//-----------begin section added to prototype
    BYTE    action;
    WORD    err;
    int     count=0;
    char    buf[80];
    PTCB    tcb;

    tcb = Get_Initial_Thread_Handle(hVM);
    dprintf(buf, "SysVMInit: VM=%x tcb=%x\r\n", hVM, tcb );

    fh = RO_OpenCreateFile(FALSE, "vxdskel.log",
                           0x0002, 0x0000, 0x12, 0x00,
                           &err, &action);
    if (!fh)
    {
        dprintf(buf, "Error %x opening file %s\n", err, "vxdskel.log" );
    }
    else
    {
        sprintf(buf, "SysVMInit: VM=%x tcb=%x\r\n", hVM, tcb );
        count = RO_WriteFile(FALSE, fh, buf, strlen(buf), filepos, &err);
        filepos += count;
    }
//-----------end section added to prototype
    return TRUE;
}

VOID OnSysVmTerminate(VMHANDLE hVM)
{
//-----------begin section added to prototype
    WORD    err;
    int     count=0;
    char    buf[80];
    PTCB    tcb;

    tcb = Get_Initial_Thread_Handle(hVM);
    dprintf( buf, "SysVmTerminate VM=%x tcb=%x\r\n", hVM, tcb );
    sprintf( buf, "SysVmTerminate VM=%x tcb=%x\r\n", hVM, tcb );
    count = RO_WriteFile(FALSE, fh, buf, strlen(buf), filepos, &err);
    filepos += count;
    RO_CloseFile( fh, &err );
//-----------end section added to prototype
}
```

```
BOOL OnCreateVm(VMHANDLE hVM)
{
//-----------begin section added to prototype
    PTCB    tcb;
    WORD    err;
    int     count=0;
    char    buf[80];

    tcb = Get_Initial_Thread_Handle(hVM);
    dprintf(buf, "Create_VM: VM=%x, tcb=%x\r\n", hVM, tcb);
    sprintf(buf, "Create_VM: VM=%x, tcb=%x\r\n", hVM, tcb);
    count = RO_WriteFile(FALSE, fh, buf, count, filepos, &err);
    filepos += count;
//-----------end section added to prototype
    return TRUE;
}

VOID OnDestroyVm(VMHANDLE hVM)
{
//-----------begin section added to prototype
    WORD    err;
    int     count=0;
    char    buf[80];
    PTCB    tcb;

    tcb = Get_Initial_Thread_Handle(hVM);
    dprintf(buf, "Destroy_VM: VM=%x tcb=%x\r\n", hVM, tcb );
    sprintf(buf, "Destroy_VM: VM=%x tcb=%x\r\n", hVM, tcb );
    count = RO_WriteFile(FALSE, fh, buf, count, filepos, &err);
    filepos += count;
//-----------end section added to prototype
}

BOOL OnCreateThread(PTCB tcb)
{
//-----------begin section added to prototype
    WORD    err;
    int     count=0;
    char    buf[80];

    dprintf(buf, "Create_Thread: VM=%x, tcb=%x\r\n", tcb->TCB_VMHandle, tcb);
    sprintf(buf, "Create_Thread: VM=%x, tcb=%x\r\n", tcb->TCB_VMHandle, tcb);
    count = RO_WriteFile(FALSE, fh, buf, count, filepos, &err);
    filepos += count;
//-----------end section added to prototype
    return TRUE;
}
```

```
VOID OnDestroyThread(PTCB tcb)
{
//----------begin section added to prototype
    WORD    err;
    int     count=0;
    char    buf[80];

    dprintf( buf, "Destroy_Thread VM=%x, tcb=%x\r\n", tcb->TCB_VMHandle, tcb );
    sprintf( buf, "Destroy_Thread VM=%x, tcb=%x\r\n", tcb->TCB_VMHandle, tcb );
    count = RO_WriteFile(FALSE, fh, buf, count, filepos, &err);
    filepos += count;
//----------end section added to prototype
}
```

This code looks similar to the DDK-only version of SKELETON.C. In fact, the individual message handler functions are almost indistinguishable from their DDK-only counterparts. The VtoolsD version uses dprintf and sprintf, whereas the DDK version uses the DPRINTF macro and _Sprintf. The VToolsD version of IFSMgr services is slightly different, using RO_ instead of IFSMgr_Ring0_.

The advantage of VToolsD is not in the C code you write in your VxD. The advantage is that you don't have to write anything *other* than C code. No assembly language modules are required. The DDB and Device Control Procedure are generated by the wizard and placed in the same C file, relying on a number of clever macros in the VToolsD header files to produce a mixture of C and embedded assembly. More importantly, the VToolsD library contains *all* of the VMM and standard VxD services. With VToolsD you will probably never have to write a service wrapper in assembly. VToolsD also throws in most ANSI C run-time functions, including sprintf.

A Windows 3.x Version of SKELETON

Structurally, VxDs for Windows 3.x are the same as Windows 95 VxDs. However, Windows 95 contains a number of new VMM services and a number of new VxD services not available under Windows 3.x. A VxD that doesn't use any Windows 95-specific services will run unchanged under Windows 95. But a VxD that uses Windows 95-specific services will cause an "Unsupported Service" run-time error when run under Windows 3.x.

SKELETON.VXD does use a number of Windows 95-specific services. Windows 3.x doesn't have threads, so a Windows 3.x version of SKELETON wouldn't contain the two thread message handlers or any calls to Get_Initial_Thread_Handler in the VM message handlers. Windows 3.x doesn't have an IFSMgr VxD either, so file I/O must be done in V86 mode using Exec_VxD_Int instead of at Ring 0 with IFSMgr. (Exec_VxD_Int will be covered in Chapter 12.) Last, the VMM in Windows 3.x doesn't offer the _Sprintf service, so formatted output of message strings would have to be done in the SKELETON VxD itself (unless you use VToolsD, which provides sprintf in the run-time library). This VxD doesn't really need the full power offered by _Sprintf, and a simpler method that converts a DWORD to a hex string could be used instead.

VToolsD sells a version of their toolkit for Windows 3.x, and if you plan to write a Windows 3.x VxD in C, you need VToolsD. As this chapter showed, writing a Windows 95 VxD in C without VToolsD is possible but painful. Writing VxD in C for Windows 3.x without VToolsDs is more than painful. It's so much trouble that you might as well stick to assembly.

If you do choose Windows 3.x and C without VToolsD, here's what you're up against. While the Windows 95 DDK provided only a partial VMM wrapper library, the Windows 3.x DDK doesn't provide one at all. This means each and every VMM or VxD service called by your VxD will require you to write an assembly language wrapper function and to create an appropriate function prototype in your own VMM header file. Also, the Windows 3.x DDK doesn't provide a VMM.H, so you'll have to use the one from the Windows 95 DDK, being careful not to use any services not present in Windows 3.x.

Summary

Even with its limited functionality, SKELETON.VXD illustrates many critical issues of VxD development, requiring correct use of structure, interface, and tools. Using the wrappers supplied by WRAPPERS.CLB, you can code most of a Windows 95 VxD directly in C, even if you have only the DDK development tools. VToolsD makes the process even easier by supplying a more complete set of wrappers and convenient access to Ring 0 versions of many standard library functions. If you are developing a Windows 3.x VxD, VToolsD is more than a convenience. Whichever tool set you are using, you are now ready to write a VxD that does something — one that actually manipulates the hardware. The next chapter will explain the issues involved in manipulating basic hardware resources from Ring 0 code.

Listing 5.1   *SKELETON.C* **(DDK-only version)**

```
#include <basedef.h>
#include <vmm.h>
#include <debug.h>
#include "vxdcall.h"
#include <vxdwraps.h>
#include <wrappers.h>
#include "intrinsi.h"

#ifdef DEBUG
#define DPRINTF(buf, fmt, arg1, arg2) _Sprintf(buf, fmt, arg1, arg2 ); \
        Out_Debug_String( buf )
#else
#define DPRINTF(buf, fmt, arg1, arg2)
#endif

typedef struct tcb_s *PTCB;

BOOL OnSysVmInit(VMHANDLE hVM);
VOID OnSysVmTerminate(VMHANDLE hVM);
BOOL OnCreateVm(VMHANDLE hVM);
VOID OnDestroyVm(VMHANDLE hVM);
BOOL OnCreateThread(PTCB hThread);
VOID OnDestroyThread(PTCB hThread);

#pragma VxD_LOCKED_DATA_SEG

DWORD filepos = 0;
HANDLE fh;
char  buf[80];

#pragma VxD_LOCKED_CODE_SEG


BOOL OnSysVmInit(VMHANDLE hVM)
{
    BYTE   action;
    WORD   err;
    int       count=0;
    PTCB   tcb;

    tcb = Get_Initial_Thread_Handle(hVM);
    DPRINTF(buf, "SysVMInit: VM=%x tcb=%x\r\n", hVM, tcb );

    fh = IFSMgr_RingO_OpenCreateFile(FALSE, "vxdskel.log",
                         0x0002, 0x0000, 0x12, 0x00,
                         &err, &action);
    if (!fh)
    {
        DPRINTF(buf, "Error %x opening file %s\n", err, "vxdskel.log" );
    }
    else
    {
        _Sprintf(buf, "SysVMInit: VM=%x tcb=%x\r\n", hVM, tcb );
        count = IFSMgr_RingO_WriteFile(FALSE, fh, buf, strlen(buf), filepos, &err);
        filepos += count;
    }
    return TRUE;
}
```

Listing 5.1 (continued) *SKELETON.C* **(DDK-only version)**

```
VOID OnSysVmTerminate(VMHANDLE hVM)
{
    WORD  err;
    int   count=0;
    PTCB  tcb;

    tcb = Get_Initial_Thread_Handle(hVM);
    DPRINTF( buf, "SysVmTerminate VM=%x tcb=%x\r\n", hVM, tcb );
    _Sprintf( buf, "SysVmTerminate VM=%x tcb=%x\r\n", hVM, tcb );
    count = IFSMgr_Ring0_WriteFile(FALSE, fh, buf, strlen(buf), filepos, &err);
    filepos += count;
    IFSMgr_Ring0_CloseFile( fh, &err );
}

BOOL OnCreateVm(VMHANDLE hVM)
{
    PTCB  tcb;
    WORD  err;
    int   count=0;

    tcb = Get_Initial_Thread_Handle(hVM);
    DPRINTF(buf, "Create_VM: VM=%x, tcb=%x\r\n", hVM, tcb);
    _Sprintf(buf, "Create_VM: VM=%x, tcb=%x\r\n", hVM, tcb);
    count = IFSMgr_Ring0_WriteFile(FALSE, fh, buf, strlen(buf), filepos, &err);
    filepos += count;
    return TRUE;
}

VOID OnDestroyVm(VMHANDLE hVM)
{
    WORD  err;
    int   count;
    PTCB  tcb;

    tcb = Get_Initial_Thread_Handle(hVM);
    DPRINTF(buf, "Destroy_VM: VM=%x tcb=%x\r\n", hVM, tcb );
    _Sprintf(buf, "Destroy_VM: VM=%x tcb=%x\r\n", hVM, tcb );
    count = IFSMgr_Ring0_WriteFile(FALSE, fh, buf, strlen(buf), filepos, &err);
    filepos += count;
}

BOOL OnCreateThread(PTCB tcb)
{
    WORD  err;
    int       count;

    DPRINTF(buf, "Create_Thread: VM=%x, tcb=%x\r\n", tcb->TCB_VMHandle, tcb);
    _Sprintf(buf, "Create_Thread: VM=%x, tcb=%x\r\n", tcb->TCB_VMHandle, tcb);
    count = IFSMgr_Ring0_WriteFile(FALSE, fh, buf, strlen(buf), filepos, &err);
    filepos += count;
    return TRUE;
}

VOID OnDestroyThread(PTCB tcb)
{
    WORD  err;
    int   count;

    DPRINTF( buf, "Destroy_Thread VM=%x, tcb=%x\r\n", tcb->TCB_VMHandle, tcb );
    _Sprintf( buf, "Destroy_Thread VM=%x, tcb=%x\r\n", tcb->TCB_VMHandle, tcb );
    count = IFSMgr_Ring0_WriteFile(FALSE, fh, buf, strlen(buf), filepos, &err);
    filepos += count;
}
```

Listing 5.2   SKELCTRL.ASM *(DDK-only version)*

```
.386p

;****************************************************************************
;               I N C L U D E S
;****************************************************************************

    include vmm.inc
    include debug.inc

;==========================================================================
;        V I R T U A L   D E V I C E   D E C L A R A T I O N
;==========================================================================

DECLARE_VIRTUAL_DEVICE      SKELETON, 1, 0, ControlProc, UNDEFINED_DEVICE_ID, \
                            UNDEFINED_INIT_ORDER

VxD_LOCKED_CODE_SEG

;==========================================================================
;
;   PROCEDURE: ControlProc
;
;   DESCRIPTION:
;    Device control procedure for the SKELETON VxD
;
;   ENTRY:
;    EAX = Control call ID
;
;   EXIT:
;    If carry clear then
;        Successful
;    else
;        Control call failed
;
;   USES:
;    EAX, EBX, ECX, EDX, ESI, EDI, Flags
;
;==========================================================================

BeginProc ControlProc
    Control_Dispatch SYS_VM_INIT, _OnSysVmInit, cCall, <ebx>
    Control_Dispatch SYS_VM_TERMINATE, _OnSysVmTerminate, cCall, <ebx>
    Control_Dispatch CREATE_VM, _OnCreateVm, cCall, <ebx>
    Control_Dispatch DESTROY_VM, _OnDestroyVm, cCall, <ebx>
    Control_Dispatch CREATE_THREAD, _OnCreateThread, cCall, <edi>
    Control_Dispatch DESTROY_THREAD, _OnDestroyThread, cCall, <edi>

    clc
    ret

EndProc ControlProc

VxD_LOCKED_CODE_ENDS

    END
```

Listing 5.3 *SKELETON.MAK* **(DDK-only version)**

```
CVXDFLAGS = -Zdp -Gs -c -DIS_32 -Z1 -DDEBLEVEL=1 -DDEBUG -DWANTVXDWRAPS
AFLAGS    = -coff -DBLD_COFF -DIS_32 -W2 -Zd -c -Cx -DMASM6 -DDEBLEVEL=1 -DDEBUG

all: skeleton.vxd

skeleton.obj: skeleton.c
        cl $(CVXDFLAGS) -Fo$@ %s

skelctrl.obj: skelctrl.asm
        ml $(AFLAGS) -Fo$@ %s

skeleton.vxd: skelctrl.obj skeleton.obj ..\..\wrappers\vxdcall.obj
              ..\..\wrappers\wrappers.clb skeleton.def
        echo >NUL @<<skeleton.crf
-MACHINE:i386 -DEBUG -DEBUGTYPE:MAP -PDB:NONE
-DEF:skeleton.def -OUT:skeleton.vxd -MAP:skeleton.map
-VXD vxdwraps.clb wrappers.clb skelctrl.obj skeleton.obj vxdcall.obj
<<
        link @skeleton.crf
        mapsym skeleton
```

Listing 5.4 *SKELETON.DEF* **(DDK-only version)**

```
VXD SKELETON
SEGMENTS
    _LTEXT     CLASS 'LCODE'   PRELOAD NONDISCARDABLE
    _LDATA     CLASS 'LCODE'   PRELOAD NONDISCARDABLE
    _TEXT      CLASS 'LCODE'   PRELOAD NONDISCARDABLE
    _DATA      CLASS 'LCODE'   PRELOAD NONDISCARDABLE
    _LPTEXT    CLASS 'LCODE'   PRELOAD NONDISCARDABLE
    _CONST     CLASS 'LCODE'   PRELOAD NONDISCARDABLE
    _BSS       CLASS 'LCODE'   PRELOAD NONDISCARDABLE
    _TLS       CLASS 'LCODE'   PRELOAD NONDISCARDABLE
    _ITEXT     CLASS 'ICODE'   DISCARDABLE
    _IDATA     CLASS 'ICODE'   DISCARDABLE
    _PTEXT     CLASS 'PCODE'   NONDISCARDABLE
    _PDATA     CLASS 'PCODE'   NONDISCARDABLE
    _STEXT     CLASS 'SCODE'   RESIDENT
    _SDATA     CLASS 'SCODE'   RESIDENT
    _MSGTABLE  CLASS 'MCODE'   PRELOAD NONDISCARDABLE IOPL
    _MSGDATA   CLASS 'MCODE'   PRELOAD NONDISCARDABLE IOPL
    _IMSGTABLE CLASS 'MCODE'   PRELOAD DISCARDABLE IOPL
    _IMSGDATA  CLASS 'MCODE'   PRELOAD DISCARDABLE IOPL
    _DBOSTART  CLASS 'DBOCODE' PRELOAD NONDISCARDABLE CONFORMING
    _DBOCODE   CLASS 'DBOCODE' PRELOAD NONDISCARDABLE CONFORMING
    _DBODATA   CLASS 'DBOCODE' PRELOAD NONDISCARDABLE CONFORMING
    _16ICODE   CLASS '16ICODE' PRELOAD DISCARDABLE
    _RCODE     CLASS 'RCODE'
EXPORTS
    SKELETON_DDB @1
```

Listing 5.5   *SKELETON.C (VToolsD version)*

```
// SKELETON.c - main module for VxD SKELETON          e

#define   DEVICE_MAIN
#include  "skeleton.h"
#undef    DEVICE_MAIN

DWORD va_arg_list[2];
DWORD filepos = 0;
HANDLE fh;

Declare_Virtual_Device(SKELETON)


DefineControlHandler(SYS_VM_INIT, OnSysVmInit);
DefineControlHandler(SYS_VM_TERMINATE, OnSysVmTerminate);
DefineControlHandler(CREATE_VM, OnCreateVm);
DefineControlHandler(DESTROY_VM, OnDestroyVm);
DefineControlHandler(CREATE_THREAD, OnCreateThread);
DefineControlHandler(DESTROY_THREAD, OnDestroyThread);

BOOL __cdecl ControlDispatcher(
    DWORD dwControlMessage,
    DWORD EBX,
    DWORD EDX,
    DWORD ESI,
    DWORD EDI,
    DWORD ECX)
{
    START_CONTROL_DISPATCH

        ON_SYS_VM_INIT(OnSysVmInit);
        ON_SYS_VM_TERMINATE(OnSysVmTerminate);
        ON_CREATE_VM(OnCreateVm);
        ON_DESTROY_VM(OnDestroyVm);
        ON_CREATE_THREAD(OnCreateThread);
        ON_DESTROY_THREAD(OnDestroyThread);

    END_CONTROL_DISPATCH

    return TRUE;
}
```

Listing 5.5 (continued) *SKELETON.C* **(VToolsD version)**

```c
BOOL OnSysVmInit(VMHANDLE hVM)
{
    BYTE    action;
    WORD    err;
    int     count;
    char    buf[80];
    PTCB    tcb;

    tcb = Get_Initial_Thread_Handle(hVM);
    dprintf("SysVmInit: VM=%x, tcb=%x\r\n", hVM, tcb);

     fh = RO_OpenCreateFile(FALSE, "vxdskel.log", 0x0002, 0x0000, 0x12, 0x00,
                            &err, &action);
    if (!fh)
        dprintf("Error %x opening file\n", err );
    else
    {
        count = sprintf(buf, "SysVmInit: VM=%x tcb=%x\r\n", hVM, tcb );
        if (count)
        {
            count = RO_WriteFile(FALSE, fh, buf, count, filepos, &err);
            filepos += count;
        }
    }
    return TRUE;
}

VOID OnSysVmTerminate(VMHANDLE hVM)
{
    WORD    err;
    int     count;
    char    buf[80];
    PTCB    tcb;

    tcb = Get_Initial_Thread_Handle(hVM);
    dprintf("SysVmTerminate: VM=%x, tcb=%x\r\n", hVM, tcb);
    count = sprintf(buf, "SysVmTerminate: VM=%x tcb=%x\r\n", hVM, tcb );
    if (count)
    {
        count = RO_WriteFile(FALSE, fh, buf, count, filepos, &err);
        filepos += count;
    }

    RO_CloseFile( fh, &err );
}
```

Listing 5.5 (continued)   *SKELETON.C* **(VToolsD version)**

```c
BOOL OnCreateVm(VMHANDLE hVM)
{
    WORD    err;
    PTCB    tcb;
    int     count;
    char    buf[80];

    tcb = Get_Initial_Thread_Handle(hVM);
    dprintf("Create_VM: VM=%x, tcb=%x\r\n", hVM, tcb);

    count = sprintf(buf, "Create_VM: VM=%x, tcb=%x\r\n", hVM, tcb );
    if (count)
    {
        count = RO_WriteFile(FALSE, fh, buf, count, filepos, &err);
        filepos += count;
    }
    return TRUE;
}

VOID OnDestroyVm(VMHANDLE hVM)
{
    char    buf[80];
    int     count;
    WORD    err;

    dprintf("Destroy_VM: VM=%x tcb=%x\n", hVM );
    count = sprintf(buf, "Destroy_VM: VM=%x\r\n", hVM );
    if (count)
    {
        count = RO_WriteFile(FALSE, fh, buf, count, filepos, &err);
        filepos += count;
    }
}

BOOL OnCreateThread(THREADHANDLE hThread)
{
    PTCB    tcb = (PTCB)hThread;
    char    buf[80];
    int     count;
    WORD    err;

    dprintf("Create_Thread: VM=%x, tcb=%x\r\n", tcb->TCB_VMHandle, tcb);

    count = sprintf(buf, "Create_Thread: VM=%x, tcb=%x\r\n", tcb->TCB_VMHandle, tcb );
    if (count)
    {
        count = RO_WriteFile(FALSE, fh, buf, count, filepos, &err);
        filepos += count;
    }
    return TRUE;
}
```

Listing 5.5 (continued)   `SKELETON.C` **(VToolsD version)**

```
VOID OnDestroyThread(THREADHANDLE hThread)
{
    PTCB    tcb = (PTCB)hThread;
    char    buf[80];
    int     count;
    WORD    err;

    dprintf("Destroy_Thread: VM=%x, tcb=%x\r\n", tcb->TCB_VMHandle, tcb);

    count = sprintf(buf, "Destroy_Thread: VM=%x, tcb=%x\r\n", tcb->TCB_VMHandle, tcb );
    if (count)
    {
        count = RO_WriteFile(FALSE, fh, buf, count, filepos, &err);
        filepos += count;
    }
}
```

Listing 5.6   `SKELETON.H` **(VToolsD version)**

```
// SKELETON.h - include file for VxD SKELETON

#include <vtoolsc.h>

#define SKELETON_Major          1
#define SKELETON_Minor          0
#define SKELETON_DeviceID       UNDEFINED_DEVICE_ID
#define SKELETON_Init_Order     UNDEFINED_INIT_ORDER
```

Listing 5.7   `SKELETON.MAK` **(VToolsD version)**

```
# SKELETON.mak - makefile for VxD SKELETON

DEVICENAME = SKELETON
FRAMEWORK = C
DEBUG = 1
OBJECTS = skeleton.OBJ

!include $(VTOOLSD)\include\vtoolsd.mak
!include $(VTOOLSD)\include\vxdtarg.mak

skeleton.OBJ:    skeleton.c skeleton.
```

Listing 5.8   *SKELETON.DEF* **(VToolsD version)**

```
VXD SKELETON
SEGMENTS
    _LTEXT      CLASS 'LCODE'   PRELOAD NONDISCARDABLE
    _LDATA      CLASS 'LCODE'   PRELOAD NONDISCARDABLE
    _TEXT       CLASS 'LCODE'   PRELOAD NONDISCARDABLE
    _DATA       CLASS 'LCODE'   PRELOAD NONDISCARDABLE
    _LPTEXT     CLASS 'LCODE'   PRELOAD NONDISCARDABLE
    _CONST      CLASS 'LCODE'   PRELOAD NONDISCARDABLE
    _BSS        CLASS 'LCODE'   PRELOAD NONDISCARDABLE
    _TLS        CLASS 'LCODE'   PRELOAD NONDISCARDABLE
    _ITEXT      CLASS 'ICODE'   DISCARDABLE
    _IDATA      CLASS 'ICODE'   DISCARDABLE
    _PTEXT      CLASS 'PCODE'   NONDISCARDABLE
    _PDATA      CLASS 'PCODE'   NONDISCARDABLE
    _STEXT      CLASS 'SCODE'   RESIDENT
    _SDATA      CLASS 'SCODE'   RESIDENT
    _MSGTABLE   CLASS 'MCODE'   PRELOAD NONDISCARDABLE IOPL
    _MSGDATA    CLASS 'MCODE'   PRELOAD NONDISCARDABLE IOPL
    _IMSGTABLE  CLASS 'MCODE'   PRELOAD DISCARDABLE IOPL
    _IMSGDATA   CLASS 'MCODE'   PRELOAD DISCARDABLE IOPL
    _DBOSTART   CLASS 'DBOCODE' PRELOAD NONDISCARDABLE CONFORMING
    _DBOCODE    CLASS 'DBOCODE' PRELOAD NONDISCARDABLE CONFORMING
    _DBODATA    CLASS 'DBOCODE' PRELOAD NONDISCARDABLE CONFORMING
    _16ICODE    CLASS '16ICODE' PRELOAD DISCARDABLE
    _RCODE      CLASS 'RCODE'
EXPORTS
    _The_DDB @1
```

Chapter 6

VxD Talks to Hardware

The last two chapters introduced the basic structure of a VxD and demonstrated a skeleton VxD that processed a few messages and did some debug output. In this chapter, I'll show you how a VxD communicates with a hardware device. This chapter will cover talking to I/O-mapped, memory-mapped, and DMA/bus-master devices. I'll save a related subject, interrupt handling, for the next chapter.

I/O-mapped versus Memory-mapped

A hardware device on a PC can be located in one of two separate address spaces: memory or I/O. A device in the memory address space, called memory-mapped, is accessed exactly like memory. It can be accessed via any of the many instructions that take a memory reference, such as MOV, ADD, OR, etc. From a high-level language, memory-mapped devices are accessed through a pointer. By contrast, a device in I/O address space (I/O-mapped) can be accessed with only a few instructions: IN, OUT, and their derivatives. There is no high-level language construct for an I/O-mapped device, although many compilers do add support via run-time library functions like inp and outp.

Another difference between the two address spaces is that I/O address space is much smaller than memory space. While the 80386 and above processors support a 4Gb memory address space, I/O address space is only 64Kb on all 80x86 processors.

Talking to an I/O-mapped Device

To communicate with an I/O-mapped device, a VxD directly executes the appropriate IN or OUT instructions or their high-level language equivalents. The processor won't trap a VxD that executes these instructions, because VxDs run at Ring 0.

If you're writing in assembly, use an IN or OUT instruction with an appropriately sized operand. For example, this code fragment writes the byte A5h to the port 300h.

```
MOV AL, 0A5h
MOV DX, 300h
OUT DX, AL
```

If you're writing in C, it's easiest to use the C run-time equivalents of IN and OUT — as long as these functions are supported by your compiler and you make sure the compiler uses the intrinsic, or inline, version. When asked to use the "intrinsic" form of a function, the compiler inserts actual IN and OUT instructions instead of making a call to the run-time library. It is important to avoid calling the library version, because few vendors supply a VxD-callable run-time library. (VToolsD is an exception).

Microsoft's 32-bit compilers support _inp and _outp for byte access, _inpw and _outpw for word access. You can force the compiler to generate intrinsic code instead of a call to the run-time by using either the -Oi compiler flag, or by using the intrinsic pragma in your code. The following code fragment writes the byte 0 to the port 300h, and uses the intrinsic pragma to guarantee inline code:

```
#pragma intrinsic(_outp)
_outp( 0x300, 0 );
```

Borland doesn't support IN and OUT equivalents (called _inpb, _inpw, _outpb, and _outpw) when generating 32-bit code. If you develop your VxD with Borland's compiler, you should use embedded assembly for input and output operations.

The run-time functions listed above are only for byte- and word-sized port accesses. Some devices, particularly recent PCI devices, support dword-sized (32-bit) accesses (IN EAX, DX and OUT DX, EAX). Neither Microsoft nor Borland provides a run-time equivalent for the dword version of IN and OUT instructions. To exploit dword I/O operations, you'll have to use embedded assembly.

By default, the Borland compiler uses its built-in assembler to translate embedded assembly, which contains only 16-bit instructions. The compiler will automatically call the stand-alone TASM32 assembler if the embedded assembly contains any 32-bit instructions. Therefore, to use IN EAX, DX and OUT DX, EAX, you must have TASM32.EXE.

Talking to a Memory-mapped Device

To access a memory-mapped device, a VxD must manipulate a specific physical address. Unfortunately, manipulating an address in the memory system's physical address space isn't as straightforward as manipulating a port in the I/O system's port address space. Even though VxDs run at Ring 0 (where they see a flat memory model), they still manipulate only linear addresses; all VxD memory accesses go through the page tables for linear-to-physical address translation. Thus, before a VxD can access a particular memory-mapped device, it must configure the page tables to assure that the device's physical address corresponds to a linear address. The resulting linear address may then be used as a "plain old pointer". Even though this mapping information comes from the page tables, a VxD should never directly manipulate the page tables; it should use VMM services instead.

This procedure takes one of two forms, depending upon whether the device can be dynamically reconfigured. Older ISA devices with jumper-selected addresses are "statically configured"; they are guaranteed to reside at the same address for the life of the Windows session. Many modern devices, however, can be reconfigured at run time (for example, PCI, PCMCIA, and ISAPNP devices). As part of its support for Plug and Play, Windows 95 may move these dynamically configurable devices around (both in I/O and memory space) as the devices are started and stopped. (See Chapters 9 and 10 for a full discussion of Windows 95 Plug and Play.) A Windows 95 VxD obtains a linear address for a statically configured device with a single call to MapPhysToLinear — just as in Windows 3.x. If the device is run-time configurable, though, the process is more complicated and not Windows 3.x compatible. I'll discuss each situation separately.

Statically Configured Memory-mapped Devices

A VxD for an ISA device can obtain a linear address for its device by calling _MapPhysToLinear. Given a physical address and the region size, _MapPhysToLinear returns a linear address that maps to that physical address region. (Both VToolsD and the DDK C wrapper libraries contain a wrapper for _MapPhysToLinear.)

The Calling Interface for _MapPhysToLinear

```
DWORD _MapPhysToLinear(DWORD PhysAddr, DWORD nBytes, DWORD Flags);
PhysAddr: physical address to be mapped
nBytes: size of region to be mapped, in bytes
Flags: must be zero
Returns: linear address of region; this linear address is in the
         system arena, and so is valid no matter which Win32 process
         is current
```

In the following code fragment, the VxD accesses a device mapped to the 256Kb region starting at 16Mb:

```
BYTE *lin;
DWORD phys, size;

phys = 0x01000000L;     // 16Mb
size = 256*1024;        // 256Kb
lin = (BYTE *)_MapPhysToLinear(phys, size, 0L) ;
*lin = 0xA5;            // write out to device
if (*lin != 0xA5)      // read back from same location
    return 0;          // error
```

Dynamically Configurable Devices

A VxD for a Plug and Play device shouldn't use _MapPhysToLinear because a Plug and Play device may change its physical address while Windows 95 is running. Moreover, because the linear address returned by _MapPhysToLinear maps to the same physical address for the life of Windows, calling the service multiple times would waste page table entries. The VMM provides no "unmap" service.

Instead of calling _MapPhysToLinear, a VxD for a device with a dynamically reconfigurable physical address must divide the "map" process into multiple steps. Each step calls a VMM service that can be reversed:

- _PageReserve, to allocate a block of linear address space. This is really a set of page table entries.
- _PageCommitPhys, to map the linear address range to the device's physical address (by setting the physical address field of the allocated page table entries).
- _LinPageLock, to prevent the Virtual Memory Manager from swapping out the pages, thus making the linear address usable during interrupt time.

Here's a function, MyMapPhysToLinear, which performs this three-step mapping and returns a linear address:

```
DWORD MyMapPhysToLinear( DWORD phys, DWORD size )
{
    DWORD lin;
    DWORD nPages = size / 4096;

    lin = _PageReserve( PR_SYSTEM, nPages, 0 );
    if (lin == -1)
     return 0;
    if (!_PageCommitPhys( lin, nPages, phys, PC_INCR | PC_WRITEABLE))
     return 0;
    if (!_LinPageLock( lin, nPages, 0 ))
     return 0;
}
```

This function uses the _PageReserve, _PageCommitPhys, and _LinPageLock services. Let's examine each call in detail.

MyMapPhysToLinear passes PR_SYSTEM into ipage when calling _PageReserve (see next page), so that the linear address is valid for any address context, regardless of the current Win32 process and current VM. (Note that PR_SHARED would have the same effect, since that arena is also valid for all processes and VMs.) MyMapPhysToLinear doesn't use any of the predefined values for flags. If the call to _PageReserve fails, MyMapPhysToLinear immediately returns zero (failure) to its caller.

The Calling Interface for _PageReserve

```
PVOID _PageReserve (DWORD ipage, DWORD npages, DWORD flags) ;
ipage: determines which arena the linear address will be in
       PR_PRIVATE to allocate the linear address in the private arena
       PR_SHARED to allocate the linear address in the shared arena
       PR_SYSTEM to allocate the linear address in the system arena
nPages: number of pages to allocate
flags: PR_FIXED prevents PageReallocate from moving pages
       PR_STATIC forces future calls to commit, decommit and frees
           this linear address to also specify PR_STATIC
       PR_4MEG forces linear address on a 4Mb boundary
```

MyMapPhysToLinear passes the linear page number given by _PageReserve (linear address shifted right by 12) and the caller's physical address to _PageCommitPhys. Calling with the PC_INCR flag causes an "incremental" wrapping: i.e. the first page in the linear address range maps the first page of the physical region; the second page in the linear address range maps the second page of the physical region; etc.

The Calling Interface for _PageCommitPhys

```
BOOL _PageCommitPhys(DWORD ipage, DWORD npages, DWORD physpage, DWORD flags);
ipage: first page number of linear range to be mapped
npages: number of pages to commit
physpage: first physical address to be mapped, as a page number (linear >> 12)
flags: PC_INCR maps linear pages to successive contiguous physical pages
       (if not set, all linear pages in range are mapped to same physpage)
       PC_USER marks all pages as accessible to Ring 3
       PC_WRITEABLE marks all pages as writeable (else write will page-fault)
```

Finally, the same linear page and number of pages is passed to _LinPageLock. No flags are specified because the pages should be locked regardless of the type of swap device; the device doesn't use DMA (it's memory-mapped instead), and the linear address was already allocated from the system arena.

The Calling Interface for _LinPageLock

```
DWORD _LinPageLock(DWORD LinPgNum, DWORD nPages, DWORD Flags);
LinPgNum: page to lock
nPages: number of pages to lock
Flags: PAGELOCKEDIFDP locks pages only if swap device uses DOS or BIOS services
       PAGEMAPGLOBAL returns an alias linear address in the system arena so
           region can be accessed regardless of current context
       PAGEDIRTY marks dirty bit in page table entry. Use if DMA device will
           be writing to pages, because processor won't know pages are dirty
```

To undo the mapping, call _LinPageUnlock, _PageDecommit and _PageFree. Each of these calls undoes the work of its counterpart which was called earlier. That's all I will say about these services, because the parameters are all self-explanatory and no special flags are required.

Another Data Transfer Method: DMA

When the CPU transfers individual bytes to a device through an I/O port or a memory location, the processor must fetch one or more instructions and generate target addresses for every single byte of data transferred to or from the device. An alternative method, DMA (Direct Memory Access), can significantly reduce bus traffic during a transfer. In a DMA transfer, the device itself takes over the bus from the processor and transfers the data, eliminating the instruction fetches associated with a CPU-driven transfer.

There are two types of DMA: system DMA and bus-master DMA. In system DMA, the system DMA controller (every PC has two of these) and the device work together to take over the bus from the processor and transfer the data. The Sound-Blaster card is the best known system DMA device. In bus-master DMA, the device itself acts as "master" of the bus, requiring no help from the system DMA controller or the processor. Bus-master DMA is common for PCI devices.

A device that uses DMA as its data transfer method still needs I/O-mapped or memory-mapped control ports. By writing to the control ports, the processor can tell the device where to find the system memory buffer, how large the buffer is, and when to start the transfer. A buffer to be used in a DMA transaction must meet a number of allocation requirements, which I'll explain in detail later in this chapter.

VxDs that use DMA — either system or bus-master — for their data transfer method should use the services provided by the VDMAD. The VDMAD (Virtual DMA Controller) does more than virtualize the PC's two system DMA controllers. It also provides services useful to VxDs that are performing system DMA or bus-master transfers.

Using System DMA

For system DMA, a VxD uses VDMAD services to claim usage of one of the seven DMA channels supported by the PC and to request that VDMAD issue appropriate instructions to the DMA controller. Note that because VxDs run at Ring 0, there is nothing to prevent a VxD from interacting with the controller directly. However, doing so could interfere with DMA transfers on other channels, because of the way the registers on the DMA controller are laid out. (Specifically, because there is only a single mode and a single mask register, not one set for each channel, so the VDMAD must be aware of all reads and writes to/from the controller in order to correctly virtualize DMA transfers.)

In addition to using VDMAD services to program the controller, a VxD must also allocate a buffer suitable for DMA and obtain the buffer's physical address. A buffer used for a system DMA transfer must meet several strict requirements. The DMA buffer must be

- physically contiguous,
- fixed and pagelocked,
- aligned on a 64Kb boundary, and
- located below 16Mb in physical memory.

These requirements are necessary because the system DMA controller has no knowledge of linear addresses or pages and performs no address translation. The controller is programmed with a starting physical address and simply increments (or decrements) that address with each byte transferred in order to generate the next physical address.

System DMA Buffer Requirements

The buffer must be physically contiguous because the processor views the linear address space as a series of 4Kb pages. Through the page tables, each 4Kb page can be mapped to a different location in physical address space. A buffer made up of pages that map to noncontiguous physical addresses won't work for DMA, because the DMA controller can only increment through a series of physical addresses (or decrement through a decreasing series).

To understand the requirement for fixed and pagelocked memory, consider the situation illustrated in Figure 6.1. The VxD, through VDMAD services, initializes the DMA controller with the physical address of the desired buffer and instructs the controller to begin the transfer. The controller transfers a byte to physical memory, and the processor regains control of the bus. Assuming that the buffer's pages were not fixed, the virtual memory manager may then decide to move a page — the one being used in the transfer — by copying the page contents to another location in physical memory and then updating the page's linear address in the page tables.

At some later time the DMA controller steals the bus again and continues the transfer, using the original physical address programmed during the initialization process (plus one for each byte already transferred). When the transfer completes, the VxD examines the new data using the same linear address, but the expected data is not at that linear address, because that linear address maps to a new physical address. The DMA controller stored the data at the location given by the original physical address.

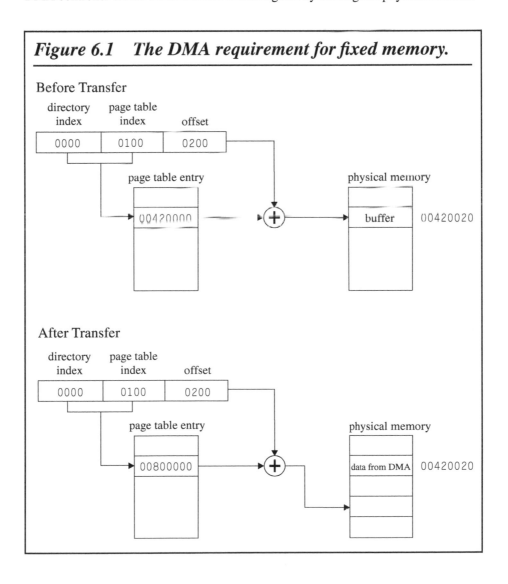

Figure 6.1   The DMA requirement for fixed memory.

Pagelocking the buffer also prevents a similar problem, where the memory manager swaps the page contents out to disk during the transfer. The DMA controller continues to store data at the original physical address. But when the VxD accesses the page after the transfer, expecting to see the new data, the memory manager swaps the page contents in from disk instead, and the VxD sees "old" data.

The two requirements for physical location below 16Mb and 64Kb alignment have nothing to do with either Windows or the processor but are a limitation of the PC architecture. The original PC used 20-bit physical addresses, but the PC's DMA controller chip had only a 16-bit address register. To make it possible to perform DMA transfers anywhere within the entire 1Mb of the PC's address space, the PC system designers added a page register external to the DMA controller to store the upper four bits of the address. They also added extra logic so that the page register put these upper 4 bits onto the address bus at the same time the DMA controller placed its 16 bits on the bus, forming a full 20-bit address for main memory.

When the PC-AT was introduced, the page registers grew to 8 bits, and again extra logic made those 8 bits appear on the address bus when the DMA controller placed the lower 16 bits on the bus. To remain compatible, today's system designers still use this 24-bit DMA scheme even though processors have a 32-bit bus. One side effect of this decision is that system DMA can only occur in the lowest 16Mb (24 bits) of memory.

How is this relevant to the 64Kb boundary requirement? Suppose you want to perform a DMA transfer of 1000h bytes to physical address 6F000h. To do this, you write the lower 16 bits (F000h) into the DMA controller's address register and the upper 4 bits (6h) to the proper page register (there is one per DMA channel). You also set the controller for a transfer count of 1000h bytes. The physical address of the very last byte is 70000h (6F000h + 01000h). But the physical address generated when the last byte is transferred is actually 60000h. The DMA controller address register correctly rolls over from FFFFh to 0000h, but the page register containing the upper 4 bits doesn't increment from 6h to 7h. Therefore, all system DMA transfers must stay on the same 64Kb "page".

A Function for Allocating a System DMA Buffer

Although the VMM provides a number of different types of memory allocation services for VxDs to use, only one will meet the requirements for a system DMA buffer. That service is _PageAllocate. (Note that _PageAllocate is one of the VMM services provided in the Windows 95 DDK VMM wrapper library.)

The Calling Interface for _PageAllocate

```
ULONG _PageAllocate(ULONG nPages, ULONG pType, ULONG VM, ULONG AlignMask,
                    ULONG minPhys, ULONG maxPhys, ULONG *PhysAddr,
                    ULONG flags);
nPages: number of 4Kb pages
pType: PG_VM (specific to VM)
       PG_SYS (valid for all VMs)
       PG_HOOKED (same as PG_VM, hold-over from Win3.x)
VM: handle of VM or zero if PG_SYS
AlignMask: used if PAGEUSEALIGN bit in Flags is set
           00h forces 4Kb,
           0Fh forces 64Kb alignment,
           1Fh forces 128Kb alignment
minPhys: minimum acceptable physical page
maxPhys: maximum acceptable physical page
*PhysAddr: pointer to DWORD where physical address will be returned
flags: zero or more of the following bits
       PAGEZEROINIT (pages are filled with zeroes)
       PAGELOCKED (pages are locked, can be unlocked with _PageUnLock)
       PAGELOCKEDIFDP (locks pages only if the virtual page swap device uses
                       MS-DOS or BIOS functions to write to the hardware)
       PAGEFIXED (pages are locked at fixed linear address,
                  can't be unlocked or moved)
       PAGEUSEALIGN: pages allocated meet AlignMask, minPhys and maxPhys
                     restrictions; ignored unless PAGEFIXED also set
       PAGECONTIG: pages allocated are physically contiguous; ignored unless
                   PAGEUSEALIGN is also set
       Note: unless one of PAGELOCKED, PAGELOCKEDIFDP or PAGEFIXED is set,
       no physical pages are allocated, only linear pages
Returns: linear address of buffer
```

The function `AllocSysDmaBuffer` (contained in the file `DMAALLOC.C` in the `\DMAALLOC` directory of the code disk) uses `_PageAllocate` with the appropriate parameters to allocate a system DMA buffer.

```
DWORD AllocSysDmaBuffer( DWORD nPages, DWORD *pPhysAddr )
{
    return _PageAllocate(nPages, PG_SYS, 0, 0x0F, 0, 0x1000,
                         pPhysAddr, PAGEFIXED | PAGEUSEALIGN |
                         PAGECONTIG );
}
```

PG_SYS allows the VxD to access the buffer at hardware interrupt time, regardless of which VM is currently executing at the time of the interrupt. The AlignMask, minPhys, and maxPhys parameters, combined with the PAGEUSEALIGN flag bit, correspond exactly to the "64Kb alignment" and "below 16Mb" requirements. (Note that the maxPhys parameter is <u>not</u> 16Mb, but 16Mb/4Kb, which is the physical address expressed as a page number.) The PAGEFIXED flag meets the fixed and pagelocked requirement. The function return value is the buffer's linear address, and the physical address is returned at PhysAddr.

In general, a VxD should only pagelock a buffer when it's absolutely necessary — in this case, only for the duration of the DMA transfer. But because of the way _PageAllocate uses the Flag parameter, the physical contiguity requirement forces AllocSysDmaBuffer to allocate a buffer that is permanently fixed and pagelocked.

A VxD cannot use a buffer allocated by a Win32 application for system DMA, because there is no way to force that buffer to meet physical contiguity and alignment requirements.

Overview of VDMAD Services

After allocating a system DMA buffer from the VMM, the VxD uses VDMAD services to program the DMA controller. The standard documentation explains the individual VDMAD services well enough, but fails to outline the overall sequence of services used to perform a transfer. Here is a summary of the overall sequence.

Before the first transfer, the VxD calls

- VDMAD_Virtualize_Channel to reserve the channel and obtain a DMA "handle" used in calls to other VDMAD services.

Then, for every transfer, the VxD calls

- VDMAD_Set_Region_Info to program the system DMA controller with the buffer's physical address and size,
- VDMAD_Set_Phys_State to program the system DMA controller's mode, and
- VDMAD_Phys_Unmask_Channel to unmask the channel on the system DMA controller.

The VMM/VxD library included with VToolsD provides C-callable wrappers for all VDMAD services. The Windows 95 DDK wrapper library doesn't have the necessary wrappers, but WRAPPERS.CLB does include all VDMAD services discussed in this chapter.

VDMAD Services in Detail

VDMAD_Virtualize_Channel can be used to virtualize a channel. If you pass in a non-null callback parameter, VDMAD will call your VxD back whenever Ring 3 code changes the state of your channel by accessing one of the DMA controller registers. By responding to this callback, your VxD can virtualize the channel itself. Or you can pass NULL for the callback parameter to tell VDMAD you're not really virtualizing the channel, you only want the DMA "handle" returned by the service, which you need for other VDMAD calls.

The Calling Interface for VDMAD_Virtualize_Channel

```
HANDLE VDMAD_Virtualize_Channel(DWORD Channel, PVOID CallbackProc );
Channel: DMA channel to virtualize/use, 0-7
CallbackProc: called to notify of Ring 3 access to DMA controller
Returns: DMA handle to be used in calls to other VDMAD services
         or zero if fail
```

The next call is VDMAD_Set_Region_Info, where "region" refers to the DMA buffer. The DMAHandle is, of course, the one returned by VDMAD_Virtualize_Channel. The BufferId parameter should be zero if you've allocated your own buffer (otherwise it refers to the buffer ID returned by the service VDMAD_Request_Buffer). The documentation says that the LockStatus parameter should be "zero if not locked, non-zero if locked". If this parameter is zero, VDMAD will send a warning message to the debugger during the next step (VDMAD_Set_Phys_State) — a gentle reminder that you probably forgot to lock. The Region parameter, containing the buffer's linear address, and the PhysAddr parameter are both provided by the initial call to _PageAllocate.

The Calling Interface for VDMAD_Set_Region_Info

```
VOID VDMAD_Set_Region_Info(HANDLE DMAHandle, BYTE BufferID,
                           BOOLEAN LockStatus, DWORD Region,
                           DWORD RegionSize, DWORD PhysAddr);
DMAHandle: handle returned by VDMAD_Virtualize_Channel
BufferID: id returned by VDMAD_Request_Buffer, or zero
LockStatus: zero if pages are not locked, non-zero if locked;
Region: Linear address of DMA buffer
RegionSize: size of DMA buffer, in bytes
PhysAddr: physical address of DA buffer
```

While `VDMAD_Set_Region_Info` gives the VDMAD information about the DMA buffer, `VDMAD_Set_Phys_State` gives VDMAD information about the transfer itself. There is no explanation of the `VMHandle` parameter in Microsoft's documentation. In fact, the VDMAD does nothing more with this parameter than see if it's a valid VM handle; if not it sends a warning message to the debugger. To avoid this warning, use the handle of the current VM, returned by `Get_Cur_VM_Handle`.

The Calling Interface for `VDMAD_Set_Phys_State`

```
void VDMAD_Set_Phys_State(HANDLE DMAHandle, HANDLE VMHandle,
                          WORD Mode, WORD ExtMode);
DMAHandle: handle returned by VDMAD_Virtualize_Channel
VMHandle: any VM handle
Mode: bitmap corresponding to system DMA controller's Mode register
```

The `Mode` parameter isn't explained in the documentation either, but it corresponds exactly to the mode register of the DMA controller, which controls transfer direction, auto-initialization, etc. VxDs should always use the flag `DMA_single_mode`, to be consistent with the way PC architecture defines system DMA bus cycles. The `VDMAD.H` provided by VToolsD provides #`defines` for these values, as does the `WRAPPERS.H` for the `WRAPPERS.CLB` library. The `ExtMode` parameter, used only in EISA and PS/2 DMA transfers, also has #`defines` in VToolsD `VDMAD.H` and in `WRAPPERS.H`.

When `VDMAD_Set_Phys_State` returns, the VDMAD has programmed the DMA controller base register, page register, count (using the address and size from the previous call to `VDMAD_Set_Region_Info`), and mode register (with the mode parameter from this call).

The final step is to enable the transfer by unmasking the channel with a call to `VDMAD_Phys_Unmask_Channel`. Once again, the call requires a VM handle, and the return value from `Get_Cur_VM_Handle` will do. This call unmasks the channel in the actual system DMA controller, which means the DMA controller is ready to begin the transfer. The transfer will actually begin when your device asserts the proper bus signals (`DMA_REQx`, `DMA_GRANTx`).

The Calling Interface for `VDMAD_Phys_Unmask_Channel`

```
void VDMAD_Phys_Unmask_Channel( HANDLE DMAHandle, HANDLE VMHandle);
DMAHandle: handle returned by VDMAD_Virtualize_Channel
VMHandle: any VM handle
```

The following code fragment combines a call to `AllocSysDmaBuf` with the VDMAD calls described above to set up a system DMA transfer on Channel 3. The transfer uses `DMA_type_write` mode, meaning the transfer "writes" to memory (from the device). This example also specifies `DMA_single_mode`, the mode used for normal system DMA bus cycles.

```
DWORD lin, size, phys;
BYTE ch;
DMAHANDLE dmaHnd;

size = 4 * 4 * 1024;
ch = 3;
lin = AllocSysDmaBuf( size/4096, &phys );
dmaHnd = VDMAD_Virtualize_Channel(ch, NULL );
VDMAD_Set_RegionInfo( dmaHnd, 0, TRUE, lin, size, phys );
VDMAD_Set_Phys_State( dmaHnd, Get_Cur_VM_Handle(),
                      DMA_type_write | DMA_single_mode, 0 );
VDMAD_Phys_Unmask_Channel( dmaHnd );
```

Using Bus-master DMA

A buffer used for a bus-master transfer has fewer restrictions than a system DMA buffer. Bus-master transfers still require fixed and pagelocked buffers, but the new buses (like PCI) that support bus-master transfers aren't limited by the old ISA 64Kb alignment and 16Mb maximum restrictions. Whether a bus-master transfer requires a physically contiguous buffer depends on whether or not the bus-master device supports a feature called "scatter-gather".

In a "scatter-gather" transfer, the DMA buffer, described by a single linear address and size, may be composed of multiple physical regions instead of a single physically contiguous region. A "scatter-gather" driver programs a bus-master device with the physical address and size of each of these regions, then the device initiates DMA transfers to/from each of the regions in turn, without any more intervention from the driver — or the processor, for that matter.

DMA buffers for devices without scatter-gather support must consist of physically contiguous pages — i.e. `_PageAllocate` must be called with the `PAGECONTIG` flag. The following function, `AllocBusMasterBuffer` (also contained in the file `DMAALLOC.C` in the `\DMAALLOC` directory of the code disk), uses `_PageAllocate` with the appropriate parameters to allocate a buffer for a bus-master *without* scatter-gather support.

```
DWORD AllocBusMasterBuffer( DWORD nPages, DWORD *pPhysAddr )
{
    return _PageAllocate(nPages, PG_SYS, 0, 0, 0, 0x100000, pPhysAddr,
                    PAGEFIXED | PAGEUSEALIGN | PAGECONTIG );
}
```

Note that alignment and maximum physical address requirements have relaxed. The AlignMask parameter now specifies 4Kb instead of 64Kb, and maxPhys now specifies the page number for 4Gb. PAGECONTIG is set to get contiguous pages; the PAGEUSEALIGN bit is set because PAGECONTIG requires it; and PAGEFIXED is set because PAGEUSEALIGN requires it. The function returns the buffer's linear address and stores the physical address at *pPhysAddr. This physical address is used to program the bus-master device with the address of the transfer.

Bus-masters that don't support scatter-gather don't require physically contiguous pages. The following function, AllocScatterGatherBuffer (also contained in the file DMAALLOC.C in the \DMAALLOC directory of the code disk), uses _PageAllocate with the appropriate parameters to allocate a buffer for a bus-master *with* scatter-gather support.

```
DWORD AllocScatterGatherBuffer( DWORD nPages, DWORD *pPhysAddr )
{
    return _PageAllocate(nPages, PG_SYS, 0, 0, 0, 0x100000,
                    pPhysAddr, 0 );
}
```

Notice that the last argument, Flags, in this call to _PageAllocate is zero. PAGECONTIG isn't set, which means PAGEALIGN doesn't need to be set, which means PAGEFIXED doesn't need to be set.

The function return value is the buffer's linear address, but the physical address returned at pPhysAddr is not valid. When PAGEFIXED is clear, _PageAllocate allocates linear pages (slots in the page tables) but marks the pages as not present in physical memory. This state is called "committed", but "not present". (Note that _PageAllocate behaves a bit differently under Windows 3.x: see the section "Windows 3.x Differences" at the end of this chapter).

The VxD can wait until the time of the actual transfer to allocate physical pages (make them "present") and meet the remaining buffer requirements — fixed and pagelocked. This strategy reduces overall system demands for physical memory, a limited commodity. When the transfer is over, the VxD can unlock the pages again, allowing the virtual memory manager the flexibility of swapping these pages to disk to free up physical memory for another use.

In addition to pagelocking the buffer before the scatter-gather transfer, a VxD needs to acquire the physical address of each page in the buffer (remember, they're not physically contiguous) in order to program the device for the transfer. The

VDMAD provides a service for just this purpose: one call to `VDMAD_Scatter_Lock` will lock all the pages in a linear address range and return the physical address of each page. Unfortunately, using this service is tricky. The documentation is incomplete, and the `VMDAD.H` header file (in both VToolsD and the Windows 95 DDK) incorrectly defines the structure it uses.

Examining Linear and Physical Addresses in the Debugger

Both SoftIce/Windows and WDEB386 let you examine memory manager data structures. I used this feature to verify the behavior of the `AllocSysDMABuf`, `AllocBusMasterBuf`, and `AllocScatterGatherBuf` functions. I used the WDEB386 command `.m` to dump all the memory manager information for the linear address range returned by `_PageAllocate`. In each case I allocated four pages, so I dumped four linear addresses.

WDEB386 shows that `AllocSysDMABuf` does meet system DMA requirements: is fixed and locked; four physical pages are contiguous; each physical page is aligned on a 4Kb boundary (implicitly meeting the requirement that the buffer not cross a 64Kb boundary); and each physical page is below 16Mb.

The Buffer Attributes After a Call to `AllocSysDMABuf`

```
.m C156D000
C156D000 committed r/w user Fixed present locked Phys=00250000 Base=C156D000
.m C156D100
C156D100 committed r/w user Fixed present locked Phys=00251000 Base=C156D000
.m C156D200
C156D200 committed r/w user Fixed present locked Phys=00252000 Base=C156D000
.m C156D300
C156D300 committed r/w user Fixed present locked Phys=00253000 Base=C156D000
```

WDEB386 shows that `AllocBusMasterBuf` does meet bus-master (no scatter-gather) requirements: is fixed and locked; four physical pages are contiguous; each physical page is aligned on a page boundary; and each page is located well above 16Mb (my system had 40Mb of physical RAM).

The Buffer Attributes After a Call to `AllocBusMasterBuf`

```
.m C156D000
C156D000 committed r/w user Fixed present locked Phys=027fc000 Base=C156D000
.m C156D100
C156D100 committed r/w user Fixed present locked Phys=027fd000 Base=C156D000
.m C156D200
C156D200 committed r/w user Fixed present locked Phys=027fe000 Base=C156D000
.m C156D300
C156D300 committed r/w user Fixed present locked Phys=027ff000 Base=C156D000
```

The Right Way to Use `VDMAD_Scatter_Lock`

The first parameter to `VDMAD_Scatter_Lock` is a VM handle parameter, and you can pass in the return value from `Get_Cur_VM_Handle` (see previous "VDMAD Services in Detail" section for an explanation of this technique). The other parameters need a lot of explanation because the available documentation is incomplete and confusing.

(Examining Linear and Physical Addresses in the Debugger — continued)

Finally, WDEB386 shows that `AllocScatterBuf` doesn't really allocate any physical pages. Though the memory manager says the pages are "committed" (have page table entries), they are marked as "not present", so no physical address is shown.

The Buffer Attributes After a Call to `AllocScatterBuf`

```
.m C1573000
C156D000 committed r/w user Swapped not-present Base=C1573000
.m C1573100
C156D100 committed r/w user Swapped not-present Base=C1573000
.m C1573200
C156D200 committed r/w user Swapped not-present Base=C1573000
.m C1573300
C156D300 committed r/w user Swapped not-present Base=C1573000
```

After a call to `VDMAD_Scatter_Lock`, the same buffer meets bus-master (scatter-gather) requirements. The pages are still "swapped" — but this really seems to mean "swappable" as opposed to "fixed". Now, however, the pages are present, locked, and have a physical address. Note that the physical addresses are not contiguous and that each is located above 16Mb.

The Buffer Attributes After a Call to `VDMAD_Scatter_Lock`

```
.m C1573000
C1573000 committed r/w user Swapped present locked Phys=0155c000 Base=C1573000
.m C1573100
C1573100 committed r/w user Swapped present locked Phys=015a9000 Base=C1573000
.m C1573200
C1573100 committed r/w user Swapped present locked Phys=0168b000 Base=C1573000
.m C15732300
C1573100 committed r/w user Swapped present locked Phys=0168f000 Base=C1573000
```

The Calling Interface for VDMAD_Scatter_Lock

```
DWORD VDMAD_Scatter_Lock(HANDLE VMHandle, DWORD Flags, PVOID pDDS,
                         PDWORD pPTEOffset);
VMHandle: any VM handle
Flags: 0: copy phys adddr and size to DDS
       1: copy raw PTE to DDS
       3: don't lock not-present pages
pDDS: pointer to DDS structure
pPTEOffset: if flags is 1 or 3, contains the 12-bit offset portion
            of the physical address for the first region.
Returns: 0 if no pages were locked
         1 if all pages were locked
         2 if some pages were locked
```

If Bit 0 of the Flags parameter is clear, the VDMAD fills in the caller's DDS structure with the physical address and size of each physical region in the buffer's linear address range. If Bit 0 is set, the VDMAD fills the DDS structure with the PTE (page table entry) for each page in the buffer. Your VxD can then derive the physical address and size of each region from the PTEs. For most VxDs, the physical address and size of each region is sufficient, so Bit 0 would be clear. A pager VxD would typically set Bit 0, because it can use the other PTE information (like the present bit and the dirty bit).

Only a pager VxD would use Bit 1 of the Flag parameter (which is ignored unless Bit 0 is also set). Setting Bit 1 tells the VDMAD to not lock, or return the address of, pages that are not present. Other VxDs usually clear Bit 1 so that the VDMAD locks pages whether or not they are marked "present". Because when used for a DMA buffer, the pages are already locked and present, Bit 1 doesn't really matter, but it's more efficient to tell the VDMAD to ignore the present/not-present attribute by clearing the bit.

According to the documentation, the second parameter should be a "pointer to the extended DDS structure". But the EXTENDED_DDS structure definition in VDMAD.H is incorrect. Here is the definition of the *correct* structure (DDS) to pass (via a pointer) to VDMAD_Scatter_Lock:

```
typedef struct
{
   EXTENDED_DDS extdds;
   union
   {
      REGION aRegionInfo[16];
      DWORD  aPte[16];
   }
} DDS;
```

```
typedef struct
{
    DWORD    PhysAddr;
    DWORD    Size;
} REGION;

typedef struct Extended_DDS_Struc
{
    DWORD    DDS_size;
    DWORD    DDS_linear;
    WORD     DDS_seg;
    WORD     RESERVED;
    WORD     DDS_avail;
    WORD     DDS_used;
} EXTENDED_DDS, *PEXTENDED_DDS;
```

The DDS and REGION structures above aren't contained in any VToolsD or Windows 95 DDK header files, but they are in WRAPPERS.H. I created them after figuring out how VDMAD_Scatter_Lock really uses the structure passed to it (by looking at the VDMAD source contained in the Windows 95 DDK). To understand this complicated set of structures within structures, it's best to step back and think about what the service is really doing.

A DMA buffer, described by a single linear address and size, can be composed of multiple physical regions, each of varying size. For example, a 16Kb buffer is always composed of four pages, 4Kb each. But this buffer can be composed of 1, 2, 3, or 4 physically contiguous regions. This is illustrated in Figure 6.2.

VDMAD_Scatter_Lock takes the buffer's linear address and size and returns either: the physical address and size of each of the physically contiguous regions (if Bit 0 of Flags is clear) or the PTE for each of the pages (if Bit 0 of Flags is set). All of this information is recorded — albeit in a most complicated manner — in the DDS structure described above.

The VxD fills in (as input) the DDS_size (size of buffer, in bytes) and DDS_linear (linear address of buffer) fields of the EXTENDED_DDS structure. VDMAD provides (as output) one of the two members of the union inside DDS: either the array of REGION structures or the array of DWORD PTEs, depending on the Flags parameter.

The call to VDMAD passes a pointer to the DDS which contains both pieces, the EXTENDED_DDS and the union. Your VxD fills in as input DDS_avail which tells the VDMAD the number of REGIONs or DWORDs in the union. The VDMAD fills in DDS_used on return, which tells your VxD how many of the REGIONs or DWORDs were filled in with physical address and size or with PTEs.

Note that in my definition, the two arrays contain 16 elements, which means the DDS structure supports a maximum DMA buffer size of 64Kb (16*4Kb). A buffer of 16 pages could consist of 16 physically discontiguous pages, in which case the VDMAD would need a REGION structure to describe each. If your VxD for a bus-master device uses more than 64Kb in a single bus-master transfer, increase this array size.

Using Events with Bus-master DMA Transfers

Commonly, DMA drivers start the first transfer in non-interrupt code, service an interrupt generated by the device when the transfer is complete, and start the next transfer directly from the interrupt handler. However, only VxD services marked specifically as asynchronous may be called at interrupt time, so it's vital to know which VDMAD services are asynchronous. According to the DDK documentation, only VDMAD_Physically_Unmask_Channel and VDMAD_Physically_Mask_Channel are asynchronous. The VDMAD source code reveals several other asynchronous services too, including VDMAD_Set_Region_Info and VDMAD_Set_Phys_State. But, VDMAD_Scatter_Lock and VDMAD_Scatter_Unlock are conspicuously missing.

A system DMA VxD can make all of its calls from its interrupt handler, because all of the VDMAD services it uses are asynchronous (even if not documented as so). But a bus-master VxD needs VDMAD_Scatter_Unlock, which can't be called at interrupt time. The synchronous services, VDMAD_Scatter_Lock and VDMAD_Scatter_Unlock, must be called outside of the interrupt handler. This is accomplished by having the interrupt handler use VMM services to schedule an event, and calling VDMAD_Scatter_Lock and VDMAD_Scatter_Unlock from the event handler. In fact, it's really just as easy to do the entire sequence of VDMAD calls in the event handler.

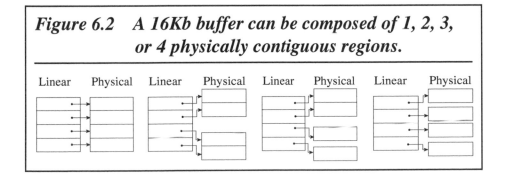

Figure 6.2   A 16Kb buffer can be composed of 1, 2, 3, or 4 physically contiguous regions.

If your VxD allocated the DMA buffer itself, you can schedule a global event, because any buffer allocated by a VxD comes from the 3Gb–4Gb system arena, visible regardless of the currently executing thread and VM. On the other hand, if your VxD didn't allocate the buffer, but instead pagelocked a buffer allocated by a Win32 process, then that buffer resides in the 2Gb–3Gb private arena and is valid only in the context of the same Win32 process that called your VxD for the page-lock. In this case, you must schedule a thread event so that your event handler runs in the correct context. Handling interrupts, as well as using thread and global events, will be covered in the next chapter.

Windows 3.x Differences

There are only minor differences in talking to hardware from a VxD when running under Win3.x.

- Accessing I/O-mapped hardware is no different at all — it works exactly as described earlier in the chapter.

- When accessing memory-mapped hardware, use _MapPhysToLinear, a simpler method than multiple VMM calls to _PageReserve/_PageCommitPhys/_LinPageLock. The simple method is sufficient because the device's physical address cannot change (no Plug and Play). Also, the other VMM services (_PageReserve, etc.) are Windows 95-specific.

- To perform system DMA, use the VDMAD services as described above. However, your VxD must allocate the DMA buffer during Sys_Critical_Init message processing because the PAGECONTIG flag passed to _PageAllocate isn't valid after initialization.

- Bus-master DMA is no different at all.

Summary

Talking to the hardware from your VxD is pretty straightforward if you only need to manipulate I/O ports. Most devices that use memory-mapped I/O are only slightly more challenging. Devices that support DMA are considerably more challenging, because they interact with physical memory in more complex patterns. Even so, with careful consideration of the paging and address translation issues involved, you can write a VxD that can manipulate the necessary physical memory.

I/O ports, memory, and DMA channels, though, are only part of the hardware a VxD needs to manipulate. VxDs aren't just called by applications — they are often invoked as asynchronous interrupt handlers. The next chapter explains how Windows virtualizes interrupts and how to register a VxD as the handler for a particular interrupt.

Listing 6.1   `DMAALLOC.C`

```c
// DMAALLOC.c - main module for VxD DMAEXAMP
#define WANTVXDWRAPS

#include <basedef.h>
#include <vmm.h>
#include <debug.h>
#include "vxdcall.h"
#include <vxdwraps.h>
#include <wrappers.h>

#ifdef DEBUG
#define DPRINTF0(buf, fmt) _Sprintf(buf, fmt ); Out_Debug_String( buf )
#define DPRINTF1(buf, fmt, arg1) _Sprintf(buf, fmt, arg1 ); Out_Debug_String( buf )
#define DPRINTF2(buf, fmt, arg1, arg2) _Sprintf(buf, fmt, arg1, arg2 ); \
    Out_Debug_String( buf )
#else
#define DPRINTF0(buf, fmt)
#define DPRINTF1(buf, fmt, arg1)
#define DPRINTF2(buf, fmt, arg1, arg2)
#endif

PVOID AllocSysDmaBuf( DWORD nPages, PVOID pPhysAddr );
PVOID AllocBusMasterBuf( DWORD nPages, PVOID pPhysAddr );
PVOID AllocScatterGatherBuf( DWORD nPages, PVOID pPhysAddr );

PVOID lin;
char buf[80];

BOOL OnSysDynamicDeviceInit(VMHANDLE hVM)
{
    BOOL  rc;
    DWORD PTEOffset;
    DWORD nPages, phys;
    DDS   myDDS;
    int i;

    DPRINTF0(buf, "DynInit\r\n" );
    nPages = 4;
    lin = AllocScatterGatherBuf( nPages, &phys );
    if (!lin)
    {
        DPRINTF0(buf, "ERR PageAlloc\r\n" );
    }
    else
    {
        DPRINTF2(buf, "Lin=%x, Phys=%x\r\n", lin, phys);
    }
```

Listing 6.1 *(continued)*   DMAALLOC.C

```
    myDDS.dds.DDS_linear = lin;
    myDDS.dds.DDS_size = 4 * 4 * 1024;
    myDDS.dds.DDS_seg = myDDS.dds.RESERVED = 0;
    myDDS.dds.DDS_avail = 16;
    rc = VDMAD_Scatter_Lock( Get_Cur_VM_Handle(), 0, &myDDS, &PTEOffset );
    DPRINTF1(buf, "Scatter_Lock rc=%x\r\n", rc);
    DPRINTF1(buf, "nRegions=%x\r\n", myDDS.dds.DDS_used);
    for (i=0; i < myDDS.dds.DDS_used; i++)
    {
        DPRINTF2(buf, "Region phys=%x size=%d\r\n", myDDS.aRegionInfo[i].PhysAddr,
                myDDS.aRegionInfo[i].Size );
    }

    return TRUE;
}


BOOL OnSysDynamicDeviceExit(void)
{
    BOOL rc;
    DPRINTF0(buf, "DynExit\r\n" );
    rc = _PageFree( lin, 0 );
    if (!rc)
        DPRINTF0(buf, "PageFree failed\n");
    return TRUE;
}


PVOID AllocSysDmaBuf( DWORD nPages, PVOID pPhysAddr )
{
    return( _PageAllocate(nPages, PG_SYS, 0, 0x0F, 0, 0x1000L, pPhysAddr,
            PAGECONTIG | PAGEUSEALIGN | PAGEFIXED ) );
}


PVOID AllocBusMasterBuf( DWORD nPages, PVOID pPhysAddr )
{
    return( _PageAllocate(nPages, PG_SYS, 0, 0, 0, 0x100000L, pPhysAddr,
            PAGECONTIG | PAGEUSEALIGN | PAGEFIXED ) );
}

PVOID AllocScatterGatherBuf( DWORD nPages, PVOID pPhysAddr )
{
    return( _PageAllocate(nPages, PG_SYS, 0, 0, 0, 0x100000L, pPhysAddr, 0 ) );
}
```

Listing 6.2   *DMADDB.ASM*

```
    .386p

;******************************************************************************
;                   I N C L U D E S
;******************************************************************************

    include vmm.inc
    include debug.inc

;=============================================================================
;          V I R T U A L   D E V I C E   D E C L A R A T I O N
;=============================================================================

DECLARE_VIRTUAL_DEVICE      DMAALLOC, 1, 0, ControlProc, UNDEFINED_DEVICE_ID, \
                            UNDEFINED_INIT_ORDER

VxD_LOCKED_CODE_SEG

;=============================================================================
;
;   PROCEDURE: ControlProc
;
;   DESCRIPTION:
;    Device control procedure for the SKELETON VxD
;
;   ENTRY:
;    EAX = Control call ID
;
;   EXIT:
;    If carry clear then
;         Successful
;    else
;         Control call failed
;
;   USES:
;    EAX, EBX, ECX, EDX, ESI, EDI, Flags
;
;=============================================================================

BeginProc ControlProc
    Control_Dispatch SYS_DYNAMIC_DEVICE_INIT, _OnSysDynamicDeviceInit, cCall, <ebx>
    Control_Dispatch SYS_DYNAMIC_DEVICE_EXIT, _OnSysDynamicDeviceExit, cCall, <ebx>

    clc
    ret

EndProc ControlProc

VxD_LOCKED_CODE_ENDS

    END
```

Listing 6.3   *DMAALLOC.MAK*

```
CFLAGS     = -DWIN32 -DCON -Di386 -D_X86_ -D_NTWIN -W3 -Gs -D_DEBUG -Zi
CVXDFLAGS  = -Zdp -Gs -c -DIS_32 -Zl -DDEBLEVEL=1 -DDEBUG
LFLAGS     = -machine:i386 -debug:notmapped,full -debugtype:cv
             -subsystem:console kernel32.lib
AFLAGS     = -coff -DBLD_COFF -DIS_32 -W2 -Zd -c -Cx -DMASM6 -DDEBLEVEL=1 -DDEBUG

all: dmaalloc.vxd

dmaalloc.obj: dmaalloc.c
        cl $(CVXDFLAGS) -Fo$@ -Fl %s

dmaddb.obj: dmaddb.asm
        ml $(AFLAGS) -Fo$@ %s

vxdcall.obj: vxdcall.c
        cl $(CVXDFLAGS) -Fo$@ %s

dmaalloc.vxd: dmaddb.obj dmaalloc.obj vxdcall.obj ..\wrappers\wrappers.clb dmaalloc.def
        echo >NUL @<<dmaalloc.crf
-MACHINE:i386 -DEBUG -DEBUGTYPE:MAP -PDB:NONE
-DEF:dmaalloc.def -OUT:dmaalloc.vxd -MAP:dmaalloc.map
-VXD vxdwraps.clb wrappers.clb vxdcall.obj dmaddb.obj dmaalloc.obj
<<
        link @dmaalloc.crf
        mapsym dmaalloc
```

Listing 6.4   `DMAALLOC.DEF`

```
VXD DMAALLOC DYNAMIC
SEGMENTS
    _LTEXT     CLASS 'LCODE'    PRELOAD NONDISCARDABLE
    _LDATA     CLASS 'LCODE'    PRELOAD NONDISCARDABLE
    _TEXT      CLASS 'LCODE'    PRELOAD NONDISCARDABLE
    _DATA      CLASS 'LCODE'    PRELOAD NONDISCARDABLE
    _LPTEXT    CLASS 'LCODE'    PRELOAD NONDISCARDABLE
    _CONST     CLASS 'LCODE'    PRELOAD NONDISCARDABLE
    _BSS       CLASS 'LCODE'    PRELOAD NONDISCARDABLE
    _TLS       CLASS 'LCODE'    PRELOAD NONDISCARDABLE
    _ITEXT     CLASS 'ICODE'    DISCARDABLE
    _IDATA     CLASS 'ICODE'    DISCARDABLE
    _PTEXT     CLASS 'PCODE'    NONDISCARDABLE
    _PDATA     CLASS 'PCODE'    NONDISCARDABLE
    _STEXT     CLASS 'SCODE'    RESIDENT
    _SDATA     CLASS 'SCODE'    RESIDENT
    _MSGTABLE  CLASS 'MCODE'    PRELOAD NONDISCARDABLE IOPL
    _MSGDATA   CLASS 'MCODE'    PRELOAD NONDISCARDABLE IOPL
    _IMSGTABLE CLASS 'MCODE'    PRELOAD DISCARDABLE IOPL
    _IMSGDATA  CLASS 'MCODE'    PRELOAD DISCARDABLE IOPL
    _DBOSTART  CLASS 'DBOCODE'  PRELOAD NONDISCARDABLE CONFORMING
    _DBOCODE   CLASS 'DBOCODE'  PRELOAD NONDISCARDABLE CONFORMING
    _DBODATA   CLASS 'DBOCODE'  PRELOAD NONDISCARDABLE CONFORMING
    _16ICODE   CLASS '16ICODE'  PRELOAD DISCARDABLE
    _RCODE     CLASS 'RCODE'
EXPORTS
    DMAALLOC_DDB @1
```

## Chapter 7

Handling Hardware Interrupts in a VxD

Windows' translation of hardware interrupts into events that can trigger execution of ISRs residing in various virtual machines is one of the most confusing and complicated parts of the virtual environment. Windows must not only make certain that any associated VxD sees the interrupt but also must assure that the appropriate virtual machines see the interrupt. This process is not only complicated but also involves a large amount of overhead — so much overhead that an ISR residing in a DOS program running under Windows can exhibit as much as 20 times more latency than the same ISR under pure DOS. (For details, see "The Tao of Interrupts," by David Long, *Microsoft Developer Network CD*.)

An interrupt can trigger activity that cascades through four levels of code:

* the processor vectors to a routine in the VMM;

* the VMM calls registered handlers in one or more VxDs;

* the VMM then (potentially) simulates the interrupt for protected mode handlers; and

* the VMM then (potentially) simulates the interrupt for V86 mode handlers.

The programmer can install an interrupt handler at any but the first of these levels. (Actually, since a VxD runs at Ring 0, it could also install at the first level, directly in the actual IDT. Microsoft strongly warns against doing this.) Handlers installed by Windows applications qualify as protected mode handlers, running in the System VM. Handlers installed by DOS applications qualify as V86 mode handlers, running in the same VM as the DOS application.

Interrupts and the VMM

Under Windows, the processor runs in three different states. The processor runs in V86 Mode when a DOS application is executing (in a V86 VM). The processor runs in Ring 3 protected mode when the System VM (Windows) is executing or when a DOS VM has switched into protected mode. The processor runs in Ring 0 protected mode when VMM or a VxD is executing.

Regardless of the processor's current state, when a hardware interrupt occurs, the processor switches to protected mode at Ring 0. The processor then finds the address of the interrupt handler in the IDT and begins executing the handler. This isn't a Windows rule — it's the way the 80x86 architecture works.

As Figure 7.1 shows, however, Windows doesn't use the IDT to vector to what one normally thinks of as a interrupt handler. Instead, Windows makes all IDT entries point to a routine in the VMM. The VMM routine figures out whether it was called as the result of an exception or an interrupt. The VMM manages exceptions itself but hands all hardware interrupts to an important VxD called the VPICD (Virtual Programmable Interrupt Controller Device.) VPICD will pass the interrupt on to another VxD for servicing if a VxD has registered for the interrupt. If not, the VPICD will pass the interrupt on to one of the VMs, a process known as reflection.

A VxD registers for a specific hardware interrupt by calling the VPICD service VPICD_Virtualize_IRQ and passing to the VPICD the address of a callback routine. Once a VxD has registered for an interrupt it may act as a true interrupt handler, servicing the interrupting device itself, or the VxD may use another VPICD service,

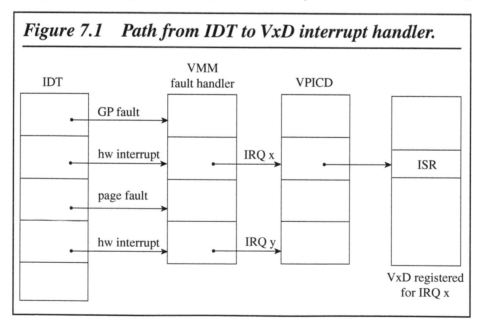

Figure 7.1   Path from IDT to VxD interrupt handler.

VPICD_Set_Int_Request, to reflect the interrupt to a VM. That is, instead of servicing the device itself, the VxD lets the VM's handler do it. The section "Virtualizing a Hardware Interrupt" in Chapter 8 will explain reflection in more detail. (See also the sidebar "Interrupt Latency under Windows".)

Using VPICD Services

This section will examine the VPICD services used by a VxD that handles a hardware interrupt. Although the VPICD exports close to two dozen services, a typical VxD uses only a few of them:

- VPICD_Virtualize_IRQ to install an interrupt handler.
- VPICD_Phys_Unmask to unmask the interrupt at the PC Interrupt Controller (PIC).
- VPICD_Phys_EOI to send an EOI to the PIC.
- VPICD_Phys_Mask to mask the interrupt at the PIC.
- VPICD_Force_Default_Behavior to uninstall an interrupt handler.

Interrupt Latency under Windows

The delay between the hardware interrupt signal and the execution of its handler is called interrupt latency. Because of the complicated reflection process involved, latency for protected mode or V86 mode handlers can be significant — times around 1 ms are not uncommon. To minimize interrupt latency, handling of hardware interrupts should be done in a VxD.

Unfortunately, not even a VxD can guarantee real-time response to an interrupt. There are several factors that make true real-time response impossible under Windows (both 95 and 3.x), including ring transitions and the multiple layers of VMM and VPICD handlers. But the factor that overwhelms all others is the abillity of applications to disable processor interrupts. When processor interrupts are disabled, not even a VxD interrupt handler can run.

The VMM allows both DOS and Windows applications (and DLLs) to turn off interrupts. (Refer to the section "Trapping Interrupts and Exceptions" in Chapter 3 for more details). Although applications could also turn off interrupts under plain DOS, the consequences are often worse under Windows simply because users typically run multiple applications, and the chances that one application will disable interrupts for a long period are increased.

The VMM/VxD library included with VToolsD provides C-callable wrappers for all VPICD services. The Windows 95 DDK wrapper library doesn't, but the WRAPPERS.CLB library does include VPICD functions discussed in this chapter as well as the others discussed in the section "Virtualizing a Hardware Interrupt" in Chapter 8.

The example VxD for this chapter, VXDISR.VXD, demonstrates a simple interrupt handler. This VxD services one of the few standard PC devices that isn't already controlled by another VxD: the Real Time Clock, which generates an interrupt on IRQ 8. The Real Time Clock is not the same as the 8254 Timer device. The timer generates an interrupt on IRQ 0, and is controlled by another VxD, the VTD.

Examining VPICD Services in Detail: *VXDIRQ.C*

The VXDISR VxD has only two message handlers, On_Sys_Vm_Init and On_Sys_-Vm_Terminate, which install and uninstall an interrupt handler, respectively. On_Sys_Vm_Init calls the service VPICD_Virtualize_IRQ to install an interrupt handler.

The Calling Interface for *VPICD_Virtualize_IRQ*

```
IRQHANDLE VPICD_Virtualize_IRQ(VPICD_IRQ_DESCRIPTOR *vid);
vid: pointer to structure which describes the interrupt to be virtualized
typedef struct
{
    USHORT VID_IRQ_Number;    // IRQ to virtualize
    USHORT VID_Options;
    // VPICD_OPT_CAN_SHARE: allow other VxDs to virtualize IRQ also
    // VPICD_OPT_REF_DATA: pass VID_Hw_Int_Ref as param to Hw_Int_Handler
    ULONG VID_Hw_Int_Proc;    // callback for hardware interrupt
    ULONG VID_Virt_Int_Proc;
    ULONG VID_EOI_Proc;
    ULONG VID_Mask_Change_Proc;
    ULONG VID_IRET_Proc;
    ULONG VID_IRET_Time_Out;
    PVOID VID_Hw_Int_Ref;     // pass this data to Hw_Int_Handler
} VPICD_IRQ_DESCRIPTOR;
Returns: handle to be used in future VPICD calls
        or zero if call failed (IRQ already virtualized or invalid IRQ)
```

This service is well behaved, i.e. it doesn't install the handler directly into the IDT, but simply registers the handler with the VPICD. This service uses only a single parameter, a pointer to a `VPICD_IRQ_DESCRIPTOR` structure. The return value is an IRQ "handle", required in calls to other VPICD services.

Because VXDISR is simply handling an interrupt, as opposed to fully virtualizing it (I'll discuss virtualization in the next chapter), it uses only a few fields in this structure. `VID_IRQ_Number` is the number of the IRQ the VxD wants to service. `VID_Hw_Int_Proc` is the address of the interrupt service routine. `VID_Options` is a bitmapped flag. `VPICD_OPT_CAN_SHARE` allows other VxDs to call `VPICD_Virtualize_IRQ` for the same IRQ. (VXDISR doesn't set this bit: the device itself must support IRQ sharing, and simply setting the option bit won't make IRQ-sharing work.) The `VPICD_OPT_REF_DATA` bit works in conjunction with the `VID_Hw_Int_Ref` parameter. If `VPICD_OPT_REF_DATA` is set, the VPICD passes `VID_Hw_Int_Ref` as a parameter when it calls the interrupt handler. `VID_Hw_Int_Ref` is used as reference data, so VXDISR passes a pointer to its device context structure.

Be sure to set the other callback fields (`VID_Virt_Int_Proc`, `VID_EOI_Proc`, `VID_Mask_Change_Proc`, and `VID_IRET_Proc`) to `NULL`. The VPICD uses these callbacks to notify a VxD of other interrupt-related events, such as when a V86 mode or protected mode handler is called. The section "Virtualizing a Hardware Interrupt" in Chapter 8 will demonstrate use of these other callbacks.

After installing its interrupt handler, `On_Sys_Vm_Init` enables the RTC interrupt in two steps. In the first it writes to an RTC register to enable the interrupt "at the device". In the second step, `On_Sys_Vm_Init` calls `VPICD_Physically_Unmask` using the same IRQ handle returned by `VPICD_Virtualize_IRQ`, which programs the PIC to recognize interrupts on IRQ 8. This second step enables the interrupt "at the PIC". A VxD should always use the VPICD service instead of writing directly to the PIC mask register.

The Calling Interface for `VPICD_Physically_Unmask`

```
void VPICD_Physically_Unmask(IRQHANDLE hnd);
hnd: IRQ handle returned by VPICD_Virtualize_IRQ
```

The `On_Sys_Vm_Terminate` function reverses the steps taken at initialization, first disabling interrupts at the device, then calling `VPICD_Physically_Mask` to disable the interrupt at the PIC, and finally uninstalling the handler with a call to `VPICD_Force_Default_Behavior`.

The Calling Interface for `VPICD_Physically_Mask`

```
void VPICD_Physically_Mask(IRQHANDLE hnd);
hnd: IRQ handle returned by VPICD_Virtualize_IRQ
```

The Calling Interface for `VPICD_Force_Default_Behavior`

```
void VPICD_Force_Default_Behavior(IRQHANDLE hnd);
hnd: IRQ handle returned by VPICD_Virtualize_IRQ
```

Assembly Thunks and C Handlers

Many VMM and VxD services require a callback function parameter. The VXDISR example in this chapter introduces a callback convention that all other VxDs in this book will follow. All registered callback functions reside in the VxD's assembly language file. The name of each registered callback function ends in "Thunk". Each callback function always transfers parameters from registers to the stack and calls an analogous function in the VxD's C file. The name of the C function is similar to the callback in the assembly function, except that the C function ends in "Handler" instead of in "Thunk".

The `HwIntProc` Callback: `DDBISR.ASM` and `VXDISR.C`

When a hardware interrupt occurs, the VPICD calls the registered `Hw_Int_Proc` callback with the handle of the current VM in `EBX` and the IRQ handle in `EAX`. Because `On_Sys_Vm_Init` set the `VPICD_OPT_REF_DATA` bit in `VID_Options` when registering the handler, `EDX` contains reference data. The registered interrupt handler is `_HwIntProcThunk` in `DDBISR.ASM` (Listing 7.2, page 122). This function does nothing more than push the current VM handle, IRQ handle, and reference data on the stack and call the real interrupt handler, `HwIntProcHandler` in `VXDISR.C` (Listing 7.1, page 117).

The first action taken by `HwIntProcHandler` is to cast the reference data to a pointer to its device context structure. The `DEVICE_CONTEXT` structure contains all the VxD needs to know about the device: its I/O address, state information, etc. `HwIntProcHandler` reads from the RTC Status C register to clear the interrupt.

The Calling Interface for `VPICD_Phys_EOI`

```
void VPICD_Phys_EOI( IRQHANDLE hnd );
hnd: handle returned by VPICD_Virtualize_IRQ
```

Immediately before returning, HwIntProcHandler calls VPICD_Phys_EOI to send an EOI (End Of Interrupt) command for IRQ 8 (see the sidebar "EOI Handling in Windows"). This EOI tells the PIC to recognize further interrupts from the RTC. A common mistake in coding an interrupt handler is to forget the EOI. The result is an interrupt handler that is called once but never again: although the device itself may be generating more interrupts, the PIC doesn't let these interrupts through to the processor until an EOI is received.

HwIntProcHandler returns a boolean indicating whether or not it serviced (cleared) the interrupt. On return, its caller, _HwIntProcThunk in DDBISR.ASM (Listing 7.2, page 122), examines this return value. If true (meaning the interrupt was serviced,) _HwIntProcThunk clears the Carry flag before returning to the VPICD, otherwise _HwIntProcThunk sets the Carry flag.

The VPICD uses this return value to support shared interrupts. If more than one VxD virtualizes the same IRQ, and both set VPICD_OPT_CAN_SHARE during registration, the VPICD keeps the registered interrupt handlers in a list. When the interrupt occurs, the VPICD calls the first handler on the list. When that handler returns, the VPICD examines the Carry flag. If Carry is set, meaning the interrupt handler did not service the interrupt, the VPICD calls the next handler in the list. This continues until one of the handlers services the interrupt.

Event Handling in VXDISR

In many cases, a VxD's interrupt handler isn't able to fully process the interrupt because the VMM or VxD services required aren't asynchronous. (See the section "Using Events with Bus-master DMA Transfers" in Chapter 6 for an explanation of synchronous and asynchronous services.) In this situation, the interrupt handler must schedule an event (which will be called later) and call the needed VMM/VxD service from the event callback. HwIntProcHandler demonstrates this technique, even though it doesn't really need it (the only VxD service it uses is VPICD_Phys_EOI, which is asynchronous).

EOI Handling in Windows

Windows uses the interrupt controller's EOI mechanism differently than does DOS. The VPICD is the first VxD to be notified of an interrupt, and the VPICD immediately sends a "specific EOI" to the controller — specifically for the level of the interrupting device. Then the VPICD masks (disables) that particular interrupt level on the controller. These two actions allow other interrupt levels to be recognized, including those of lesser priority than the interrupting level. When a VxD calls VPICD_Phys_EOI before exiting the interrupt handler, the VPICD unmasks (enables) interrupts on that same level.

The Calling Interface for `Schedule_Global_Event`

```
EVENTHANDLE Schedule_Global_Event(void *EventCallback, void *RefData );
EventCallback: pointer to callback function;
RefData: pointer to reference data to be passsed to callback function
```

`HwIntProcHandler` schedules a global event, meaning that the event callback could occur in the context of any VM. A global event is used because the actions taken in the event callback aren't specific to any one VM. The parameters to `Schedule_Global_Event` are straightforward: a pointer to the callback function and a pointer to reference data. The return value is an `EVENTHANDLE`, which is used to cancel the event.

As with the interrupt handler, the function passed to `Schedule_Global_Event` is actually a procedure in `DDBISR.ASM`, called `EventThunk`. This procedure takes the three parameters passed in by the VMM — the current VM handle in `EBX`, the reference data in `EDX` and a pointer to the Client Register Structure in `EBP` — and pushes them on the stack before calling the real event handler in `VXDISR.C`. (The Client Register Structure was introduced in Chapter 4.)

`EventHandler` is the name of the real event handler. `EventHandler` first casts the reference data to a `DEVICE_CONTEXT` pointer, then zeros out the event handle and increments the `EventCounter` field of the structure.

If your VxD ever cancels an event from an interrupt handler or timeout, the event handler must take special precautions to prevent cancellation of an already-dispatched event. Although VXDISR doesn't have cancel code, it follows this rule anyway. An event handler guards against this condition by zeroing out the event handle as its very first action. This precaution ensures that if the VxD's cancel code interrupts the event handler, the handle passed to VMM cancel service will be zero. It's permitted to pass the VMM cancel routine a handle of zero, but it's not ok to cast the handle of an event that is already in progress.

Windows 3.x Differences

There is only one minor difference in handling an IRQ in a Windows 3.x VxD as compared to a Windows 95 VxD. The VPICD doesn't support the `VPICD_OPT_REF_DATA` flag for `VPICD_Virtualize_IRQ`, so no reference data can be passed to the interrupt handler. Since the VXDISR example above used this reference data to provide a pointer to the `DEVICE_CONTEXT` associated with the interrupting device, how does a Windows 3.x handler get context information?

A Windows 3.x interrupt handler must provide its own context information. When the VxD supports only a single device instance, this is trivial. The assembly language handler pushes the hard-coded address of the one and only device context structure before calling the C handler routine.

```
EXTRN ptrDevice:DWORD      ;declared in C module, as is Device structure

BeginProc HwIntProcThunkDev
    mov    edi, ptrDevice
    cCall  _HwIntHandler, <ebx, eax, edi>
    or     eax, eax
    ret
EndProc HwIntProcThunkDev
```

By extending this concept a little further, the VxD can support multiple device instances, and thus multiple device contexts. Declare a different entry point in the assembly language module for each device context and have each entry point push the address of its own device context structure onto the stack before calling the C routine.

```
EXTRN ptrDevice1:DWORD     ;declared in C module, as is Device1 structure
EXTRN ptrDevice2:DWORD     ;declared in C module, as is Device2 structure

BeginProc HwIntProcThunkDev1
    mov    edi, ptrDevice1
    cCall  _HwIntHandler, <ebx, eax, edi>
    or     eax, eax
    ret
EndProc HwIntProcThunkDev1

BeginProc HwIntProcThunkDev2
    mov    edi, ptrDevice2
    cCall  _HwIntHandler, <ebx, eax, edi>
    or     eax, eax
    ret
EndProc HwIntProcThunkDev2
```

Of course, the initialization code that registers the interrupt handlers with the VPICD must change also. When registering a handler for Device1, HwIntProcThunkDev1 is the handler; when registering for Device2, HwIntProcThunkDev2 is the handler.

Note that for both single and multiple device instances, the real handler in the C module remains ignorant of these changes in the assembly language module. HwIntProcHandler keeps its DEVICE_CONTEXT* parameter, only this time it's provided by the HwIntProcThunk instead of the VPICD.

Summary

Once you understand the role of the VPICD with regard to hardware interrupt handlers, writing a VxD that services an interrupt isn't much harder than writing a DOS ISR. Instead of calling `DOS Set Vector`, use `VPICD_Virtualize_IRQ`. Instead of writing to the PIC directly to unmask an IRQ, use `VPICD_Unmask_IRQ`, and instead of sending an EOI to the PIC directly, use `VPICD_Phys_EOI`.

However, you may discover your VxD gets less than exciting performance. Even when implemented in a VxD, an ISR running under Windows will show substantially worse latency than a similar ISR running under DOS. The fact that a Windows or DOS application can actually disable processor interrupts for an indeterminate time means that even a VxD ISR can be delayed indefinitely.

Even so, most modern hardware has quick response time and some buffering. These factors mean that a VxD ISR may be an acceptable solution for all but applications with "hard" real-time requirements.

Listing 7.1   VXDISR.C

```c
#include <basedef.h>
#include <vmm.h>
#include <debug.h>
#include <vxdwraps.h>
#include <vpicd.h>

#include <vxdcall.h>
#include <wrappers.h>
#include <intrinsi.h>

#define RTC_IRQ      8

#define RTC_STATUSA      0xA
#define RTC_STATUSB      0xB
#define RTC_STATUSC      0xC

#define STATUSB_ENINT      0x40

#define CMOS_ADDR      0x70
#define CMOS_DATA      0x71

typedef struct
{

   VPICD_IRQ_DESCRIPTOR      descIrq;
   IRQHANDLE                 hndIrq;
   EVENTHANDLE               hEvent;
   DWORD                     EventCounter;
   BYTE                      StatusA;
   BYTE                      StatusB;
} DEVICE_CONTEXT;

DEVICE_CONTEXT      rtc;

BOOL OnDeviceInit(VMHANDLE hVM);
void OnSystemExit(VMHANDLE hVM);
BOOL _stdcall HwIntProcHandler(VMHANDLE hVM, IRQHANDLE hIRQ, void *Refdata);
VOID _stdcall EventHandler(VMHANDLE hVM, PVOID Refdata, CRS *pRegs);
void CmosWriteReg( BYTE reg, BYTE val );
BYTE CmosReadReg( BYTE reg );
```

Listing 7.1 (continued)   *VXDISR.C*

```
// functions in asm module
void EventThunk( void );
void HwIntProcThunk( void );

BOOL OnSysDynamicDeviceInit(VMHANDLE hVM)
{
   OnDeviceInit( hVM );
   return TRUE;
}

BOOL OnSysDynamicDeviceExit(void)
{
   OnSystemExit(Get_Cur_VM_Handle() );
   return TRUE;
}

BOOL OnDeviceInit(VMHANDLE hVM)
{
    rtc.descIrq.VID_IRQ_Number = RTC_IRQ;
    rtc.descIrq.VID_Options = VPICD_OPT_REF_DATA;
    rtc.descIrq.VID_Hw_Int_Ref = &rtc;
    rtc.descIrq.VID_Hw_Int_Proc = (ULONG)HwIntProcThunk;
    rtc.descIrq.VID_EOI_Proc =
    rtc.descIrq.VID_Virt_Int_Proc =
    rtc.descIrq.VID_Mask_Change_Proc =
    rtc.descIrq.VID_IRET_Proc = 0;

    rtc.descIrq.VID_IRET_Time_Out = 500;

    if (!(rtc.hndIrq = VPICD_Virtualize_IRQ(&rtc.descIrq)))
        return FALSE;

    rtc.StatusA = CmosReadReg(RTC_STATUSA);
    rtc.StatusB = CmosReadReg(RTC_STATUSB);
```

Listing 7.1 (continued)   `VXDISR.C`

```c
    // set interrupt frequency to only 2 times a sec
    CmosWriteReg(RTC_STATUSA, rtc.StatusA | 0x0F );
    // enable clock interrupts
    CmosWriteReg(RTC_STATUSB, rtc.StatusB | STATUSB_ENINT);
    // clear flags
    CmosReadReg(RTC_STATUSC);

    rtc.EventCounter = 0;

    VPICD_Physically_Unmask(rtc.hndIrq);

    return TRUE;
}

VOID OnSystemExit(VMHANDLE hVM)
{

    CmosWriteReg(RTC_STATUSA, rtc.StatusA );
    CmosWriteReg(RTC_STATUSB, rtc.StatusB );

    Cancel_Global_Event(rtc.hEvent);
    VPICD_Physically_Mask(rtc.hndIrq);
    VPICD_Force_Default_Behavior(rtc.hndIrq);
}

BOOL __stdcall HwIntProcHandler(VMHANDLE hVM, IRQHANDLE hIRQ, void *Refdata)
{
   DEVICE_CONTEXT *pRtc = (DEVICE_CONTEXT *)Refdata;

    CmosReadReg( RTC_STATUSC );

    VPICD_Phys_EOI(hIRQ);        // tell VPICD to clear the interrupt

    pRtc->hEvent = Schedule_Global_Event(EventThunk, (ULONG)pRtc );

    return TRUE;                 // thunk will clear carry
}
```

Listing 7.1 (continued) VXDISR.C

```c
VOID __stdcall EventHandler(VMHANDLE hVM, PVOID Refdata, CRS* pRegs)
{
    DEVICE_CONTEXT *rtc = (DEVICE_CONTEXT *)Refdata;

    rtc->hEvent = 0;
    rtc->EventCounter++;
}

BYTE CmosReadReg( BYTE reg )
{
    BYTE    data;

    _asm
    {
        ; disable NMI then ints
        mov    al, reg
        or     al, 80h
        cli

        ; first output reg to address port
        out    CMOS_ADDR, al
        jmp    _1

_1:
        jmp    _2
_2:
        ; then read data from data port
        in     al, CMOS_DATA
        mov    data, al
        jmp    _3
_3:
        jmp    _4
_4:
        ; reenable NMI then ints
        xor    al, al
        out    CMOS_ADDR, al
        sti
    }

    return data;
}
```

Listing 7.1 (continued)   *VXDISR.C*

```c
void CmosWriteReg( BYTE reg, BYTE val )
{
    _asm
    {
        ; disable NMI then ints
        mov   al, reg
        or    al, 80h
        cli

        ; first output reg to address port
        out   CMOS_ADDR, al
        jmp   __1
__1:
        jmp   __2
__2:

        ; then output val to data port
        mov   al, val
        out   CMOS_DATA, al
        jmp   __3
__3:
        jmp   __4
__4:

        ; reenable NMI then ints
        xor   al, al
        out   CMOS_ADDR, al
        sti
    }
}
```

Listing 7.2   DDBISR.ASM

```
.386p

;*****************************************************************************
;                    I N C L U D E S
;*****************************************************************************

    include vmm.inc
    include debug.inc

;============================================================================
;       V I R T U A L   D E V I C E   D E C L A R A T I O N
;============================================================================

DECLARE_VIRTUAL_DEVICE    VXDISR, 1, 0, ControlProc, UNDEFINED_DEVICE_ID, \
                          UNDEFINED_INIT_ORDER

VxD_LOCKED_CODE_SEG

;============================================================================
;
;   PROCEDURE: ControlProc
;
;   DESCRIPTION:
;    Device control procedure for the SKELETON VxD
;
;   ENTRY:
;    EAX = Control call ID
;
;   EXIT:
;    If carry clear then
;        Successful
;    else
;        Control call failed
;
;   USES:
;    EAX, EBX, ECX, EDX, ESI, EDI, Flags
;
;============================================================================

BeginProc ControlProc
    Control_Dispatch DEVICE_INIT, _OnDeviceInit, cCall, <ebx>
    Control_Dispatch SYSTEM_EXIT, _OnSystemExit, cCall, <ebx>
Control_Dispatch SYS_DYNAMIC_DEVICE_INIT, _OnSysDynamicDeviceInit, cCall, <ebx>
Control_Dispatch SYS_DYNAMIC_DEVICE_EXIT, _OnSysDynamicDeviceExit, cCall
    clc
    ret

EndProc ControlProc
```

Listing 7.2 (continued)   DDBISR.ASM

```
PUBLIC _HwIntProcThunk
_HwIntProcThunk PROC NEAR      ; called from C, needs underscore

    sCall HwIntProcHandler, <ebx, eax, edx>
    or    ax, ax
    jnz   clearc
    stc
    ret

clearc:
    clc
    ret

 _HwIntProcThunk ENDP

VxD_LOCKED_CODE_ENDS

VxD_CODE_SEG


BeginProc _EventThunk

    sCall EventHandler, <ebx,edx ehp>
    ret

EndProc _EventThunk

VXD_CODE_ENDS

    END
```

Listing 7.3 *VXDISR.MAK*

```
CVXDFLAGS  = -Zdp -Gs -c -DIS_32 -Zl -DDEBLEVEL=1 -DDEBUG -DWANTVXDWRAPS
AFLAGS     = -coff -DBLD_COFF -DIS_32 -W2 -Zd -c -Cx -DMASM6 -DDEBLEVEL=1 -DDEBUG

all: vxdisr.vxd

vxdisr.obj: vxdisr.c
        cl $(CVXDFLAGS) -Fo$@ %s

ddbisr.obj: ddbisr.asm
        ml $(AFLAGS) -Fo$@ -Fl %s

vxdisr.vxd: ddbisr.obj vxdisr.obj ..\wrappers\vxdcall.obj vxdisr.def
        echo >NUL @<<vxdisr.crf
-MACHINE:i386 -DEBUG -DEBUGTYPE:MAP -PDB:NONE
-DEF:vxdisr.def -OUT:vxdisr.vxd -MAP:vxdisr.map
-VXD vxdwraps.clb wrappers.clb ddbisr.obj vxdisr.obj vxdcall.obj
<<KEEP
        link @vxdisr.crf
        mapsym vxdisr
```

Listing 7.4   *VXDISR.DEF*

```
VXD VXDISR DYNAMIC
SEGMENTS
    _LTEXT      CLASS 'LCODE'   PRELOAD NONDISCARDABLE
    _LDATA      CLASS 'LCODE'   PRELOAD NONDISCARDABLE
    _TEXT       CLASS 'LCODE'   PRELOAD NONDISCARDABLE
    _DATA       CLASS 'LCODE'   PRELOAD NONDISCARDABLE
    _LPTEXT     CLASS 'LCODE'   PRELOAD NONDISCARDABLE
    _CONST      CLASS 'LCODE'   PRELOAD NONDISCARDABLE
    _BSS        CLASS 'LCODE'   PRELOAD NONDISCARDABLE
    _TLS        CLASS 'LCODE'   PRELOAD NONDISCARDABLE
    _ITEXT      CLASS 'ICODE'   DISCARDABLE
    _IDATA      CLASS 'ICODE'   DISCARDABLE
    _PTEXT      CLASS 'PCODE'   NONDISCARDABLE
    _PDATA      CLASS 'PCODE'   NONDISCARDABLE
    _STEXT      CLASS 'SCODE'   RESIDENT
    _SDATA      CLASS 'SCODE'   RESIDENT
    _MSGTABLE   CLASS 'MCODE'   PRELOAD NONDISCARDABLE IOPL
    _MSGDATA    CLASS 'MCODE'   PRELOAD NONDISCARDABLE IOPL
    _IMSGTABLE  CLASS 'MCODE'   PRELOAD DISCARDABLE IOPL
    _IMSGDATA   CLASS 'MCODE'   PRELOAD DISCARDABLE IOPL
    _DBOSTART   CLASS 'DBOCODE' PRELOAD NONDISCARDABLE CONFORMING
    _DBOCODE    CLASS 'DBOCODE' PRELOAD NONDISCARDABLE CONFORMING
    _DBODATA    CLASS 'DBOCODE' PRELOAD NONDISCARDABLE CONFORMING
    _16ICODE    CLASS '16ICODE' PRELOAD DISCARDABLE
    _RCODE      CLASS 'RCODE'
EXPORTS
    VXDISR_DDB @1
```

Chapter 8

VxDs for Virtualization

Earlier chapters explained how to write a "driver" VxD, that is a VxD that interfaces to and controls a hardware device. Topics included interfacing to I/O mapped, memory-mapped, and DMA devices, as well as hardware interrupts. This chapter will focus on a different aspect of VxD functionality: how to virtualize a hardware device (I/O-mapped or memory-mapped) and how to virtualize a hardware-generated interrupt.

Windows virtualizes physical devices because with multitasking, there is always the possibility of two processes attempting to use a device simultaneously. Virtualization wouldn't be required if every process went through the same driver to access the device; in that case, the driver could serialize the access.

Unfortunately, some applications (especially DOS applications) attempt to manipulate the hardware directly, instead of calling the operating system's driver. Because VxDs rely upon the 80x86's port-trapping and page-trapping hardware instead of an explicit call to a device driver, the VxD can intercept any VM's attempt to access a device. This includes even direct manipulations by a DOS application. Thus, the VxD can reliably detect when multiple VMs are trying to access the same device.

Note that Windows does *not* rely on VxDs to detect conflicts between multiple Windows applications trying to access the same device. The port-trapping and page-trapping features work on a per-VM basis, and all Windows applications live in the same VM. It is the job of a Windows driver DLL to serialize access to the device by multiple Windows applications.

Thus, a VxD that virtualizes a device is responsible for detecting and resolving conflicts between multiple VMs that want to use the same device. The VxD "resolves" the conflict by enforcing a particular "arbitration policy". In the Windows environment, the most common policies are:

- Allowing one VM to access the physical device and ignoring the other VMs. The VPD (Virtual Printer Device) uses this, the simplest form of virtualization.

- Allowing one VM to access the physical device and virtualizing the device for the other VMs. The VKD (Virtual Keyboard Device) takes this approach. The VKD assigns one VM to have the input focus, and that VM gets access to the physical keyboard, which includes keyboard interrupts. The VKD also makes sure the other VMs see an empty keyboard buffer

- Allowing multiple VMs to share the same physical device while maintaining the illusion, from the VM point of view, of exclusive access. The VDD (Virtual Display Device) behaves this way. Each windowed DOS VM writes directly to what it thinks is display memory, while the VDD remaps this memory to another buffer, which appears in a window.

- Allowing one VM to access the virtual device while the VxD independently controls the physical device. The VCD (Virtual Com Device) uses this, perhaps the most complicated form of virtualization. The VCD buffers incoming serial data, and transparently "feeds" it to a VM by reflecting the interrupt and then, when the VM interrupt handler reads the serial port data register, substituting an already-received byte from the buffer.

Like physical devices, hardware-generated interrupts must also be virtualized. Hardware-generated interrupts have no knowledge of VMs. Interrupts are virtualized to assure that each interrupt is visible to every VM that needs it, regardless of which VM was running when the interrupt was generated.

This chapter presents two example VxDs, PORTTRAP and PAGETRAP, that illustrate the techniques involved in virtualizing both port-mapped and memory-mapped devices. A third example, REFLECT, virtualizes a hardware interrupt. All of these VxDs use the simplest arbitration policy to resolve access conflicts. Avoiding unnecessary complexity in the arbitration policy emphasizes the basic techniques that are core to all virtualization VxDs: port-trapping, page-trapping, and interrupt reflection.

VMM and Processor Exceptions

At Windows startup, the VMM installs handlers in the IDT for all processor exceptions, including faults, traps, and interrupts. VxDs may then use various VMM services to register for notification from the VMM when a particular fault, trap, or interrupt occurs. The VPICD always registers with the VMM for all hardware interrupts, then other VxDs register with the VPICD to receive notification of hardware interrupts.

Although VMM provides a general purpose Hook_VMM_Fault service, which can be used to hook any type of fault, trap, or interrupt, most VxDs should register their handlers via more specialized services. The VMM offers other entry points specifically for use by port trap handlers and page fault handlers (Install_IO_Handler and Hook_V86_Page). By using these specific services, VxDs can take advantage of the pre-processing work done by the VMM fault handler, which figures out which VM caused the exception, which port or page the VM accessed, and even the specific instruction that causes the trap/fault. Similarly, VxDs should use the VPICD_-Virtualize_IRQ service to register a hardware interrupt handler rather than calling Hook_VMM_Fault.

The VxDs presented in this chapter will use the specialized VMM and VPICD services mentioned above. PORTTRAP will use Install_IO_Handler to receive callbacks on I/O port access. PAGETRAP will use Hook_V86_Page to receive callbacks on access to memory pages. REFLECT will use VPICD_Virtualize_IRQ service to get callbacks on hardware interrupts.

Device Ownership

Both PAGETRAP and PORTTRAP use a very simple algorithm for device management. Succinctly stated, the strategy is: "you touch it, you own it until you die." The first VM to access the device is declared the owner VM, and ownership is relinquished when a VM is terminated. If any other VM attempts to access the device while it is owned, the VxD may ask the user to decide which VM should be the owner.

The concept of device ownership is fundamental to a virtualization VxD. Typically the VxD disables local trapping of port I/O or of page faults to allow the owner VM direct access to the device without causing a trap, a step which improves performance. Also, if the device generates interrupts, the VxD makes sure that only the owner VM sees them.

Some VxDs allow access to specific I/O ports within a device without assigning an owner, if such accesses are benign and non-destructive. For example, the VCD (Virtual Com Device) allows any VM to configure a serial port with baud rate, parity, etc. Instead of outputting the bytes to the serial port, however, the VCD stores them in its own virtual copy of the serial port registers. Ownership is assigned when a VM accesses the serial port's interrupt or data registers. As part of assigning ownership, the VCD copies the virtual registers for that VM to the real serial port registers.

Implementing this type of behavior is more complicated and requires in-depth knowledge of how VMs are expected to access the device. If a VM accesses the device in a way that the VxD doesn't expect and, thus, doesn't handle it properly — for example, not reading a status register before writing to a register — the device won't function as the VM expects.

Virtualizing an I/O-mapped Device: The PORTTRAP Example

Writing a VxD to demonstrate I/O-mapped virtualization using port-trapping is complicated by the fact that Windows contains VxDs that virtualize most of the standard PC I/O port devices, and the VMM allows only one VxD to trap access to a given port. Rather than take over an existing device, this chapter's PORTTRAP traps the ports of an imaginary device at I/O address 300h–307h.

The PORTTRAP example (Listing 8.1, page 151) is the most elaborate of the examples in this chapter. It allocates per-VM storage in the VMM's Control Block and allows the user to resolve contention between VMs. Even so, PORTTRAP requires a very modest amount of code: only three message handlers (OnDeviceInit, OnSystemExit, and OnVmTerminate) and a port trap handler.

The Initialization Routine: OnDeviceInit

```
BOOL OnDeviceInit(VMHANDLE hVM)
{
    int i;

    for (i=0; i < device.numIoPorts; i++)
    {
        if (!Install_IO_Handler(device.IoBase+i, PortTrapThunk ))
        {
            DPRINTF1(buf, "Error installing handler for io %x\r\n", IO_BASE+i );
            return FALSE;
        }
    }
    if (device.cbOffset = _Allocate_Device_CB_Area(sizeof(DEVICE_CB), 0))
    {
        DPRINTF0("Error alloc'ing control block\r\n" );
        return FALSE;
    }

    return TRUE;
}
```

OnDeviceInit calls the VMM service Install_IO_Handler to register a port trap handler for each of the trapped ports. The VxD calls Install_IO_Handler in a loop, passing the same callback function each time (PortTrapThunk), but a different port number. Because the same callback function is used for all the ports, when the trap handler is invoked it will need to determine which port was accessed before it can act appropriately. This is an easy decision, because the port number is provided to the callback routine. An alternative method is to give each port its own callback routine.

The `Install_IO_Handler` service initially enables trapping for all VMs, current and future, which means that PORTTRAP doesn't have to take any special action when new VMs are created. A VxD could change this initial behavior by calling other VMM services: `Disable_Global_Trapping` and `Enable_Global_Trapping` change the trapping state of a specific port for all VMs; `Enable_Local_Trapping` and `Disable_Local_Trapping` change the trapping state only for a specific VM and a specific port.

The Calling Interface for `Install_IO_Handler`

```
BOOL Install_IO_Handler(DWORD PortNum, PIO_HANDLER IOCallback);
PortNum: I/O port number
IOCallback: pointer to callback function, called when VM accesses
            PortNum
```

PORTTRAP uses both device context and per-VM data structures. The device context structure, `DEVICE_CONTEXT`, includes fields for items like the I/O port base address and the handle of the owner VM. The per-VM structure, `DEVICE_CB`, consists of a single boolean field. This boolean is set whenever a user is asked to choose an owner VM from among two contending VMs. `OnDeviceInit` uses the VMM service `_Allocate_Device_CB_Area` to allocate room for a `DEVICE_CB` in the VM Control Block, then stores the returned offset in the device context.

Handling Different IO Types: *PortTrapThunk*

When a port trap occurs, the VMM calls the handler registered through `Install_IO_Handler`. As is the case with other example VxDs in this book, the actual registered callback is found in the assembly module. In this case the function is `_PortTrapThunk` in the module `PORTDDB.ASM` (Listing 8.2, page 154), and like the other example VxDs we've seen so far, `_PortTrapThunk` does minimal processing before calling the "real" callback in the C module, which is `PortTrapHandler`.

When the VMM invokes a port trap handler, the register data is set up as follows:

```
Input:
EAX=data for OUT instruction
EBX=current VM handle
ECX=IOType      //BYTE_INPUT, BYTE_OUTPUT, WORD_INPUT, WORD_OUTPUT,
                //DWORD_INPUT, DWORD_OUTPUT, STRING_IO, REP_IO,
                //ADDR_32_IO, REVERSE_IO
EDX=port number
EBP=address of Client Register Structure
Output:
EAX=data returned by IN instruction
```

PortTrapThunk passes all these parameters on to the C routine, after some initial pre-processing which involves the macro Emulate_Non_Byte_IO.

```
BeginProc PortTrapThunk

Emulate_Non_Byte_IO
cCall _PortTrap, <ebx, ecx, edx, ebp, eax>
ret

EndProc PortTrapThunk
```

The VxD is "emulating" non-byte I/O because its hardware understands only byte-sized access. Nothing prevents an application from issuing word or dword IN/OUT instructions, or even from performing "string I/O" using REP INSB/OUTSB. The VMM provides the macro Emulate_Non_Byte_IO to allow a VxD port trap handler to pass non-byte accesses back to the VMM. This macro expands to

```
;Emulate_Non_Byte_IO macro expansion
cmp ecx, BYTE_OUTPUT
jbe SHORT Byte_IO
VMMJmp  Simulate_IO
Byte_IO:
;cCall macro expansion
push eax
push ebp
push edx
push ecx
push ebx
call _PortTrap
;C routine returned with data in EAX, just return as is to VMM
ret
```

If the IOType parameter in ECX indicates a byte-sized access, the generated code falls through to the code after the macro, which pushes parameters on the stack and calls the C routine. If IOType is non-byte, then the code jumps to the VMM service Simulate_IO. This service breaks down a word access into two sequential calls back into the port trap handler, each with ECX=BYTE_INPUT or BYTE_OUTPUT. The service similarly breaks down dword and string access into multiple calls into the port trap handler.

The C routine PortTrapHandler called by _PortTrapThunk passes a return value in EAX, which _PortTrapThunk passes on to the VMM when it returns. If IOType was an IN of any size, the VMM will move the contents of EAX to the Client Register Structure EAX field. The end result is that the return value from _PortTrapThunk appears to the VM as the result of an IN instruction.

If your hardware directly supports word or dword I/O, your handler should also support these modes directly, rather than using Emulate_Non_Byte_IO.

Checking Ownership: *PortTrapHandler*

After taking care of non-byte access with the macro Emulate_Non_Byte_IO, the assembly language routine _PortTrapThunk calls the function PortTrapHandler in the C module to do the real work — to allow port access by the owner VM while preventing access from a non-owner VM.

```
DWORD _stdcall PortTrapHandler(VMHANDLE hVM, DWORD IOType, DWORD Port,
                               CLIENT_STRUCT *pcrs, DWORD Data)
{
   DEVICE_CB *pCB;
   BOOL      bThisVMIsOwner;
   VMHANDLE  newVMOwner;

   bThisVMIsOwner = TRUE;

   if (!device.VMOwner)
   {
      // device doesn't have an owner, assign this VM as owner
      SetOwner(hVM, &device);
   }
```

```
    else if (device.VMOwner && (device.VMOwner != hVM))
    {
        // device has an owner, but it's not this VM
        pCB = (DEVICE_CB *)((char *)hVM + device.cbOffset);
        if (pCB->flags & FLAGS_CONTENDED)
        {
            // this VM has already attempted to grab the device
            bThisVMIsOwner = FALSE;
        }
        else
        {
            newVMOwner = SHELL_Resolve_Contention(device.VMOwner, hVM,
                                                  device.DeviceName );
            if (newVMOwner != device.VMOwner)
            {
                bThisVMIsOwner = FALSE;
                Data = 0xFFFFFFFF;
            }
        }
    }

    if (bThisVMIsOwner)
    {
        if (IOType & BYTE_INPUT)
        {
            Data = _inp( Port );
        }
        else if (IOType & BYTE_OUTPUT)
        {
            _outp( Port, Data );
        }
    }

    return Data;
}
```

If the VMOwner field of DEVICE_CONTEXT is set to zero, then the device doesn't have an owner yet. In this case, the code calls the subroutine SetOwner to assign the VM that caused the trap as the owner. SetOwner updates the VMOwner field of DEVICE_CONTEXT and disables local trapping for the new owner VM, using the VMM service Disable_Local_Trapping. This service takes as parameters a VM handle and a port number. SetOwner calls the service in a loop, using the same VM handle (the new owner) and changing the port number each time to disable trapping on each of the device's ports. With local trapping disabled, the owner VM can now access the device without causing a fault and, thus, without interference from PORTTRAP. Access by any other VM will still cause a fault and a call to PortTrapHandler.

If the device does have an owner but it's not the VM that caused the trap, PORT-TRAP may use the `SHELL_Resolve_Contention` service to ask the user which VM should be owner: the already-assigned owner VM or the new "contender" VM. However, the VxD doesn't bother the user every time a non-owner VM accesses the device, only the very first time. The `FLAGS_CONTENDED` bit in the Flag field in the per-VM control block determines whether the VxD queries the user.

If `FLAGS_CONTENDED` is set, it means the VxD has already warned the user once that this VM is accessing the port and asked the user to assign an owner. In this case, `PortTrapHandler` simply sets the local variable `bThisVMIsOwner` to `FALSE`, which prevents code executed later in the function from performing the I/O access on behalf of the VM.

If `FLGS_CONTENDED` is clear, the VxD immediately sets it and then calls `SHELL_Resolve_Contention`, passing as parameters the VM handle of the current owner, the VM handle of the "contender" and a pointer to a device name. (See the sidebar "Why Blue Text?" for details on the `SHELL_Resolve_Contention` display.) The SHELL VxD then displays a dialog box listing the name of each VM (usually corresponding to the name of the DOS application running in the VM) and the name of the device, and the user chooses which VM should own the device.

`SHELL_Resolve_Contention` returns to `PortTrapHandler` with the handle of the chosen VM as a return value. If the user has *not* chosen the contending VM as owner, then `PortTrapHandler` sets the local variable `bThisVMIsOwner` to `FALSE`, so that code later in the function will not perform the I/O.

Why Blue Text?

Why does `SHELL_Resolve_Contention` sometimes display a blue text screen instead of a dialog box?

`SHELL_Resolve_Contention` appears to behave inconsistently, sometimes displaying a true Windows dialog box on top of the GUI, and sometimes going into full-screen mode and displaying a blue text message. Many developers think this blue screen is ugly and would like to force `SHELL_Resolve_Contention` to always display a true dialog box.

Bad news: you can't. The SHELL VxD's behavior depends on the current state of the GUI subsystem of the System VM, as well as which VM is current when `SHELL_Resolve_Contention` is called. In short, if the GUI subsystem is already "busy" when this SHELL function is called, a true dialog box cannot be displayed, so the SHELL VxD does the next best thing: switches to text mode and displays an ugly blue screen with the message on it.

At this point, PortTrapHandler has determined whether or not the VM that caused the port trap is indeed the owner VM, and thus should be allowed to access the port, and has set bThisVMIsOwner accordingly. If bThisVMIsOwner is now TRUE, PortTrapHandler carries out the I/O access on behalf of the VM, using the IOType parameter to determine whether to execute an IN or OUT and the Port parameter to determine the port address. If the access was an OUT, the Data parameter provides the output data. If the access was an IN, PortTrapHandler sets Data to the result of the IN. Finally, PortTrapHandler returns to his caller with Data as a return value. As explained in the previous section, the VMM propagates the port trap handler return value back to the VM, so the VM sees this value as the result of its IN instruction.

Processing VM_TERMINATE

Once a VM has acquired ownership of a device, it continues to own it until

- the VM terminates or
- the user selects a different owner through the Shell_Resolve_Contention service.

To detect the first case, PORTTRAP processes the VM_TERMINATE message. OnVmTerminate checks to see if the VM being destroyed is the device owner and, if so, sets VMOwner to zero to mark the device as unowned. OnVmTerminate does not need to re-enable port-trapping for the VM, because the VM itself is being destroyed.

Using PORTTRAP

I've implemented PORTTRAP as a static VxD so that it is present for the creation and destruction of all VMs. Under Windows 95, you can load a static VxD one of two ways: a device= statements in the [386Enh] section of SYSTEM.INI, or a registry entry under SYSTEM\CurrentControlSet\Services\VxD. For details on static load methods, refer to Chapter 4.

An easy way of testing PORTTRAP is to open several DOS boxes and use DEBUG to access the device through one of the ports at 300h–307h. (Use the i and o commands for input and output.) You'll see that after you access any one of the eight I/O ports that make up the imaginary device in one DOS box, the first access to the device in a different DOS box results in the "Device Contention" dialog box from the SHELL VxD. If you assign the original DOS box as owner, subsequent accesses by the second DOS box will not result in the dialog box. But if you open up a third DOS box and access the port from there, you will once again see the Device Contention dialog.

Virtualizing a Memory-mapped Device: The PAGETRAP Example

A device that is memory-mapped, as opposed to I/O-mapped, may also need a VxD to perform device arbitration. The need for such a VxD depends on where the device is mapped in memory. A device mapped above 1Mb in physical memory by definition cannot be accessed by a DOS application, and so doesn't need to be virtualized. But a device mapped below 1Mb can be accessed by a DOS application, and so may need a VxD for virtualization.

Because the only standard PC memory-mapped device is the video adapter, and the Video Device Driver (VDD) already virtualizes it, I've designed PAGETRAP (Listing 8.5, page 157) to virtualize the monochrome video adapter. If you don't have a monochrome video adapter, then PAGETRAP will still work, as PORTTRAP did, on an imaginary device.

The Initialization Routine

To intercept access to a memory-mapped device, PAGETRAP calls the following VMM services in its `Device_Init` message handler:

- `_Assign_Device_V86_Pages`, to tell the VMM that the VxD will be using a specific range of pages in V86 linear address space (i.e. below 1Mb),

- `_ModifyPageBits`, to mark the pages as not present so that VM access to the pages will cause a page fault, and

- `Hook_V86_Page`, to register a page fault handler for those pages.

Note that PAGETRAP does *not* allocate pages in physical memory, because the memory is already supplied by the device.

```
BOOL OnDeviceInit(VMHANDLE hVM)
{
   DWORD    PageNum = device.RegionPhysAddr >> 12;
   DWORD    nPages = device.RegionSize / 4096;

   if (!_Assign_Device_V86_Pages(PageNum, nPages, hVM, 0 ))
   {
      DPRINTF("Assign_Device_V86_Pages failed\r\n");
      return FALSE;
   }
```

```
if (!Hook_V86_Page(PageNum, PageFaultThunk ))
{
    DPRINTF("Hook_V86_Page failed\r\n");
    return FALSE;
}

if (!_ModifyPageBits(hVM, PageNum, nPages, ~P_AVAIL, 0, PG_HOOKED, 0 ))
{
    DPRINTF("ModifyPageBits failed\r\n");
    return FALSE;
}

return TRUE;
}
```

_Assign_Device_V86_Pages allows a VxD to claim pages in a VM's linear address space for use by a device. Later calls will associate physical address space with these linear pages. PAGETRAP uses the monochrome video adapter's physical address and size, stored in the DEVICE_CONTEXT structure, to derive the values for the VMLinrPage and nPages parameters.

The Calling Interface for _Assign_Device_V86_Pages

```
BOOL _Assign_Device_V86_Pages (DWORD VMLinrPage, DWORD nPages,
                              VMHANDLE hVM, DWORD flags);
VMLinrPage: linear page number (linear address >> 12)
nPages: number of (4 KB) pages
hVM: zero for global assignment
     non-zero VM handle for local assignment
flags: reserved; must be 0
```

A zero value for the hVM parameter means the assignment is global, that is, the pages are assigned to the device in all VMs (present and future). A non-zero value means the assignment is local; the pages are assigned to the device only in the VM identified by hVM. The VMM will return an error if one VxD has claimed a page globally and another VxD tries to claim the same page, whereas two different VxDs can both claim the same page locally without error. PAGETRAP uses zero for hVM, so that the device pages are claimed in all VMs.

Next, the OnDeviceInit routine calls Hook_V86_Page to register a page fault handler routine. PAGETRAP only hooks a single page. If you're writing a VxD for a device that spans multiple pages, you will need to call this service repeatedly — once for each page. I'll explain the page fault handler code in detail later.

The Calling Interface for Hook_V86_Page

```
BOOL Hook_V86_Page(DWORD PageNum, PV86Page_HANDLER Callback );
PageNum: linear page number
Callback: pointer to callback function,
          called when any VM causes a page-fault on PageNum
```

Last, OnDeviceInit calls _ModifyPageBits to mark the device page as not present in the System VM. Once again, the parameters hVM, VMLinPgNum, and nPages are self-explanatory. The bit-mapped values for the bitAnd and bitOr parameters match the processor's page table entry bits exactly.

The Calling Interface for _ModifyPageBits

```
BOOL _ModifyPageBits(VMHANDLE hVM, DWORD VMLinPgNum, DWORD nPages,
                     DWORD bitAnd, DWORD bitOR, DWORD pType,
                     DWORD Flags);
```

To force a page fault, PAGETRAP must clear the P_PRES, P_WRITE, and P_USER bits. The VMM.H header files has a #define for this particular combination of bits:

```
#define P_AVAIL (P_PRES | P_WRITE | P_USER)
```

To clear these three bits and leave all other bits as is, PAGETRAP uses a value of (~P_AVAIL) for the bitAND parameter and 0 for the bitOR. PAGETRAP uses a value of PG_HOOKED for the pType parameter, because the DDK documentation says that PG_HOOKED must be used if P_PRES, P_WRITE, or P_USER is being cleared.

PAGETRAP calls _ModifyPageBits with the very same parameters in its OnCreateVm message handler, so that the device pages are also marked as not present in the page tables for each new VM.

The Fault Handler Routine

PageFaultHandler [which is called by _PageFaultThunk in PAGEDDB.ASM (Listing 8.6, page 160)] has two jobs: it arbitrates access to its memory-mapped device, and it maps the owner VM's linear address to the device's physical address. PAGETRAP uses the same strategy that PORTTRAP did for device arbitration: you touch it, you own it. PAGETRAP's implementation is even simpler, though, as it doesn't ask the user to resolve contention. This means PAGETRAP uses no per-VM data and thus doesn't need to allocate space in the CB. PageFaultHandler merely watches for the first VM to access the device, and assigns that VM as owner.

```
VOID __stdcall PageFaultHandler(VMHANDLE hVM, DWORD PageNumber)
{
    if (device.VMOwner)
    {
        // device already has an owner, owner wouldn't cause a page
        // fault therefore this VM is not owner
        if (!_MapIntoV86( _GetNulPageHandle(), PageNumber, hVM,
                        PageNumber, device.RegionSize / 4096, 0, 0 ))
        {
            DPRINTF0("MapIntoV86 failed\r\n");
        }
    }
    else
    {
        device.VMOwner = hVM;
        _PhysIntoV86( PageNumber, hVM, PageNumber,
                    device.RegionSize / 4096, 0 );
    }
}
```

After an owner has been assigned, PAGETRAP causes all owner VM accesses to the memory-mapped device to go straight to the device, while all non-owner accesses are either ignored (writes) or return 0xFF (reads). To get this behavior, PAGETRAP uses the service _PhysIntoV86, which updates the VM's page tables to map a range of linear address space to a range of physical memory.

The Calling Interface for _PhysIntoV86

```
BOOL _PhysIntoV86(DWORD PhysPage, VMHANDLE hVM,
                  DWORD VMLinPgNum, DWORD nPages, DWORD Flags);
```

PageFaultHandler uses the handle of the faulting VM (provided by the caller, _PageFaultThunk) for hVM. Both PhysPage and VMLinPgNum are set equal to device.RegionPhysAddr >> 12 and nPages is set to device.RegionSize/4096. These values make linear page 0xB0 in the faulting VM map to physical page 0xB0. After this call, reads and writes by the VM to the device's linear address go directly to the device, without page-faulting.

This is the action taken by PageFaultHandler if the device had no owner. On the other hand, if the device already has an owner — a VM already accessed the pages and was assigned ownership — then PAGETRAP must take another action. Ideally, PAGETRAP would make it seem as if the device isn't present at that address, perhaps by returning 0xFF as a result of the VM's read of this address. But unlike a port trap handler, a VxD page fault handler doesn't have a return value that it can use to return 0xFF for a particular memory read access.

PAGETRAP has two options to trick the non-owner VM into seeing no device at physical address 0xB0000. One is to _PageAllocate a region of physical memory, fill it with 0xFF, and, when the page fault occurs, map the VM's pages to the allocated page. The VM will then see a region of memory that initially reads 0xFFs (although the page can be written to and read back with a new value). Presumably the device region would not read 0xFFs if the device was actually present at that page, and the VM would then determine the device wasn't present and would not attempt further access.

The other option achieves the same result with less work. Instead of mapping to a target page of 0xFFs, the VxD can map the VM's pages to a special page already allocated by the VMM called the "null page". The null page is mapped to different locations at different times, so the contents are random. This behavior should also cause the VM to determine that the device isn't present and not attempt further access. In my experience, the null page often maps to non-existent RAM, which does result in reading 0xFF.

PAGETRAP uses the null page approach. If the device is already owned, PageFaultHandler first calls _GetNulPageHandle to return the memory handle of the null page. Then PageFaultHandler calls the VMM service _MapIntoV86 to map the VM's linear address space to this null page.

The Calling Interface for _MapIntoV86

```
BOOL _MapIntoV86(MEMHANDLE hMem, VMHANDLE hVM,
                 DWORD VMLinPageNumber, DWORD nPages, DWORD PageOff,
                 DWORD Flags);
```

PAGETRAP uses the handle returned by _GetNulPageHandle for the hMem parameter and the VM handle of the faulting VM for hVM. Once again, VMLinPgNum is device.RegionPhysAddr >> 12 and nPages is set to device.RegionSize/4096. This service has an additional parameter, PageOff, which PAGETRAP sets to 0 so the first page of the linear region is mapped into the first page of the physical (null page) region. After the call to _MapIntoV86 with these parameters, reads and writes by the VM to the device's linear address go directly to the null page without page-faulting.

Processing VM_Terminate

PAGETRAP also processes the VM_TERMINATE message. OnVmTerminate checks to see if the VM being destroyed is the device owner, and if so, sets VMOwner in the device context to zero to mark the device as unowned. It is not necessary to do anything with the VM's page tables since the VM is being destroyed.

Using PAGETRAP

You can test PAGETRAP by opening several DOS prompt windows and using DEBUG to read and write to the monochrome adapter at B000:0000h. If you have an adapter installed, you should be able to read and write to it via DEBUG in the first DOS window that was opened, but you should see random data in the window when reading and writing to it from subsequent DOS windows. If you don't have an adapter at all, you'll read only 0xFFs from the first DOS window and random data from the other DOS windows.

Virtualizing a Hardware Interrupt

When virtualizing a device that generates interrupts, a VxD may virtualize the interrupt by "reflecting" it to a VM for servicing instead of servicing it in the VxD. A VxD reflects an interrupt — causes the interrupt handler in a VM to execute — by using VPICD services. A VxD can reflect an interrupt to any VM it chooses, but most VxDs assign VM ownership of a device through port-trapping or page-trapping, or even through an API, and then reflect all interrupts to the owner VM.

Because hardware interrupts occur asynchronously, any VM could be executing at the time a VxD calls the VPICD service for reflection. As the first step in reflection, the VPICD must force the desired VM to be scheduled. The VPICD forces the scheduling change by calling the VMM service Call_Priority_VM_Event with the highest priority, Time_Critical_Boost.

The VPICD provides a callback with this service, so the VMM may notify the VPICD when the target VM has been scheduled. The VPICD responds to the callback by using another VMM service, Simulate_Int, to modify the VM's execution environment. Simulate_Int changes the VM's state information so that it appears to execute an INT instruction: the VM's CS, IP, and flags registers are pushed onto the VM's stack; and the VM's new CS and IP values are fetched from the VM's IVT (location 0000:0000h in the VM's address space). In addition, the VPICD also clears the VM's interrupt flag because it's really simulating a hardware interrupt, not a software interrupt. When the VPICD returns from this callback and the VMM switches back to V86 mode, the VM immediately executes the interrupt handler for the hardware interrupt that was originally fielded by the VPICD.

Which VM?

The VPICD itself will reflect a hardware interrupt that is not claimed by any other VxD. Although the VPICD doesn't know about any other hardware devices besides its own (the PIC), it must still decide which VM gets the interrupt. In making this decision, the VPICD differentiates between local interrupts and global interrupts. A local interrupt is one that was disabled (in the physical PIC) at Windows startup. A global interrupt is one that was enabled at Windows startup. Note that since a global interrupt is enabled, a global interrupt must already have an interrupt handler installed in the BIOS, in a DOS driver, or in a TSR when Windows begins. We'll explore the importance of this statement shortly.

After Windows initializes, a VM may install an interrupt handler and then enable it in the PIC. By definition, that's a local interrupt. The VPICD now considers the VM that enabled the interrupt to be its owner, and from this point on the VPICD will always reflect this interrupt to the owner VM. This policy makes sense because the VM interrupt handler exists only in the installing VM; reflecting the interrupt to any other VM would result in calling an invalid address.

Global interrupts, on the other hand, do not have owners, but are reflected to whatever VM happens to be executing at the time the interrupt occurred. This works because a global interrupt was enabled when Windows started, and therefore had a handler installed when Windows started, which in turn means that the "global" handler exists in all VMs. Thus, it really doesn't matter to which VM the VPICD reflects a global interrupt — each has an IVT that points to the same handler. The difference between global and local interrupts is illustrated in Figure 8.1.

Once the VPICD has chosen a VM for reflection, it must make another choice: whether to call the protected mode or V86 mode handler. As Chapter 4 explained, all VMs start in V86 mode, and thus, have a V86 component; some VMs later switch to protected mode, and thus have a PM component also. One or both of these two components may install an interrupt handler. V86 interrupt handlers are those installed by a VM's V86-mode component, which includes the BIOS and DOS. PM interrupt handlers are those installed by a VM's PM component — usually a Windows DLL, but possibly a DOS-extended application using DPMI.

The VPICD always calls the protected mode handler, if one is installed. Only if no protected mode handler has been installed does the VPICD call the V86 mode handler. VPICD maintains a pseudo-IDT, which is updated when a protected mode application installs an interrupt handler through `DOS Set Vector` or `DPMI Set Protected Mode Vector`. This pseudo-IDT is used to get the address of the protected mode handler. Similarly, VPICD maintains a pseudo-IVT, which is updated when a DOS application installs an interrupt handler (or when a Windows application calls `DPMI Set Real Mode Vector`), and this pseudo-IVT provides the address of the V86 mode handler.

The above describes the VPICD's default behavior when no VxD has registered for the interrupt. If a VxD has registered for the interrupt and plans to reflect it to a VM, then it is the VxD's responsibility to choose the appropriate VM and direct the interrupt to the correct handler (protected mode or V86 mode). Typically, a VxD tracks ownership of a device and reflects the interrupt to the owner VM. The VxD passes the owner's VM handle to the VPICD as part of the call to `VPICD_Set_Int_Request`. (This service will be described in detail later in this chapter.)

A VxD for Hardware Interrupt Reflection

The REFLECT VxD (Listing 8.9, page 163) illustrates how to reflect an interrupt to an owner VM. The example code virtualizes the Real Time Clock interrupt, IRQ 8 (not to be confused with the timer interrupt on IRQ 0), but can be easily modified to work with any IRQ.

REFLECT virtualizes IRQ 8 during `Device_Init` processing with a call to `VPICD_Virtualize_IRQ`, passing a pointer to its `VPICD_IRQ_DESCRIPTOR` structure, `IrqDesc`. `VPICD_Virtualize_IRQ` returns an IRQ handle, which REFLECT stores in its device context. This handle will be used later when calling other VPICD services.

Figure 8.1 **VPICD associates an owner with each interrupt so that it can force the scheduling of the appropriate VM when a local interrupt is received. Global interrupts go to whichever VM is currently executing.**

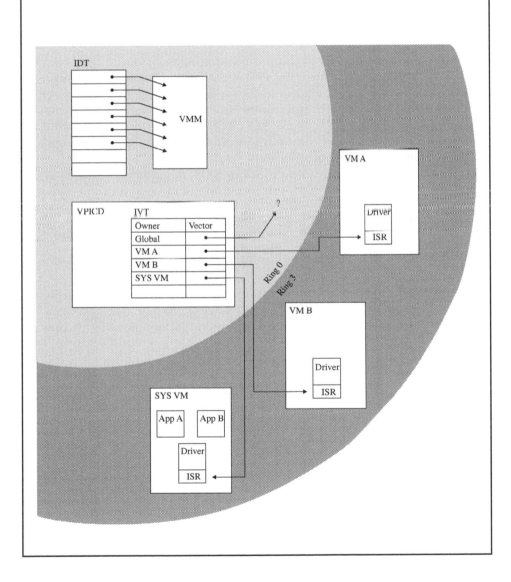

```
struct VPICD_IRQ_Descriptor {
   USHORT VID_IRQ_Number;
   USHORT VID_Options;
   ULONG  VID_Hw_Int_Proc;
   ULONG  VID_Virt_Int_Proc;
   ULONG  VID_EOI_Proc;
   ULONG  VID_Mask_Change_Proc;
   ULONG  VID_IRET_Proc;
   ULONG  VID_IRET_Time_Out;
   PVOID  VID_Hw_Int_Ref;
}
```

The VPICD_IRQ_DESCRIPTOR structure contains pointers to five callback functions, which the VPICD uses to notify the VxD of changes to the state of the physical and the virtualized IRQ. These callbacks are the key to reflecting an IRQ to an owner VM. The VXDISR VxD discussed in a previous chapter used this same structure but filled in only the VID_Hw_Int_Proc field. REFLECT fills in all five fields. VXDISR needed only one callback because it actually serviced the interrupt; REFLECT is only reflecting the interrupt to a VM for servicing.

The VPICD_IRQ_DESCRIPTOR structure used by REFLECT is statically initialized as follows:

```
VPICD_IRQ_DESCRIPTOR IrqDesc = { RTC_IRQ, VPICD_OPT_REF_DATA,
                                 HwIntThunk, VirtIntThunk, EOIThunk,
                                 MaskChangeThunk, IRETThunk, 500,
                                 &device };
```

REFLECT follows the same framework as the other VxDs in this book: all registered callback functions reside in the assembly language module. The C function always ends in the name "Handler". In the sections below, I'll talk only about the handler functions in the C module.

Callbacks: *MaskChangeHandler*

```
VOID MaskChangeHandler(VMHANDLE hVM, IRQHANDLE hIRQ, BOOL bMasking)
{
   if (!bMasking)
   {
      if (!device.VMOwner)
      {
         device.VMOwner = hVM;
      }
```

```
    else
    {
        if (device.VMOwner != hVM))
        {
            device.VMOwner = SHELL_Resolve_Contention(device.VMOwner, hVM
                                                        device.DeviceName );
        }
    }
    VPICD_Physically_Unmask( hIRQ );
    }
    else
    {
        device.VMOwner = 0;
        VPICD_Physically_Mask( hIRQ );
    }
}
```

When a VM masks or unmasks IRQ 8 in the interrupt controller, the VPICD calls MaskChangeHandler. REFLECT is more interested in unmasking than masking. REFLECT's rule for ownership is: "you enable the interrupt in the PIC, you own it". So if bMasking is FALSE, the function examines the VMOwner field in the device context to see any VM owns the IRQ. If no VM currently owns IRQ 8, MaskChangeHandler assigns the current VM as owner by setting VMOwner to the VM that is doing the unmasking, hVM.

If a VM already owns the IRQ, but a different (non-owner) VM is attempting the unmask, then MaskChangeHandler uses the SHELL_Resolve_Contention service to ask the user which VM should own the device. (See the earlier discussion of PORTTRAP for details on SHELL_Resolve_Contention.)

After determining the owner VM, MaskChangeHandler calls VPICD_Physically_Unmask to unmask the interrupt in the actual interrupt controller, then returns to the VPICD. Physically unmasking the interrupt is an important step. If no VxD has virtualized the IRQ, the VPICD traps all INs and OUTs to the interrupt controller and will unmask the interrupt on the VM's behalf. But once a VxD has virtualized an interrupt, the VPICD gets out of the way and the VxD must unmask the interrupt on the VM's behalf. The unmask service requires an IRQHANDLE parameter so MaskChangeHandler supplies the handle stored in the device context (the one returned by VPICD_Virtualize_IRQ).

If the VM is masking (disabling) the IRQ, REFLECT sets VMOwner to 0, then passes the mask request on to the VPICD with a call to the service VPICD_Physically_Mask and exits. It's not strictly necessary to set the owner to "none" in response to a mask because the interrupt can't even get to the processor while masked. However, the only other time the VxD could realistically set the owner to "none" would be in response to a VM_Terminate message. Setting the owner to "none" in response to a mask is more useful, because many applications will disable (mask) the interrupt as soon as they've finished with the device (as opposed to waiting until the user exits the program). By unassigning ownership at this time, the VxD can let another VM use the device.

Callbacks: HwIntHandler

```
BOOL _stdcall HwIntHandler(VMHANDLE hVM, IRQHANDLE hIRQ)
{
    if (device.VMOwner && !device.bVMIsServicing)
    {
        VPICD_Set_Int_Request( device.VMOwner, hIRQ );
    }
    else
    {
        EatInt();
    }
    return TRUE;
}
```

The actual reflection process occurs in HwIntHandler, which the VPICD calls whenever an interrupt occurs on IRQ 8. HwIntHandler then reflects, or simulates, an interrupt into the owner VM, but only under certain conditions:

- current (interrupted) VM is the device owner, and
- current VM's handler isn't servicing the device interrupt.

REFLECT uses the flag bVMIsServicing in the device context to prevent an interrupt from being simulated to the VM while the VM is still handling a previous interrupt. If the VM is overwhelmed with too many simulated interrupts, the interrupts will nest and the VM interrupt handler's stack will overflow. This flag is set and cleared in the VirtIntHandler and IRETHandler routines, which will be discussed shortly.

If the two conditions are met, REFLECT reflects the interrupt to the owner VM by calling VPICD_Set_Int_Request. This service requires two parameters, an IRQHANDLE and a VMHANDLE. HwIntHandler uses the IRQHANDLE field of the device context for the first, and the VMOwner field for the second. Note that when this service returns to HwIntHandler, the VM interrupt handler has not been called, the VPICD has only scheduled an event to take action later. However, HwIntHandler has done its duty, and now returns.

If HwIntHandler does not reflect the interrupt because conditions aren't right, it must service the interrupt itself. It does so by calling the subroutine EatInt. Clearing the interrupt in the device is an important step. If the interrupt is not cleared at the device, then the IRQ will remain asserted and the VPICD will never see another interrupt from that device because IRQs for ISA devices are edge-triggered.

```
void EatInt( void )
{
    unsigned char temp;

    temp = CmosReadReg( RTC_STATUSC );
    VPICD_Phys_EOI( device.IrqHandle );
}
```

The behavior of EatInt is specific to the RTC device: it clears the pending device interrupt by reading a status register. Because the interrupt was actually serviced, if only to be discarded, EatInt also calls VPICD_Phys_EOI to tell the VPICD to EOI the controller. Finally, EatInt returns to its caller, HwIntHandler.

HwIntHandler always returns TRUE to its caller, _HwIntThunk. This return causes _HwIntThunk to clear the Carry flag before returning to the VPICD. Carry clear on return informs the VPICD that the IRQ was processed by the VxD, and so the VPICD should not call the next VxD in the sharing chain. As written, REFLECT does not share interrupts, because the RTC hardware can't share its interrupt with other devices.

If your device does properly support sharing IRQs, you can easily enhance the VxD. Your HwIntHandler should first ask the device if it has an interrupt pending and if not, return with FALSE. The _HwIntThunk would then set the Carry flag, so that the VPICD calls the next VxD handler in the chain.

Callbacks: EOIHandler

```
void _stdcall EOIHandler(VMHANDLE hVM, IRQHANDLE hIRQ)
{
    VPICD_Phys_EOI( hIRQ );
    VPICD_Clear_Int_Request( device.VMOwner, hIRQ );
}
```

EOIHandler is called whenever the VM interrupt handler — executed eventually as a result of REFLECT's call to VPICD_Set_Int — issues an EOI to the interrupt controller. EOIHandler first calls VPICD_Phys_EOI on behalf of the VM that attempted to issue an EOI. The only parameter expected by VPICD_Phys_EOI is the IRQ handle. Last, EOIHandler calls VPICD_Clear_Int_Request, supplying the handle of the owner VM as the hVM parameter.

This call to VPICD_Clear_Int_Request clears the request set by HwIntHandler's call to VPICD_Set_Int_Request. Without this step, the VPICD would again reflect the interrupt to the VM handler some time after EOIHandler returned to the VPICD.

Callbacks: *VirtIntHandler* and *IRETHandler*

```
void VirtIntHandler(VMHANDLE hVM, IRQHANDLE hIRQ)
{
    device.bVMIsServicing = TRUE;

}
```

VirtIntHandler is called each time the VPICD begins simulating the interrupt into a VM. That is, it marks the beginning of the execution of the VM's interrupt handler. VirtIntHandler sets the bVMIsServicing flag, which prevents HwIntHandler from reflecting further interrupts into the VM until the VM handler has returned with an IRET.

```
void _stdcall IRETHandler(VMHANDLE hVM, IRQHANDLE hIRQ)
{
    device.bVMIsServicing = FALSE;
}
```

REFLECT knows when the VM handler has returned because another callback, IRETHandler, is called at that time. IRETHandler clears the bVMIsServicing flag, which allows HwIntHandler to reflect an interrupt once again.

Summary

Writing a VxD to virtualize a device is very different than writing a VxD to control a device, because it requires a completely different set of VMM and VxD services. Many VxDs today don't virtualize at all, because they are written for newer devices and there are no DOS or Windows applications that use this hardware directly.

If you do need to virtualize an I/O-mapped or memory-mapped device, trapping port or memory accesses is actually pretty easy. Virtualizing an interrupt is more complicated, simply because the process of interrupt reflection under Windows is itself complicated.

The last three chapters have talked about controlling hardware in a VxD and virtualizing hardware in a VxD. The next two chapters deal with another hardware aspect, discovering a device's configuration: I/O address, IRQ, etc.

Listing 8.1   `PORTTRAP.C`

```c
#include <basedef.h>
#include <vmm.h>
#include <debug.h>
#include <vxdwraps.h>

#include <vxdcall.h>
#include <wrappers.h>
#include <intrinsi.h>

#ifdef DEBUG
#define DPRINTF0(buf)  Out_Debug_String( buf )
#define DPRINTF1(buf, fmt, arg1)  _Sprintf(buf, fmt, arg1 ); Out_Debug_String( buf )
#else
#define DPRINTF0(buf)
#define DPRINTF1(buf, fmt, arg1 )
#endif


#define IO_BASE        0x300
#define NUM_IO_PORTS   8
#define FLAGS_CONTENDED 0x0001
typedef struct
{
    WORD        numIoPorts;
    WORD        IoBase;
    VMHANDLE    VMOwner;
    DWORD       cbOffset;
    char        DeviceName[8];
} DEVICE_CONTEXT;

typedef struct
{
    WORD        flags;
} DEVICE_CB;

DEVICE_CONTEXT device = { NUM_IO_PORTS, IO_BASE, NULL, 0,
                          {'P','O','R','T','T','R','A','P'} };

char buf[80];

BOOL OnDeviceInit(VMHANDLE hVM);
void OnSystemExit(VMHANDLE hVM);
void OnVmTerminate(VMHANDLE hVM);
void SetOwner( VMHANDLE newVMOwner, DEVICE_CONTEXT *dev );
DWORD _stdcall PortTrapHandler(VMHANDLE hVM, DWORD IOType, DWORD Port,
                          CLIENT_STRUCT *pcrs, DWORD Data);

// functions in asm module
void PortTrapThunk( void );
```

Listing 8.1 *(continued)*   *PORTTRAP.C*

```
BOOL OnDeviceInit(VMHANDLE hVM)
{
   int i;

   for (i=0; i < device.numIoPorts; i++)
   {
      if (!Install_IO_Handler(device.IoBase+i, PortTrapThunk ))
      {
         DPRINTF1(buf, "Error installing handler for io %x\r\n", IO_BASE+i );
         return FALSE;
      }
   }
   if (device.cbOffset = _Allocate_Device_CB_Area(sizeof(DEVICE_CB), 0))
   {
      DPRINTF0("Error alloc'ing control block\r\n" );
      return FALSE;
   }

   return TRUE;
}

VOID OnSystemExit(VMHANDLE hVM)
{
   int i;

   for (i=0; i < device.numIoPorts; i++)
   {
     if (!Remove_IO_Handler(device.IoBase+i))
     {
        DPRINTF1( buf, "Error removing handler for io %x\r\n", device.IoBase+i);
        break;
     }
   }
   if (device.cbOffset)
      _Deallocate_Device_CB_Area( device.cbOffset, 0 );
}

VOID OnVmTerminate(VMHANDLE hVM)
{
   if (hVM == device.VMOwner)
   {
      device.VMOwner = 0;
   }
}
```

Listing 8.1 (continued)   `PORTTRAP.C`

```c
DWORD _stdcall PortTrapHandler(VMHANDLE hVM, DWORD IOType, DWORD Port,
                              CLIENT_STRUCT *pcrs, DWORD Data)
{
    DEVICE_CB *pCB;
    BOOL      bThisVMIsOwner;
    VMHANDLE  newVMOwner;

    bThisVMIsOwner = TRUE;

    if (!device.VMOwner)
    {
        // device doesn't have an owner, assign this VM as owner
        SetOwner(hVM, &device);
    }

    else if (device.VMOwner && (device.VMOwner != hVM))
    {
        // device has an owner, but it's not this VM
        pCB = (DEVICE_CB *)((char *)hVM + device.cbOffset);
        if (pCB->flags & FLAGS_CONTENDED)
        {
            // this VM has already attempted to grab the device
            bThisVMIsOwner = FALSE;
        }
        else
        {
            newVMOwner = SHELL_Resolve_Contention(device.VMOwner, hVM, device.DeviceName );
            if (newVMOwner != device.VMOwner)
            {
                bThisVMIsOwner = FALSE;
                Data = 0xFFFFFFFF;
            }
        }
    }

    if (bThisVMIsOwner)
    {
        if (IOType & BYTE_INPUT)
        {
            Data = _inp( Port );
        }
        else if (IOType & BYTE_OUTPUT)
        {
            _outp( Port, Data );
        }
    }

    return Data;
}
```

Listing 8.1 (continued)   PORTTRAP.C

```c
void SetOwner( VMHANDLE newVMOwner, DEVICE_CONTEXT *dev )
{
    int i;

    for (i=0; i < dev->numIoPorts; i++)
    {
        Disable_Local_Trapping( dev->VMOwner, dev->IoBase+i );
        Enable_Local_Trapping( newVMOwner, dev->IoBase+i);
    }
    dev->VMOwner = newVMOwner;
}
```

Listing 8.2   PORTDDB.ASM

```asm
.386p

;*****************************************************************************
;                    I N C L U D E S
;*****************************************************************************

    include vmm.inc
    include debug.inc

;=============================================================================
;        V I R T U A L   D E V I C E   D E C L A R A T I O N
;=============================================================================

DECLARE_VIRTUAL_DEVICE    PORTTRAP, 1, 0, ControlProc, UNDEFINED_DEVICE_ID, \
                          UNDEFINED_INIT_ORDER

VxD_LOCKED_CODE_SEG

;=============================================================================
;
;   PROCEDURE: ControlProc
;
;   DESCRIPTION:
;    Device control procedure for the SKELETON VxD
;
;   ENTRY:
;    EAX = Control call ID
;
;   EXIT:
;    If carry clear then
;        Successful
;    else
;        Control call failed
;
;   USES:
;    EAX, EBX, ECX, EDX, ESI, EDI, Flags
;
;=============================================================================
```

Listing 8.2 (continued)   *PORTDDB.ASM*

```
BeginProc ControlProc
    Control_Dispatch DEVICE_INIT, _OnDeviceInit, cCall, <ebx>
    Control_Dispatch SYSTEM_EXIT, _OnSystemExit, cCall, <ebx>
    Control_Dispatch VM_TERMINATE, _OnVmTerminate, CCall, <ebx>
Control_Dispatch SYS_DYNAMIC_DEVICE_INIT, _OnSysDynamicDeviceInit, cCall, <ebx>
Control_Dispatch SYS_DYNAMIC_DEVICE_EXIT, _OnSysDynamicDeviceExit, cCall
    clc
    ret

EndProc ControlProc


VxD_LOCKED_CODE_ENDS

VxD_CODE_SEG


PUBLIC _PortTrapThunk
_PortTrapThunk PROC NEAR ; called from C, needs underscore
    Emulate_Non_Byte_IO155
    sCall PortTrapHandler, <ebx, ecx, edx, ebp, eax>
    ret

_PortTrapThunk ENDP

VXD_CODF_ENDS

    END
```

Listing 8.3   *PORTTRAP.MAK*

```
CVXDFLAGS  = -Zdp -Gs -c -DIS_32 -Z1 -DDEBLEVEL=1 -DDEBUG -DWANTVXDWRAPS
AFLAGS     = -coff -DBLD_COFF -DIS_32 -W2 -Zd -c -Cx -DMASM6 -DDEBLEVEL=1 -DDEBUG

all: porttrap.vxd

porttrap.obj: porttrap.c
        cl $(CVXDFLAGS) -Fo$@ %s

portddb.obj: portddb.asm
        ml $(AFLAGS) -Fo$@ %s

porttrap.vxd: portddb.obj porttrap.obj ..\wrappers\vxdcall.obj porttrap.def
        echo >NUL @<<porttrap.crf
-MACHINE:i386 -DEBUG -DEBUGTYPE:MAP -PDB:NONE
-DEF:porttrap.def -OUT:porttrap.vxd -MAP:porttrap.map
-VXD vxdwraps.clb wrappers.clb portddb.obj porttrap.obj vxdcall.obj
<<
        link @porttrap.crf
        mapsym porttrap
```

Listing 8.4   `PORTTRAP.DEF`

```
VXD VXDISR DYNAMIC
SEGMENTS
    _LTEXT      CLASS 'LCODE'    PRELOAD NONDISCARDABLE
    _LDATA      CLASS 'LCODE'    PRELOAD NONDISCARDABLE
    _TEXT       CLASS 'LCODE'    PRELOAD NONDISCARDABLE
    _DATA       CLASS 'LCODE'    PRELOAD NONDISCARDABLE
    _LPTEXT     CLASS 'LCODE'    PRELOAD NONDISCARDABLE
    _CONST      CLASS 'LCODE'    PRELOAD NONDISCARDABLE
    _BSS        CLASS 'LCODE'    PRELOAD NONDISCARDABLE
    _TLS        CLASS 'LCODE'    PRELOAD NONDISCARDABLE
    _ITEXT      CLASS 'ICODE'    DISCARDABLE
    _IDATA      CLASS 'ICODE'    DISCARDABLE
    _PTEXT      CLASS 'PCODE'    NONDISCARDABLE
    _PDATA      CLASS 'PCODE'    NONDISCARDABLE
    _STEXT      CLASS 'SCODE'    RESIDENT
    _SDATA      CLASS 'SCODE'    RESIDENT
    _MSGTABLE   CLASS 'MCODE'    PRELOAD NONDISCARDABLE IOPL
    _MSGDATA    CLASS 'MCODE'    PRELOAD NONDISCARDABLE IOPL
    _IMSGTABLE  CLASS 'MCODE'    PRELOAD DISCARDABLE IOPL
    _IMSGDATA   CLASS 'MCODE'    PRELOAD DISCARDABLE IOPL
    _DBOSTART   CLASS 'DBOCODE'  PRELOAD NONDISCARDABLE CONFORMING
    _DBOCODE    CLASS 'DBOCODE'  PRELOAD NONDISCARDABLE CONFORMING
    _DBODATA    CLASS 'DBOCODE'  PRELOAD NONDISCARDABLE CONFORMING
    _16ICODE    CLASS '16ICODE'  PRELOAD DISCARDABLE
    _RCODE      CLASS 'RCODE'
EXPORTS
    VXDISR_DDB @1
```

Listing 8.5   PAGETRAP.C

```c
// PAGETRAP.c - main module for VxD PAGETRAP

#include <basedef.h>
#include <vmm.h>
#include <debug.h>
#include <vxdwraps.h>

#include <vxdcall.h>
#include <wrappers.h>
#include <intrinsi.h>

#ifdef DEBUG
#define DPRINTF0(buf)  Out_Debug_String( buf )
#define DPRINTF1(buf, fmt, arg1)  _Sprintf(buf, fmt, arg1); Out_Debug_String( buf )
#else
#define DPRINTF0(buf)
#define DPRINTF1(buf, fmt, arg1)
#endif

#define DEVICE_PHYS_ADDR     0xB0000L
#define DEVICE_REGION_SIZE   4096

typedef struct
{
    DWORD        RegionSize;
    DWORD        RegionPhysAddr;
    VMHANDLE     VMOwner;
    DWORD        linAddr;
} DEVICE_CONTEXT;


DEVICE_CONTEXT device = { DEVICE_REGION_SIZE, DEVICE_PHYS_ADDR };

char buf[80];

BOOL OnDeviceInit(VMHANDLE hVM);
void OnSystemExit(VMHANDLE hVM);
BOOL OnCreateVm(VMHANDLE hVM);
void OnVmTerminate(VMHANDLE hVM);
DWORD _stdcall PageTrapHandler(VMHANDLE hVM,DWORD PageNumber);

// functions in asm module
void PageFaultThunk( void );

BOOL OnSysDynamicDeviceInit(VMHANDLE hVM)
{
   OnDeviceInit(hVM);
   return TRUE;
}

BOOL OnSysDynamicDeviceExit(void)
{
   OnSystemExit(Get_Cur_VM_Handle());
   return TRUE;
}
```

Listing 8.5 (continued)   `PAGETRAP.C`

```c
BOOL OnDeviceInit(VMHANDLE hVM)
{
    DWORD    PageNum = device.RegionPhysAddr >> 12;
    DWORD    nPages = device.RegionSize / 4096;

    if (!_Assign_Device_V86_Pages(PageNum, nPages, hVM, 0 ))
    {
        DPRINTF0("Assign_Device_V86_Pages failed\r\n");
        return FALSE;
    }

    if (!Hook_V86_Page(PageNum, PageFaultThunk ))
    {
        DPRINTF0("Hook_V86_Page failed\r\n");
        return FALSE;
    }

    if (!_ModifyPageBits(hVM, PageNum, nPages, ~P_AVAIL, 0, PG_HOOKED, 0 ))
    {
        DPRINTF0("ModifyPageBits failed\r\n");
        return FALSE;
    }

    return TRUE;
}

VOID OnSystemExit(VMHANDLE hVM)
{
    DWORD    PageNum = device.RegionPhysAddr >> 12;
    DWORD    nPages = device.RegionSize / 4096;

    if (!Unhook_V86_Page(PageNum, PageFaultThunk ))
    {
        DPRINTF0("Unhook_V86_Page failed\r\n");
    }
    if (!_DeAssign_Device_V86_Pages( PageNum, nPages, hVM, 0))
    {
        DPRINTF0("DeAssign_Device_V86_Pages failed\r\n");
    }
}

BOOL OnCreateVm(VMHANDLE hVM)
{
    if (!_ModifyPageBits(hVM, device.RegionPhysAddr >> 12, device.RegionSize / 4096,
                    ~P_AVAIL, 0, PG_HOOKED, 0 ))
    {
        DPRINTF0( "ModifyPageBits failed\r\n");
        return FALSE;
    }

    return TRUE;
}
```

Listing 8.5 (continued)   `PAGETRAP.C`

```
VOID OnVmTerminate(VMHANDLE hVM)
{
    if (hVM == device.VMOwner)
    {
        device.VMOwner = 0;
    }
}


VOID __stdcall PageFaultHandler(VMHANDLE hVM, DWORD PageNumber)
{
    if (device.VMOwner)
    {
        // device already has an owner, owner wouldn't cause a page fault
        // therefore this VM is not owner
        if (!_MapIntoV86( _GetNulPageHandle(), PageNumber, hVM,
                          PageNumber, device.RegionSize / 4096, 0, 0 ))
        {
            DPRINTF0("MapIntoV86 failed\r\n");
        }
    }
    else
    {
        device.VMOwner = hVM;
        _PhysIntoV86( PageNumber, hVM, PageNumber, device.RegionSize / 4096, 0 );
    }
}
```

Listing 8.6   `PAGEDDB.ASM`

```
.386p

;****************************************************************************
;                     I N C L U D E S
;****************************************************************************

    include vmm.inc
    include debug.inc

;==========================================================================
;        V I R T U A L   D E V I C E   D E C L A R A T I O N
;==========================================================================

DECLARE_VIRTUAL_DEVICE    PAGETRAP, 1, 0, ControlProc, UNDEFINED_DEVICE_ID, \
                          UNDEFINED_INIT_ORDER

VxD_LOCKED_CODE_SEG

;==========================================================================
;
;   PROCEDURE: ControlProc
;
;   DESCRIPTION:
;    Device control procedure for the SKELETON VxD
;
;   ENTRY:
;    EAX = Control call ID
;
;   EXIT:
;    If carry clear then
;        Successful
;    else
;        Control call failed
;
;   USES:
;    EAX, EBX, ECX, EDX, ESI, EDI, Flags
;
;==========================================================================

BeginProc ControlProc
    Control_Dispatch DEVICE_INIT, _OnDeviceInit, cCall, <ebx>
    Control_Dispatch SYSTEM_EXIT, _OnSystemExit, cCall, <ebx>
    Control_Dispatch VM_TERMINATE, _OnVmTerminate, CCall, <ebx>
Control_Dispatch SYS_DYNAMIC_DEVICE_INIT, _OnSysDynamicDeviceInit, cCall, <ebx>
Control_Dispatch SYS_DYNAMIC_DEVICE_EXIT, _OnSysDynamicDeviceExit, cCall
    clc
    ret

EndProc ControlProc


VxD_LOCKED_CODE_ENDS
```

Listing 8.6 (continued)   *PAGEDDB.ASM*

```
VxD_CODE_SEG

PUBLIC _PageFaultThunk
_PageFaultThunk PROC NEAR ; called from C, needs underscore

    sCall PageFaultHandler, <eax, ebx>
        ret

_PageFaultThunk ENDP

VXD_CODE_ENDS

    END
```

Listing 8.7   *PAGETRAP.MAK*

```
CVXDFLAGS  =  -Zdp -Gs -c -DIS_32 -Z1 -DDEBLEVEL=1 -DDEBUG -DWANTVXDWRAPS
AFLAGS     =  -coff -DBLD_COFF -DIS_32 -W2 -Zd -c -Cx -DMASM6 -DDEBLEVEL=1 -DDEBUG

all: pagetrap.vxd

pagetrap.obj: pagetrap.c
        cl $(CVXDFLAGS) -Fo$@ %s

pageddb.obj: pageddb.asm
        ml $(AFLAGS) -Fo$@ %s

pagetrap.vxd: pageddb.obj pagetrap.obj ..\wrappers\vxdcall.obj pagetrap.def
        echo >NUL @<<pagetrap.crf
-MACHINE:i386 -DEBUG -DEBUGTYPE:MAP -PDB:NONE
-DEF:pagetrap.def -OUT:pagetrap.vxd -MAP:pagetrap.map
-VXD vxdwraps.clb wrappers.clb pageddb.obj pagetrap.obj vxdcall.obj
<<
        link @pagetrap.crf
        mapsym pagetrap
```

Listing 8.8   PAGETRAP.DEF

```
VXD PAGETRAP DYNAMIC
SEGMENTS
    _LTEXT      CLASS 'LCODE'    PRELOAD NONDISCARDABLE
    _LDATA      CLASS 'LCODE'    PRELOAD NONDISCARDABLE
    _TEXT       CLASS 'LCODE'    PRELOAD NONDISCARDABLE
    _DATA       CLASS 'LCODE'    PRELOAD NONDISCARDABLE
    _LPTEXT     CLASS 'LCODE'    PRELOAD NONDISCARDABLE
    _CONST      CLASS 'LCODE'    PRELOAD NONDISCARDABLE
    _BSS        CLASS 'LCODE'    PRELOAD NONDISCARDABLE
    _TLS        CLASS 'LCODE'    PRELOAD NONDISCARDABLE
    _ITEXT      CLASS 'ICODE'    DISCARDABLE
    _IDATA      CLASS 'ICODE'    DISCARDABLE
    _PTEXT      CLASS 'PCODE'    NONDISCARDABLE
    _PDATA      CLASS 'PCODE'    NONDISCARDABLE
    _STEXT      CLASS 'SCODE'    RESIDENT
    _SDATA      CLASS 'SCODE'    RESIDENT
    _MSGTABLE   CLASS 'MCODE'    PRELOAD NONDISCARDABLE IOPL
    _MSGDATA    CLASS 'MCODE'    PRELOAD NONDISCARDABLE IOPL
    _IMSGTABLE  CLASS 'MCODE'    PRELOAD DISCARDABLE IOPL
    _IMSGDATA   CLASS 'MCODE'    PRELOAD DISCARDABLE IOPL
    _DBOSTART   CLASS 'DBOCODE'  PRELOAD NONDISCARDABLE CONFORMING
    _DBOCODE    CLASS 'DBOCODE'  PRELOAD NONDISCARDABLE CONFORMING
    _DBODATA    CLASS 'DBOCODE'  PRELOAD NONDISCARDABLE CONFORMING
    _16ICODE    CLASS '16ICODE'  PRELOAD DISCARDABLE
    _RCODE      CLASS 'RCODE'
EXPORTS
    PAGETRAP_DDB @1
```

Listing 8.9   REFLECT.C

```c
// REFLECT.c - main module for VxD REFLECT
#include <basedef.h>
#include <vmm.h>
#include <debug.h>
#include <vxdwraps.h>
#include <vpicd.h>

#include <vxdcall.h>
#include <wrappers.h>
#include <intrinsi.h>

#define RTC_IRQ     8

#define RTC_STATUSA    0xA
#define RTC_STATUSB    0xB
#define RTC_STATUSC    0xC

#define STATUSB_ENINT    0x40

#define CMOS_ADDR    0x70
#define CMOS_DATA    0x71

typedef struct
{
    IRQHANDLE          IrqHandle;
    VMHANDLE        VMOwner;
    char            DeviceName[8];
    BOOL            bVMIsServicing;
} DEVICE_CONTEXT;

// functions in asm module
void HwIntThunk( void );
void VirtIntThunk( void );
void EOIThunk( void );
void MaskChangeThunk( void );
void IRETThunk( void );

DEVICE_CONTEXT    device = { 0, 0, {'R','E','F','L','E','C','T'} };

VPICD_IRQ_DESCRIPTOR IrqDesc = { RTC_IRQ, VPICD_OPT_REF_DATA,
                                 HwIntThunk, VirtIntThunk, EOIThunk,
                                 MaskChangeThunk, IRETThunk, 500,
                                 &device };


BOOL OnDeviceInit(VMHANDLE hVM);
void OnSystemExit(VMHANDLE hVM);
BOOL _stdcall HwIntHandler(VMHANDLE hVM, IRQHANDLE hIRQ);
void _stdcall EOIHandler(VMHANDLE hVM, IRQHANDLE hIRQ);
void _stdcall VirtIntHandler(VMHANDLE hVM, IRQHANDLE hIRQ);
void _stdcall IRETHandler(VMHANDLE hVM, IRQHANDLE hIRQ);
void _stdcall MaskChangeHandler(VMHANDLE hVM, IRQHANDLE hIRQ, BOOL bMasking);
void EatInt( void );
void CmosWriteReg( BYTE reg, BYTE val );
BYTE CmosReadReg( BYTE reg );
```

Listing 8.9 (continued)   REFLECT.C

```c
BOOL OnSysDynamicDeviceInit(VMHANDLE hVM)
{
    OnDeviceInit( hVM );
    return TRUE;
}

BOOL OnSysDynamicDeviceExit(void)
{
    OnSystemExit(Get_Cur_VM_Handle() );
    return TRUE;
}

BOOL OnDeviceInit(VMHANDLE hVM)
{
    if (!(device.IrqHandle = VPICD_Virtualize_IRQ(&IrqDesc)))
        return FALSE;

    return TRUE;
}

VOID OnSystemExit(VMHANDLE hVM)
{
    VPICD_Force_Default_Behavior(device.IrqHandle);
}

BOOL _stdcall HwIntHandler(VMHANDLE hVM, IRQHANDLE hIRQ)
{
    if (device.VMOwner && !device.bVMIsServicing)
    {
        VPICD_Set_Int_Request( device.VMOwner, hIRQ );
    }
    else
    {
        EatInt();
    }
    return TRUE;
}

void EatInt( void )
{
    unsigned char temp;

    temp = CmosReadReg( RTC_STATUSC );
    VPICD_Phys_EOI( device.IrqHandle );
}

void _stdcall EOIHandler(VMHANDLE hVM, IRQHANDLE hIRQ)
{
    VPICD_Phys_EOI( hIRQ );
    VPICD_Clear_Int_Request( device.VMOwner, hIRQ );
}

void _stdcall VirtIntHandler(VMHANDLE hVM, IRQHANDLE hIRQ)
{
    device.bVMIsServicing = TRUE;
}
```

Listing 8.9 (continued)   REFLECT.C

```c
void _stdcall IRETHandler(VMHANDLE hVM, IRQHANDLE hIRQ)
{
    device.bVMIsServicing = FALSE;
}

void _stdcall MaskChangeHandler(VMHANDLE hVM, IRQHANDLE hIRQ, BOOL bMasking)
{
    if (!bMasking)
    {
        if (!device.VMOwner)
        {
            device.VMOwner = hVM;
        }
        else
        {
            if (device.VMOwner != hVM)
            {
                device.VMOwner = SHELL_Resolve_Contention(device.VMOwner,
                                                  hVM, device.DeviceName );
            }
        }
        VPICD_Physically_Unmask( hIRQ );
    }
    else
    {
        device.VMOwner = 0;
        VPICD_Physically_Mask( hIRQ );
    }
}

BYTE CmosReadReg( BYTE reg )
{
    BYTE    data;

    _asm
    {
        ; disable NMI then ints
        mov     al, reg
        or      al, 80h
        cli

        ; first output reg to address port
        out     CMOS_ADDR, al
    jmp     _1
_1:
    jmp     _2
_2:
        ; then read data from data port
        in      al, CMOS_DATA
        mov     data, al
    jmp     _3
_3:
    jmp     _4
```

Listing 8.9 (continued)   REFLECT.C

```
_4:
        ; reenable NMI then ints
        xor     al, al
        out     CMOS_ADDR, al
        sti
    }

    return data;
}

void CmosWriteReg( BYTE reg, BYTE val )
{
    _asm
    {
        ; disable NMI then ints
        mov     al, reg
        or      al, 80h
        cli

        ; first output reg to address port
        out     CMOS_ADDR, al
    jmp     __1
__1:
    jmp     __2
__2:

        ; then output val to data port
        mov     al, val
        out     CMOS_DATA, al
    jmp     __3
__3:
    jmp     __4
__4:

        ; reenable NMI then ints
        xor     al, al
        out     CMOS_ADDR, al
        sti
    }
}
```

Listing 8.10   REFLDDB.ASM

```
.386p

;*****************************************************************************
;                I N C L U D E S
;*****************************************************************************

    include vmm.inc
    include debug.inc

;===========================================================================
;         V I R T U A L   D E V I C E   D E C L A R A T I O N
;===========================================================================

DECLARE_VIRTUAL_DEVICE    REFLECT, 1, 0, ControlProc, UNDEFINED_DEVICE_ID, \
                          UNDEFINED_INIT_ORDER

VxD_LOCKED_CODE_SEG

;===========================================================================
;
;   PROCEDURE: ControlProc
;
;   DESCRIPTION:
;    Device control procedure for the SKELETON VxD
;
;   ENTRY:
;    EAX = Control call ID
;
;   EXIT:
;    If carry clear then
;        Successful
;    else
;        Control call failed
;
;   USES:
;    EAX, EBX, ECX, EDX, ESI, EDI, Flags
;
;===========================================================================

BeginProc ControlProc
    Control_Dispatch DEVICE_INIT, _OnDeviceInit, cCall, <ebx>
    Control_Dispatch SYSTEM_EXIT, _OnSystemExit, cCall, <ebx>
Control_Dispatch SYS_DYNAMIC_DEVICE_INIT, _OnSysDynamicDeviceInit, cCall, <ebx>
Control_Dispatch SYS_DYNAMIC_DEVICE_EXIT, _OnSysDynamicDeviceExit, cCall
    clc
    ret

EndProc ControlProc
```

Listing 8.10 (continued)   *REFLDDB.ASM*

```
PUBLIC _HwIntThunk
_HwIntThunk PROC NEAR ; called from C, needs underscore

    sCall HwIntHandler, <ebx, eax>
    or      ax, ax
    jnz     clearc
    stc
    ret

clearc:
    clc
    ret

 _HwIntThunk ENDP

VxD_LOCKED_CODE_ENDS

VxD_CODE_SEG

PUBLIC _VirtIntThunk
_VirtIntThunk PROC NEAR ; called from C, needs underscore

    sCall VirtIntHandler, <ebx, eax>
    ret

 _VirtIntThunk ENDP


PUBLIC _EOIThunk
_EOIThunk PROC NEAR ; called from C, needs underscore

    sCall EOIHandler, <ebx, eax>
    ret

 _EOIThunk ENDP


PUBLIC _IRETThunk
_IRETThunk PROC NEAR ; called from C, needs underscore

    sCall IRETHandler, <ebx, eax>
    ret

 _IRETThunk ENDP


PUBLIC _MaskChangeThunk
_MaskChangeThunk PROC NEAR ; called from C, needs underscore

    sCall MaskChangeHandler, <ebx, eax, ecx>
    ret

 _MaskChangeThunk ENDP

VXD_CODE_ENDS

    END
```

Listing 8.11   *REFLECT.MAK*

```
CVXDFLAGS = -Zdp -Gs -c -DIS_32 -Zl -DDEBLEVEL=1 -DDEBUG -DWANTVXDWRAPS
AFLAGS    = -coff -DBLD_COFF -DIS_32 -W2 -Zd -c -Cx -DMASM6 -DDEBLEVEL=1 -DDEBUG

all: reflect.vxd

reflect.obj: reflect.c
        cl $(CVXDFLAGS) -Fo$@ %s

reflddb.obj: reflddb.asm
        ml $(AFLAGS) -Fo$@ %s

reflect.vxd: reflddb.obj reflect.obj ..\wrappers\vxdcall.obj reflect.def
        echo >NUL @<<reflect.crf
-MACHINE:i386 -DEBUG -DEBUGTYPE:MAP -PDB:NONE
-DEF:reflect.def -OUT:reflect.vxd -MAP:reflect.map
-VXD vxdwraps.clb wrappers.clb reflddb.obj reflect.obj vxdcall.obj
<<KEEP
        link @reflect.crf
        mapsym reflect
```

Listing 8.12   *REFLECT.DEF*

```
VXD REFLECT DYNAMIC
SEGMENTS
    _LTEXT     CLASS 'LCODE'   PRELOAD NONDISCARDABLE
    _LDATA     CLASS 'LCODE'   PRELOAD NONDISCARDABLE
    _TEXT      CLASS 'LCODE'   PRELOAD NONDISCARDABLE
    _DATA      CLASS 'LCODE'   PRELOAD NONDISCARDABLE
    _LPTEXT    CLASS 'LCODE'   PRELOAD NONDISCARDABLE
    _CONST     CLASS 'LCODE'   PRELOAD NONDISCARDABLE
    _BSS       CLASS 'LCODE'   PRELOAD NONDISCARDABLE
    _TLS       CLASS 'LCODE'   PRELOAD NONDISCARDABLE
    _ITEXT     CLASS 'ICODE'   DISCARDABLE
    _IDATA     CLASS 'ICODE'   DISCARDABLE
    _PTEXT     CLASS 'PCODE'   NONDISCARDABLE
    _PDATA     CLASS 'PCODE'   NONDISCARDABLE
    _STEXT     CLASS 'SCODE'   RESIDENT
    _SDATA     CLASS 'SCODE'   RESIDENT
    _MSGTABLE  CLASS 'MCODE'   PRELOAD NONDISCARDABLE IOPL
    _MSGDATA   CLASS 'MCODE'   PRELOAD NONDISCARDABLE IOPL
    _IMSGTABLE CLASS 'MCODE'   PRELOAD DISCARDABLE IOPL
    _IMSGDATA  CLASS 'MCODE'   PRELOAD DISCARDABLE IOPL
    _DBOSTART  CLASS 'DBOCODE' PRELOAD NONDISCARDABLE CONFORMING
    _DBOCODE   CLASS 'DBOCODE' PRELOAD NONDISCARDABLE CONFORMING
    _DBODATA   CLASS 'DBOCODE' PRELOAD NONDISCARDABLE CONFORMING
    _16ICODE   CLASS '16ICODE' PRELOAD DISCARDABLE
    _RCODE     CLASS 'RCODE'
EXPORTS
    REFLECT_DDB @1
```

Chapter 9

Plug and Play:
The Big Picture

Plug and Play is Microsoft's strategy to make new hardware devices easier to install and configure. Plug and Play requires both hardware support (devices that can identify themselves and can be configured via standard software interfaces instead of jumpers or proprietary interfaces) and software support (an operating system that can assign system resources like I/O addresses and IRQs and drivers that obtain these resource settings from the operating system). Microsoft has provided the operating system piece in Windows 95, and Windows 95 also provides the interfaces that drivers use to retrieve resources assigned to their hardware.

In Windows 95, there are two categories of hardware devices: Plug and Play devices and Legacy devices. Plug and Play devices are those that can identify themselves, declare their resource requirements, and accept run-time resource assignments. Any device for one of the newer expansion buses — PCI, EISA, PCMCIA, etc. — is by definition a Plug and Play device. Each of these buses meets the above Plug and Play requirements. Some newer ISA cards include specific support for Plug and Play (PNP). These cards, known as Plug and Play ISA or PNPISA, are also considered Plug and Play devices. Legacy devices are those older ISA cards that do not support new Plug and Play features. A Legacy device cannot be dynamically configured; its resources are either fixed in the hardware or configured by switches or jumpers. Legacy devices also fail to support any vendor-independent method of positively identifying themselves.

Plug and Play Components

The heart of Windows 95 Plug and Play support is a VxD called the Configuration Manager. The Configuration Manager relies on other VxDs to do much of the real work, including: enumerators, arbitrators, device loaders, and device drivers. Both the Configuration Manager and the enumerators make use of a system-wide database called the registry to permanently store information about devices and their drivers.

Enumerators are VxDs that run at boot and determine which hardware devices are currently installed and what resource they require. Each bus type has its own enumerator: PCI, EISA, PCMCIA, SCSI, etc. Arbitrator VxDs are specific to a type of resource: I/O address, memory address, IRQ, DMA channel. The Configuration Manager gives an arbitrator information about a set of devices that all need a resource, say an IRQ, and the arbitrator comes up with a conflict-free set of assignments, taking into account which IRQs are supported by each device and whether or not each device can share the IRQ with another. Device Loaders are VxDs that load other VxDs. Windows 95 relies on Device Loaders because many devices are managed by several layers of drivers. The Device Loader knows enough about the layering to load each driver at the right time and in the right order.

The component of greatest concern to a developer is the Plug and Play Device Driver VxD. Enumerators, arbitrators, and device loaders are provided by Microsoft with the OS, so developers only need to understand how these component VxDs fit into the overall picture, not how to write one. A Plug and Play Device Driver VxD is a normal VxD that uses Configuration Manager services to obtain its resource assignments, instead of using private methods like INI-file settings or hard-coded values — nothing more mysterious than that. A PNP Device Driver still uses VMM and other VxD (VPICD, etc.) services to do its real job, which is acting as a driver for its device.

Figure 9.1   *Registry keys, subkeys, values, and data.*

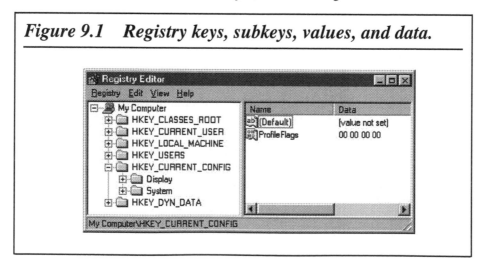

If you're supporting a Plug and Play device you should most definitely write a true Plug and Play Device Driver VxD instead of using low-level bus-specific methods in your VxD (like PCIBIOS, PCMCIA Socket services, etc.). But it's important to understand that Plug and Play device drivers aren't just for Plug and Play hardware. Plug and Play drivers are also meant for Legacy — standard ISA — hardware. For example, standard serial ports are Legacy devices, with a fixed I/O address and IRQ, so a serial port driver could be hard-coded to use those settings. But Microsoft's Windows 95 driver for standard serial ports, SERIAL.VXD, is a Plug and Play device driver, obtaining its settings from the Configuration Manager. (The Configuration Manager did, however, retrieve the settings from the registry, rather than from the device itself. Settings for Legacy devices are put in the registry by the Device Installer.) Microsoft encourages developers to write Plug and Play device drivers for all devices, including Legacy devices.

Plug and Play support is a new feature of Windows 95, not present in Windows 3.x. Therefore the information in this chapter and the next applies only to VxDs written specifically for Windows 95. A VxD for Windows 3.x must use other methods to obtain information about the resources used by its device. Other methods include querying the BIOS (e.g. COM1 and COM2 I/O address), reading SYSTEM.INI entries, or using hard-coded values.

The rest of this chapter will present an overview of Plug and Play, discussing the role of Configuration Manager and the enumerator, device loader, arbitrator, and device driver VxDs that it uses to actually implement the Plug and Play feature. The next chapter will explain in more detail the specifics of writing a Plug and Play Device Driver VxD, including a sample device driver VxD.

This chapter discusses how Plug and Play works at two times: installation and boot. Understanding how the Plug and Play components interact during operating system and driver installation is important for understanding the overall Plug and Play picture, because

The Windows 95 Registry

The registry is a binary database, accessible to the Windows 95 VMM, VxDs, and applications. The Windows 95 Plug and Play components use the registry to store and retrieve information about devices and their drivers, such as possible device configurations, device manufacturer, and the driver revision number. The registry is hierarchically structured, like a tree, where each node is called a key (Figure 9.1). One or more pieces of data, called values, can be associated with each key. A key (node) can also have subkeys, where each subkey is itself a tree with its own values and subkeys.

During installation, Windows 95 creates two keys at the root of the Windows 95 registry and several subkeys. The two root keys are HKEY_SYSTEM and HKEY_LOCAL_MACHINE (which is usually abbreviated HKLM). Most Plug and Play components other than the Configuration Manager use only two subkeys under those root keys, HKLM\ENUM and HKLM\SYSTEM\CurrentControlSet\System\Class.

the modifications made to the registry at installation time literally drive the Plug and Play boot sequence. (See the sidebar "The Windows 95 Registry" for information on how the Plug and Play components store and retrieve information about devices and their drivers.)

Plug and Play Components During Windows 95 Installation

When Windows 95 is first installed on a system, the Configuration Manager VxD identifies all the hardware devices in the system, using bus-specific modules called enumerators and detectors. Each enumerator positively identifies devices on a particular bus using bus-specific methods: the PCI enumerator reads PCI configuration space, PNPISA uses the Plug and Play isolation procedure, PCMCIA uses the Card Information Structure, etc. To find Legacy devices, the Configuration Manager uses detection modules instead of enumerators. Because of the limits of the ISA bus, detection modules must use less certain methods, such as examining hard-coded I/O locations for expected values, to detect standard ISA system hardware like the keyboard controller, interrupt controller, etc.

After an enumerator or detector has identified a new device, a module called the Device Installer creates a new hardware subkey for the device in the registry. This new key is of the form

```
HKLM\ENUM\<enumerator>\<device ID>\<instance ID>
```

The `<enumerator>` portion is either the bus name of the enumerator (PCI, SCSI, PCMCIA, etc.), or `Root` for Legacy devices found by detectors. The exact format of the `<device ID>` portion is enumerator-specific, but usually includes a combination of vendor id and adapter id, two identifiers supported by all Plug and Play buses. The `<instance ID>` uniquely identifies a particular instance of the device, and may be a serial number (as in PNPISA) or just an increasing number like `0000`, `0001`, etc.

After creating this new hardware key, the Device Installer adds registry subkeys under the hardware key, using information from either the device's information (INF) file, supplied by the vendor with the device, or from the device itself. Table 9.1 shows the values in a typical hardware key. The Device Installer always adds values called `DeviceDesc` and `Class` — two strings that describe the device and its type (network adapter, CD-ROM, etc.). For a Legacy device, the Device Installer also adds information about the device's current configuration (resource assignments). For a Plug and Play device, the Device Installer adds information about possible configurations, but not current configuration, because enumerated devices are always configured after boot. The Device Installer extracts this possible configuration information from the device itself in most cases (PNPISA, PCMCIA), or in some cases from nonvolatile system RAM (EISA).

The Device Installer always adds one more value, called Driver, under the hardware key. The data for Driver comes from the INF file. The Driver value has a misleading name, because it is not the name of the driver for the device. Instead it "points" to a software key for the device, which is always found under HKLM\SYSTEM\CURRENTCONTROLSET\SERVICES\CLASS. For example, if the Driver entry was Ports, the software key would be HKLM\SYSTEM\CURRENTCONROLSET\SERVICES\CLASS\PORTS.

The software key contains values describing the software associated with the device. Table 9.2 shows a typical software key. The enumerator or detector always adds a DevLoader value. DevLoader names the VxD that will act as a "device loader" for the driver for the device. Surprisingly, the software key does not contain a standardized value representing the driver name. But the software key does contain enough information to allow the device loader to *determine* the device driver name — more about this later in the discussion of device loaders and the boot process.

Table 9.1   Typical hardware key.

Value	Data
Class	"Display"
CompatibleIDs	"PCI/CC_0300"
ConfigFlags	00 00 00 00
DeviceDesc	"S3 Inc. Trio32/64 PCI"
Driver	"Display/0001"
HardwareID	"PCI/VEN_ 5333&DEV_8811/BUS_00&DEV_10&FUNC_00"
HWRevision	"067"
Mfg	"S3"

Table 9.2   Typical software key.

Value	Data
DevLoader	"*vdd"
DriverDesc	" S3 Inc. Trio32/64 PCI"
InfPath	"OEM1.INF"
InfSection	"S3_2"
Ver	"4.0"

Plug and Play Components
During Device Installation

The process for installation of a new device and its associated driver after initial Windows 95 installation is similar. Newly installed Plug and Play devices are discovered by an enumerator at the next boot. Newly installed Legacy devices are discovered when the user runs the Add New Hardware Wizard.

Whether the new device is Plug and Play or Legacy, the Device Installer knows it's a newly installed device because the device has no hardware key in the registry. When a new device is discovered, the Device Installer looks for the device's associated INF file, asking the user to specify its location if the file can't be found. Once the INF file is located, the enumerator creates a registry hardware key and software key and copies the driver from the installation disk, just as during the original Windows 95 installation. Once added to the registry, the "new" device becomes an "installed" device; on subsequent boots, it will be treated just like all the other installed devices.

Plug and Play Components During Boot

During installation, Windows 95 is interested only in identifying the system's devices and the drivers needed to run them. During the Windows 95 boot process, the operating system does more than identify devices and drivers, it also loads the drivers and configures the devices. The Configuration Manager VxD is the brains behind this boot process, orchestrating enumerators, arbitrators, device loaders, and the drivers themselves.

During the boot process, the Configuration Manager uses enumerator VxDs to discover devices, device loader VxDs to load driver VxDs for the devices, and arbitrator VxDs to assign conflict-free configurations to all the devices. As a last step, the Configuration Manager informs each device driver VxD of the configuration assigned to its device. The following pseudo-code shows the boot process. The following sections will explain in more detail the role of each of these types of VxDs in the boot process.

Pseudo-code for the Plug and Play Boot Process

```
For each enumerator
    CM calls enumerator to enumerate all devices on its bus
        For each device:
            Enumerator finds device, calls CM
                to create DevNode from Device ID
            if no hardware key in registry,
                device is new and must be installed
            CM sends DevLoader a PNP_New_DevNode message
                DevLoader loads a Driver VxD
                CM sends Driver VxD a PNP_New_DevNode message
                Driver VxD calls CM_Register_Device_Driver
                    to register a configuration callback
            CM returns to the enumerator
        Enumerator returns to CM
    CM links devnodes into a hardware tree
CM uses arbitrators to assign conflict-free configurations
CM traverses hardware tree, beginning at root. For each node:
    calls each Driver VxD's registered configuration function
Driver VxD calls CM_Get_Alloc_Log_Conf to discover assigned resources
```

Plug and Play Components During Boot: Enumerators

During boot, the Configuration Manager runs the same enumerators that were used during installation, one for each bus. But instead of running detectors as during installation, at boot the Configuration Manager runs the Root enumerator. The Root enumerator is different than other enumerators in that it doesn't attempt to identify any hardware, it just relies on the information already placed in the registry (in HKLM\Root) by detectors at installation.

After identifying each device, an enumerator creates a device node, a data structure containing basic information about an identified device. The device node contains fields for possible configurations, current configuration, status information (disabled, configured, etc.), and the driver for the device. The enumerator fills in these fields from values stored in the device's hardware key or from information provided by the device itself.

Device nodes serve as the basic unit of "currency" between Plug and Play components (Configuration Manager, enumerators, arbitrators, device drivers). In other words, device nodes are passed around from one component to another to identify the target device. Note that while some of the information in a device node is also found in the registry, a device node is different from a registry entry in two ways. One, the device node is in memory, not on disk, allowing much faster access. Two, the device node represents a device that is physically present on the system, whereas registry entries stay even after a device is removed to make device reinstallation easier.

As each enumerator creates a device node, it reports the new device node to the Configuration Manager. The Configuration Manager then initiates a long sequence that eventually results in the driver for the device being loaded. The enumerator then proceeds to the next device. When it has processed all device nodes, the Configuration Manager calls the next enumerator, which repeats the sequence for its own devices. When all enumerators have finished, the Configuration Manager has connected the device nodes to form a hierarchical structure called the hardware tree, an in-memory representation of the system's hardware devices.

Plug and Play Components During Boot: Device Loaders

As each device node is discovered by an enumerator, the Configuration Manager attempts to load a device driver for the device node. The Configuration Manager uses the `DevLoader` value in the device's software key (pointed to by the `Driver` value in the device's hardware key), which names the VxD responsible for loading the "real" device driver. The Configuration Manager sends a `PNP_New_DevNode` message to the VxD named as `DevLoader`, informing the VxD that a new device node has been created and that the VxD is to act as the device loader for this new device.

Two parameters are associated with a `PNP_New_DevNode` message: a pointer to the device node and a reason code describing the action the VxD should take. In this initial message to the device loader, the Configuration Manager uses the `DLVXD_LOAD_DEVLOADER` reason code. The name for this reason code is a bit confusing: the Configuration Manager is really telling the VxD to load the *driver* for the device, not to load the *device loader* for the device. `DLVXD_LOAD_DEVLOADER` really tells a VxD that "you are the device loader".

The Configuration Manager relies on device loaders instead of loading all drivers itself because some devices are managed by several layers of drivers. The device loader knows enough about the layering to load each driver at the right time and in the right order. The device loader also knows which value in the software key contains the actual driver name. For example, SCSI devices use the `MiniPortDriver` value to store the driver name, but COM ports use `PortDriver`, and network devices use `DynamicVxD`.

For those device classes that do separate the device loader VxD from the device driver VxD, the device loader must respond to the `DLVXD_LOAD_DEVLOADER` reason code by finding and loading the appropriate driver. Device loaders don't do this work themselves, but rely on two Configuration Manager services. A device loader uses `CM_Read_Registry_Value` to obtain the driver name from the appropriate entry in the software key and then `CM_Load_DLVxDs` to actually load the device driver VxD. Its job finally done, the device loader VxD returns from `PNP_New_DevNode` processing, back to the Configuration Manager.

The Configuration Manager now sends a second PNP_New_DevNode message, this one directed to the newly loaded driver VxD. The same device node parameter is used since it's still processing the same device, but this time the reason code is DLVXD_LOAD_DRIVER. Once again, the name is a bit confusing, it doesn't mean "load the driver", it means "you are the driver". The driver VxD should respond to this reason code by calling CM_Register_Device_Driver.

In cases where layering is not used, the device loader VxD and the device driver VxD are one and the same. In this simple case, when the VxD gets the DLVXD_LOAD_DEVLOADER reason code it doesn't load another VxD. Instead, the combination device loader/device driver VxD tells the Configuration Manager that it is the driver for the device by calling the Configuration Manager's CM_Register_Device_Driver function during DLVXD_LOAD_DEVLOADER processing.

One of the parameters that a device driver VxD passes to CM_Register_Device_Driver is a pointer to a callback function. The Configuration Manager calls this driver VxD function later to inform the device driver VxD of configuration events. The driver callback function and the configuration events that it processes will be covered in more detail in a later section.

Plug and Play Components During Boot: Arbitrators

At this point in the boot process, all hardware devices have been identified and drivers have been loaded. Before the drivers can access their devices, the arbitrators must find a conflict-free set of configurations for all devices.

There are four built-in arbitrators, one for each type of system resource: I/O ports, memory ranges, IRQs, and DMA channels. The I/O port arbitrator takes a list of device nodes and assigns to each the number of I/O ports it requires. The arbitrator must select ports that don't conflict with the port assignments for any other device node in the list. The other three arbitrators do exactly the same thing, each with their own resource type.

Arbitrators must handle "fussy" devices that support only a single resource assignment — e.g. a Legacy device that only supports I/O ports 200h–220h — as well as "flexible" devices — e.g. a PCI device that supports any 32-byte block within the entire 64Kb range of I/O space. The IRQ arbitrator handles an additional twist as well, because some devices support sharing an IRQ with another device, while others do not. An arbitrator returns either a success or failure code to the Configuration Manager, indicating whether or not it was successful in finding a set of allocations that worked.

The arbitration process would be fairly simple if resources were always independent of each other — if, for example, the choice for I/O port had no effect on the choice for IRQ. However, resource dependencies are common among Legacy devices: consider the standard serial port with choices of I/O 3F8h plus IRQ 4 or I/O 2F8h plus IRQ 3. To handle resource dependencies, the Configuration Manager uses the arbitrators in an iterative manner, calling each with the ARB_TEST_ALLOC reason code, which asks the arbitrator to make a trial allocation. This ARB_TEST_ALLOC allocation may occur several times. Later, when all arbitrators have returned a success code, the Configuration Manager calls each again with the ARB_SET_ALLOC reason code, telling the arbitrators to make this allocation permanent.

An example will make this process more clear. Consider a list of two devices. One is a Legacy mouse that supports any I/O port in the 200h–3FFh range, but only IRQ 4. (This resource combination is considered a single "logical configuration".) The other is a Legacy serial port that supports either I/O 3F8h plus IRQ 4 or I/O 2F8h plus IRQ 3. (Thus, the serial port is associated with two logical configurations.) To configure these two devices, the Configuration Manager must choose a logical configuration for each device, using the arbitrators to ensure that the resources that make up each chosen configuration don't conflict with each other.

Before calling the arbitrators, the Configuration Manager makes one of the logical configurations the "current" configuration. The arbitrators consider only the resources in this current configuration when making allocations — they are unaware of the resources available in any other logical configuration. In our example, suppose the Configuration Manager chose 3F8h/IRQ 4 as the current configuration for the serial port, and 200h/IRQ 4 as the current (and only) configuration for the mouse. It calls the port arbitrator first and then the IRQ arbitrator, using the ARB_TEST_ALLOC reason code for both.

In this scenario, the I/O port arbitrator can easily identify a set of non-conflicting assignments and returns TRUE, but the IRQ arbitrator cannot (both devices want IRQ 4) and returns FALSE. So the Configuration Manager makes 2F8h/IRQ 3 the current configuration for the serial port and tries again, still using the ARB_TEST_ALLOC reason code for both arbitrators. This time there is no IRQ conflict (the serial port wants IRQ 3 and the mouse wants IRQ 4) so both arbitrators return TRUE. Now the Configuration Manager calls each arbitrator again with the same current configuration as last time, but now with the ARB_SET_ALLOC reason code. When the arbitrators return, both devices have been allocated a set of non-conflicting resources: the mouse with 200h/IRQ 4 and the serial port with 2F8h/IRQ 3.

Plug and Play Components During Boot: Device Driver VxDs

After the Configuration Manager has assigned all devices a conflict-free set of resources, it must inform each driver VxD of the configuration assigned to its device. The Configuration Manager does this through the callback function registered by each driver during its call to CM_Register_Device_Driver.

The Configuration Manager passes a reason code to the configuration callback. The CONFIG_START code notifies the driver VxD that a configuration has been assigned: CONFIG_START means "start using your device's assigned configuration". A Plug and Play device driver isn't supposed to use any of its device's resources until it gets this notification. Whereas a Windows 3.x VxD usually installed I/O port handlers and virtualized an IRQ during system initialization, the rules are different under Windows 95. A Windows 95 Plug and Play device driver VxD may be loaded early in the boot process, but shouldn't do anything with system resources until explicitly notified by the Configuration Manager in this CONFIG_START message.

At the time of a CONFIG_START message, the Configuration Manager has already assigned the resources, so the driver VxD simply retrieves that assignment. (The Configuration Manager could have made it easy on the VxD by passing the resource assignments as a parameter to the configuration callback function — but it doesn't.) The VxD must make yet another call to the Configuration Manager, this time to CM_Get_Alloc_Log_Conf (Alloc stands for allocated, log stands for logical). This call returns with all configuration information in a single CMCONFIG structure: memory ranges, I/O ports, IRQs, DMA channels. Now that the device driver VxD finally knows which resources its device will be using, it can call VPICD to install an interrupt handler, call VDMAD to register a DMA channel, etc.

Summary

This chapter has introduced the component VxDs that make up Plug and Play in Windows 95 and explained how these components interact to identify devices, assign resources, and load drivers. The next chapter will focus specifically on the Plug and Play device driver VxD, and you will learn exactly what a driver VxD must do to support a Plug and Play device: which messages and callbacks it must handle and which Configuration Manager services it must call.

Chapter 10

Plug and Play Device Driver VxDs

The last chapter provided an overview of Windows 95 Plug and Play, introducing the different kinds of Plug and Play VxDs (Configuration Manager, enumerators, device loaders, arbitrators, device drivers) and the role played by each in the Windows 95 installation and boot processes. This chapter will focus on the Plug and Play Device Driver VxD, which I'll define as a VxD that interfaces to a hardware device and that obtains a device's configuration using methods that conform to Plug and Play rules.

This chapter will first explain the steps required to install a Plug and Play Device Driver VxD in the Windows 95 environment. Next, you'll see how a Plug and Play Device Driver VxD participates in the Windows 95 boot and initialization processes and how it handles other Plug and Play configuration scenarios such as device removal. The final sections will discuss in detail the code for a sample Plug and Play Device Driver VxD.

Plug and Play VxD Installation

Windows 3.x offered no standardized procedure for installing device drivers, so different vendors provided different solutions. Some vendors provided an application — sometimes a Windows program or sometimes a DOS program — that copied the driver file and made modifications to system files. Others provided only instructions and required the user to do the installation. The Plug and Play support in Windows 95 addresses this installation deficiency. Windows 95 standardizes the device installation process, both from the user perspective and from the driver vendor's perspective.

To install a new piece of hardware for Windows 95, the user first physically installs the card then boots up Windows 95. If the new hardware is Plug and Play, an enumerator automatically identifies the new device and the Device Installer prompts the user for a device installation disk. If the new hardware is Legacy, the system cannot identify the new device, and the user is required to run the Add New Hardware Wizard. This "wizard" guides the user, step-by-step, through the installation process, prompting the user for a device installation disk when required.

The device installation disk (created by the vendor) includes a device driver, a Device Information (INF) file, and optional utility or diagnostic programs used with the device. The INF file is an important piece of the Windows 95 Plug and Play standard. It provides the Device Installer with a device description for display to the user and an "installation script" to install the device driver. The installation script includes items like the name of the driver on the installation disk, the directory the driver should be copied to, and any registry entries that must be created or modified during driver installation.

Introducing the INF File

The INF text file resembles a Windows 3.x INI file. The INF is divided into sections, where each section contains one or more items. Each section relates to one step of the installation process: one section for files to be copied, another for registry entries to be added, etc. As a developer, you can create INF files with any text editor. However, Microsoft also provides an INFEDIT tool with the DDK, which allows you to navigate and edit the file in a hierarchical manner — sort of like the outline view in a word processor. Because the sections in an INF file are arranged in a hierarchy, the INFEDIT tool is very useful. (See the DDK for an explanation of how to use INFEDIT.)

The INF file can support complicated installation scenarios, but most developers will only need to handle the basics. A basic driver installation scenario includes:

- identifying the device;
- copying the driver file from the driver disk;
- identifying the device's resource requirements; and
- adding a `DevLoader` registry entry to load the driver when the device is enumerated.

Tables 10.1 and 10.2 detail the INF file sections, and items within those sections, that are required to cover this basic installation. For an actual INF file with these sections and items, as well as more details about what they mean, see the section "A Sample Plug and Play Driver VxD" later in this chapter.

Section Type	Item Name	Item Description
Version	Signature	Must be $CHICAGO$
	Class	Choose from list in Table 10.2
	Provider	Creator of INF file, typically same as vendor name
Manufacturer	Manufacturer Name	Vendor name
	Device Description	Device Manager and Add New Hardware Wizard show this string to user
	Device ID	ASCII identifier created by hardware vendor: Consists of * followed by 3-letter (EISA format) company ID then four hex digit device ID
	Install Section Name	Names later section containing installation instructions
CopyFiles	(None)	Destination file name, optional source file name
AddReg	(None)	Registry root, optional subkey, value name, and value: All drivers require one of these to specify device loader; for example, HKLM,,DevLoader,0,myvxd
LogConfig	IOConfig	Describes I/O addresses supported: minimum, maximum, size USE ONLY IF LEGACY DEVICE
	IRQConfig	Describes IRQs supported and whether or not sharable USE ONLY IF LEGACY DEVICE
Install	Copyfiles	Name of CopyFiles section in this INF file
	AddReg	Name of AddReg section in this INF file
	LogConfig	Name of LogConfig section in this INF file

Table 10.1 Standard sections for an INF file.

Table 10.2 *Device classes supported by configuration manager.*

Class Name in INF File	Class Name in Device Installer	Description
Adapter	CD-ROM controllers	Non-SCSI CD-ROM controller
DiskDrive	Disk drives	
Keyboard	Keyboard	
System	System devices	Motherboard device (PIC, PCI bridge, etc.)
MEDIA	Sound, video, and game controllers	Multimedia
Modem	Modem	
MultiFunction	Multi-function adapters	e.g. Combination modem and network adapter
Monitor	Monitor	
CDROM	CD-ROM	
Display	Display	
fdc	Floppy disk controllers	
hdc	Hard disk controllers	
Mouse	Mouse	
Ports	Ports (COM & LPT)	Serial and parallel
Printer	Printer	
MTD	Memory Technology Drivers	PCMCIA memory card
Net	Network adapters	
nodriver		Device that requires no driver
PCMCIA	PCMCIA socket	
SCSIAdapter	SCSI controllers	SCSI host adapter
Unknown	Other devices	

Plug and Play Boot Process

Driver VxD Load Sequence

Though it's not an absolute requirement, almost all Plug and Play Driver VxDs are dynamically loadable. Dynamic loading is preferred because it allows the Configuration Manager to unload a driver when its associated device is removed, either physically removed in the case of "hot insertion" devices such as PCMCIA, or logically removed when the Configuration Manager detects a device conflict or the user chooses Remove in the Device Manager.

The dynamic load procedure for a Plug and Play VxD is a convoluted process. The process begins at boot, when an enumerator identifies a particular device. The enumerator passes the Configuration Manager the Device ID and asks the Configuration Manager to create a "devnode" (device node) for the device. The Configuration Manager forms the device's hardware key by prepending HKLM\ENUM to the ASCII Device ID. This hardware key contains a Driver value that points to the software key under HKLM\SYSTEM\CURRENTCONTROLSET\SERVICES\CLASS. That software key contains a DevLoader value.

The Configuration Manager then dynamically loads the VxD specified by the DevLoader value. As a result, the VxD receives a Sys_Dynamic_Device_Init message. Most driver VxDs do minimal processing in the Sys_Dynamic_Device_Init handler, perhaps doing some one-time initialization and returning TRUE (Carry clear) from the handler to indicate success. A driver VxD does not usually call any Configuration Manager services, deferring this until the PNP_New_DevNode message (the next step in the Plug and Play sequence). A driver VxD must never access its device or install interrupt or port trap handlers during Sys_Dynamic_Device_Init handling, because the device hasn't yet been assigned an I/O address or an IRQ.

When building a dynamically loaded driver, you must specify the DYNAMIC keyword on the VXD line in your VxD's DEF file.

In the simplest case, the VxD loaded through DevLoader *is* the driver VxD that interfaces to the enumerated device. However, in some cases the VxD loaded by the DevLoader statement isn't the real driver VxD, but is simply a *device loader* for the driver VxD. This capability is used for some of Windows 95's layered subsystems: the IOS VxD loads all block device driver VxDs, the NDIS VxD loads all network driver VxDs, and the VCOMM VxD loads all port driver VxDs.

Sys_Dynamic_Device_Init processing for a device loader VxD is the same as for a true driver VxD: no interaction with the Configuration Manager; nothing that requires an I/O address or IRQ. Like a driver VxD, the device loader VxD will receive a PNP_New_DevNode message after returning TRUE from Sys_Dynamic_Device_Init.

PNP_New_DevNode *Processing*

After loading the VxD specified by the DevLoader registry value, the Configuration Manager tells the VxD which devnode caused it to be loaded, by sending the VxD a PNP_New_DevNode message. This message has two associated parameters: the devnode (passed in EBX) and a reason code (passed in EAX). The reason code must be either DL_LOAD_DEVLOADER, DL_LOAD_DRIVER, or DL_LOAD_ENUMERATOR. The PNP_New_DevNode message and its associated reason codes are one of the most confusing aspects of adding Plug and Play support in a VxD.

In the simple case, where the VxD loaded by DevLoader is really the driver VxD, the VxD's PNP_New_DevNode message handler will first receive a reason code of DL_LOAD_DEVLOADER — because the Configuration Manager knows only that this VxD is the device loader. In response to this reason code, the VxD should call CM_Register_Device_Driver to let the Configuration Manager know that this VxD is really the device driver as well as the device loader.

In the more complicated case, where the device loader VxD and the driver VxD are separate, the device loader VxD will be loaded first and will then receive the PNP_New_DevNode message with a DL_LOAD_DEVLOADER reason code. In response, a true device loader VxD uses a Configuration Manager service to load the real driver VxD. After loading the driver VxD, the Configuration Manager then sends the driver VxD its own PNP_New_DevNode message, this time with a DL_LOAD_DRIVER message, and the driver VxD responds by calling CM_Register_Device_Driver.

In both cases, the driver VxD for a particular devnode ends up calling CM_Register_Device_Driver. The driver VxD calls this function to trigger the final step in the Plug and Play process, receiving configuration notifications, which I'll address in the next section.

The Calling Interface for CM_Register_Device_Driver

```
DWORD CONFIGMG_Register_Device_Driver(DEVNODE node,
                CMCONFIGHANDLER handler,
                DWORD refData, DWORD flags);
node: registering as device driver for this node;
      provided along with PNP_New_DevNode message
handler: callback function inside the driver VxD
        which will receive configuration notifications
refData: this value will be passed
        as a parameter to the callback function
flags: CM_REGISTER_DEVICE_DRIVER_STATIC:
                device cannot be reconfigured at run-time
        CM_REGISTER_DEVICE_DRIVER_DISABLEABLE:
                device can be disabled at run-time
        CM_REGISTER_DEVICE_DRIVER_REMOVEABLE:
                device can be removed from hardware tree
```

VxDs for Plug and Play hardware should set both DISABLEABLE and REMOVEABLE. This combination of flags allows the Configuration Manager to reconfigure the device to accomodate a newly arrived Plug and Play device. A VxD for a Legacy device should set STATIC, because a Legacy device does not support reconfiguration. If your VxD does not set these flags, the Configuration Manager will never attempt to reconfigure your device — it will never send another CONFIG_START message after the initial one. In addition, the debug version of the Configuration Manager will output a warning message to the debugger, "Device does not allow rebalance and removal".

A VxD may also allocate devnode-specific, or "instance", data during PNP_New_DevNode. Commonly, a single-driver VxD will support multiple instances of the same device, for example COM1, COM2, etc. Such a driver will receive multiple PNP_New_DevNode messages (one for each physical device), and will call _Heap_Allocate during PNP_New_DevNode processing to dynamically allocate a structure for device-specific context information. A typical COM1/COM2 driver, for example, would typically allocate a structure to store the port's I/O base, IRQ, receive buffer, and transmit buffer, etc.

ConfigHandler *Processing*

After it has loaded all driver VxDs for Plug and Play devices, the Configuration Manager invokes arbitrators to assign resources to all Plug and Play devices. Once the arbitrators have made these assignments, the Configuration Manager notifies each driver VxD that it may start using the device's assigned configuration.

A VxD receives this notification through its configuration handler function, registered earlier in a call to CM_Register_Device_Driver. The VxD's configuration handler must conform to this interface:

The Calling Interface for a Configuration Callback

```
CONFIGRET CM_HANDLER ConfigHandler(CONFIGFUNC cfFunc, SUBCONFIGFUNC scfSubFunc,
                DEVNODE dnDevNode, DWORD dwRefData, ULONG ulFlags);
cfFunc: function identifier
scfSubFunc: subfunction identifier
dnDevNode: devnode handle
dwRefData: value passed as ulRefData parameter to CM_Register_Device_Driver
ulFlags: always zero
```

When notifying a VxD of a newly assigned configuration, the Configuration Manager sets the cfFunc parameter to CONFIG_START, meaning "start using your assigned configuration". When processing CONFIG_START, a VxD discovers this assigned configuration with another call to the Configuration Manager, this time to CM_Get_Alloc_Log_Conf.

The Calling Interface for CM_Get_Alloc_Log_Conf

```
CONFIGRET CM_Get_Alloc_Log_Conf(PCMCONFIG pccBuffer,
                      DEVNODE dnDevNode, ULONG ulFlags);
pccBuffer: pointer to CMCONFIG structure to receive configuration
dnDevNode: requesting configuration for this devnode
ulFlags: CM_GET_ALLOC_LOG_CONF_ALLOC to get currently allocated configuration
         CM_GET_ALLOC_LOG_CONF_BOOT_ALLOC to get boot configuration
```

This function retrieves either the currently allocated configuration or the boot configuration, depending on the value of the ulFlags parameter. When processing CONFIG_START, a VxD wants the current configuration, and so uses CM_GET_ALLOC_LOG_CONF_ALLOC. The configuration is returned in a CM_CONFIG structure that summarizes the system resources assigned to the device: memory address, I/O address, IRQ, and/or DMA channel. This structure can be confusing at first glance and isn't documented well in either the DDK or VToolsD. The following code shows how the CMCONFIG structure is defined.

```
struct Config_Buff_s                          // CM_CONFIG is typedef'ed
                                              // to struct Config_Buff_s
{
WORD    wNumMemWindows;                        // Num memory windows
DWORD   dMemBase[MAX_MEM_REGISTERS];           // Memory window base
DWORD   dMemLength[MAX_MEM_REGISTERS];         // Memory window length
WORD    wMemAttrib[MAX_MEM_REGISTERS];         // Memory window Attrib
                                              // fMD_ROM or fMD_RAM
        // fMD_24 or fMD_32 is number of address lines that device decodes
WORD    wNumIOPorts;                           // Num IO ports
WORD    wIOPortBase[MAX_IO_PORTS];             // I/O port base
WORD    wIOPortLength[MAX_IO_PORTS];           // I/O port length
WORD    wNumIRQs;                              // Num IRQ info
BYTE    bIRQRegisters[MAX_IRQS];               // IRQ list
BYTE    bIRQAttrib[MAX_IRQS];                  // IRQ Attrib list
        // fIRQD_Share if shared with another device
WORD    wNumDMAs;                              // Num DMA channels
BYTE    bDMALst[MAX_DMA_CHANNELS];             // DMA list
WORD    wDMAAttrib[MAX_DMA_CHANNELS];          // DMA Attrib list
        // fDD_BYTE if byte size channel
        // fDD_WORD if word size channel
        // fDD_DWORD if dword size channel
BYTE    bReserved1[3];                         // Reserved
};
```

The fields in the CMCONFIG structure can be partitioned into four groups: the first group describes memory resources; the second describes I/O resources; the third IRQs; and the last, DMA resources. Each of these groups conforms to a common pattern. The first field in the group (wNumMemWindows, wNumIOPorts, wNumIRQs, wNumDMAs) tells how many assignments of that type were made, and consequently, which entries in the related arrays are filled in.

For example, a zero in wNumMemWindows means no memory range was assigned, so none of the entries in the three memory-related arrays (dMemBase, dMemLength, wMemAttrib) are valid. A value of 2 for wNumIOPorts means two different I/O ranges were assigned, and the first range is described by the first entry in the two wIOPort arrays (wIOPortBase, wIOPortLength). The second range is described by the second entry in each of the two arrays. In other words, wIOPortBase[0] and wIOPortLength[0] describe the first I/O range; wIOPortBase[1] and wIOPortLength[1] describe the second I/O range.

After decoding the device's assigned resources from the CMCONFIG structure, a VxD's CONFIG_START handler should perform basic device initialization. Other VxDs do this during Sys_Init or Sys_Dynamic_Device_Init, but a Plug and Play driver VxD, although loaded early in the boot process, is unable to access the device until CONFIG_START. The driver VxD may access an I/O-mapped device with nothing more than an inp or outp to the I/O port range specified in the CMCONFIG, but access to a memory-mapped device requires calls to one or more VMM services to obtain a linear address that maps to the device's physical address in CM_CONFIG. Also at this time, a VxD will typically register an interrupt handler for the device's IRQ by calling VPICD_Virtualize_IRQ.

In some cases, the Configuration Handler will also want to process the CONFIG_FILTER message. Before choosing a logical configuration and sending the CONFIG_START message, the Configuration Manager always sends a CONFIG_FILTER message. (Note that the VxD will receive a CONFIG_FILTER message before every CONFIG_START, even if the CONFIG_START was not sent as part of the boot process.) The CONFIG_FILTER message allows the driver an opportunity to examine and modify any of the logical configurations before the Configuration Manager commits to a configuration. For example, a device that doesn't *require* page-alligned memory resources might specify an unaligned memory resource in the INF file. By responding to the CONFIG_FILTER message, the device's VxD could still attempt to optimize the transfer by changing (filtering) each logical configuration to use a page-aligned buffer instead.

Other Plug and Play Configuration Scenarios

The previous sections describe how a Plug and Play driver VxD handles boot-related configuration events. Not all configuration events, though, relate just to the boot process. The VxD's Configuration Handler function must also handle notifications triggered by such user actions as shutting down, adding devices, and removing devices. Table 10.3 summarizes the sequence of configuration events in each of these scenarios (the boot sequence is also included for completeness).

Shutdown

When a user shuts down Windows 95, each VxD Configuration Handler receives a CONFIG_SHUTDOWN notification. The DDK documentation recommends that the driver VxD "free system resources and shutdown its device". But it's interesting to note that many of the drivers whose source is in the DDK don't follow either of those instructions. It really doesn't matter if a VxD frees its system resources by unvirtualizing its IRQ and unhooking its I/O port trap handlers, because the system is shutting down anyway. As for "shutting down" your device, the action taken really depends on the kind of device. For example, an audio playback driver might stop playback on the device, or a modem might hang up a connection.

Table 10.3	Plug and Play configuration events.	
Process	**Function**	**Description**
Shutdown	CONFIG_PRESHUTDOWN	system about to shut down
	CONFIG_SHUTDOWN	system shutting down
Boot	CONFIG_START	start using assigned configuration
	CONFIG_FILTER	driver may filter logical configurations
New Configuration Assigned	CONFIG_STOP	stop using assigned configuration
	CONFIG_FILTER	driver may filter logical configurations
	CONFIG_START	start using (new) assigned configuration
Device Removal (Windows 95 knows ahead of time)	CONFIG_TEST	ok for device to be removed? return CR_SUCCESS (ok) or CR_REMOVE_VETOED (not ok)
	CONFIG_TEST_SUCCEEDED	devnode and all its children returned ok to CONFIG_TEST, device will be removed
	CONFIG_PREREMOVE	prepare for device removal
	CONFIG_PREREMOVE2	prepare for device removal
	CONFIG_REMOVE	device has been removed
Device Removal (Windows 95 knows after the fact)	CONFIG_REMOVE	device has been removed

New Configuration

Sometimes when a new device is added while the system is running (e.g. by inserting a PCMCIA card), that new device requires a resource already assigned to another device. In this case, the Configuration Manager may shuffle the resource assignments of already-present devices to satisfy the new device. If the Configuration Manager does reassign a device's resources, that device's Configuration Handler receives a CONFIG_STOP notification followed by a CONFIG_START notification. CONFIG_STOP tells the driver to stop using its allocated configuration; CONFIG_START tells the driver to start using the (newly) allocated configuration.

To stop using the device resource, the CONFIG_STOP handler may need to "undo" system calls. If the device uses an IRQ, it should be unvirtualized. If the device was memory-mapped, the linear-to-physical mapping requested by the CONFIG_START handler should be released by unlocking, decommitting, and freeing the device's linear address. Review "Talking to a Memory-mapped Device" in Chapter 6 for an explanation of these steps.

It may seem inefficient to free the linear address during CONFIG_STOP if the VxD will turn around and allocate a linear address again during the following CONFIG_START; however, there is at least one situation where a CONFIG_START does not follow a CONFIG_STOP. If the Configuration Manager attempts to reassign resources after boot because a new device was added, and the attempt results in an unresolvable conflict, the Device Manager will ask the user to choose a device to kill. This device will receive a CONFIG_STOP message and nothing else.

In most cases, the Configuration Manager follows a CONFIG_STOP with a CONFIG_START notification for the newly assigned configuration. The VxD's CONFIG_START handler acts exactly as it does during boot: it first calls CM_Get_Alloc_Log_Conf and then starts using the assigned resources returned in the CM_CONFIG structure. No special code is needed in the CONFIG_START handler to distinguish reassignment from initial boot-time assignment.

Device Removal

There are two kinds of device removal: those where the operating system knows ahead of time that the user is planning to remove the device, and those where the operating system learns of the removal after the fact. In the first case, the system can warn the device's VxD of the impending removal. The system will have advance warning, for example, when the user chooses Remove from Device Manager and when the user undocks his laptop from its docking station. In cases like these, the VxD Configuration Handler for the "about-to-be-removed" device receives a CONFIG_TEST notification before the removal. The VxD can grant its permission for the removal to proceed by returning CR_SUCCESS, or can deny permission if the device isn't ready to be removed by returning CR_REMOVE_VETOED.

If the `CONFIG_TEST` handler returns `CR_SUCCESS`, the Configuration Manager follows up with a `CONFIG_TEST_SUCCEEDED` notification, which requires no handling at all by the driver VxD. Finally, after the device is removed, the Configuration Manager sends a `CONFIG_REMOVE` notification. The `CONFIG_REMOVE` handler should halt use of the device resources (unvirtualize the IRQ, etc.). On this event, the driver should also free any devnode-specific data. (See the earlier section "Plug and Play Boot Process: `PNP_New_DevNode` Processing" for discussion of allocating devnode-specific data.)

The second class of removal happens when the operating system doesn't find out about the removal until after the fact, for example when a PCMCIA card is removed. In this case, the VxD for the just-removed device receives a `CONFIG_REMOVE` notification after the fact. Once again, a `CONFIG_REMOVE` handler should stop using device resources and free any devnode-specific data.

A Sample Plug and Play Driver VxD: `TRICORD.VxD`

The remainder of this chapter will discuss an example Plug and Play Driver VxD, `TRICORD.VXD`, and its accompanying INF file, `TRICORD.INF`. `TRICORD.VXD` is the Plug and Play Device Driver VxD for an imaginary Tricorder device produced by an imaginary vendor, the XYZ1234 Corp. `TRICORD.VXD` also acts as its own Plug and Play Device Loader, a common scenario.

While the TRICORD VxD isn't a fully functional device driver — it doesn't talk to any real hardware — it *is* a fully functional Plug and Play Driver VxD — it interacts with the Configuration Manager as required to find out what system resources the Tricorder device is using. If you already have a driver VxD and you want to add Plug and Play support, TRICORD shows you what pieces to add to your existing VxD. Or, if you are writing a Plug and Play Driver VxD from scratch, you can use TRICORD as a starting point and add device-specific functionality.

Before running TRICORD for the first time, you must run the Add New Hardware Wizard. The wizard will use TRICORD's INF file to add several registry entries and copy the VxD file. If TRICORD was a true Plug and Play device, an enumerator would automatically recognize it as a new device when first added to the system, and the Device Installer would automatically be invoked to process its INF file. But like a real Legacy device, the imaginary TRICORD device isn't automatically recognized, so you as a developer must explicitly tell Windows 95 about the new device.

TRICORD. INF **Details**

TRICORD.INF (Listing 10.5, page 213) performs a basic installation scenario as discussed earlier in this chapter. TRICORD.INF contains

- a Version section which describes the OS version and the device class (type);
- a Manufacturer section which describes the device;
- a CopyFiles section which copies TRICORD.VXD from the installation disk to the hard disk;
- an AddReg section which adds a single DevLoader entry to the device's software subtree in the registry;
- a LogConfig section which describes the resources (I/O port and IRQ) used by the device; and
- an Install section which names the CopyFiles, AddReg, and LogConfig sections.

The TRICORD.INF file is shown in the following code.

```
[Version]
Signature=$CHICAGO$
Class=OtherDevices
Provider=%String0%

[DestinationDirs]
DefaultDestDir=30,BIN

[Manufacturer]
%String0%=SECTION_0

[SECTION_0]
%String1%  = XYZ1234.Install,*XYZ1234

[XYZ1234.Install]
Copyfiles=CopyFiles_XYZ1234
AddReg=AddReg_XYZ1234
LogConfig=LogConfig_XYZ1234

[CopyFiles_XYZ1234]
tricord.vxd

[AddReg_XYZ1234]
HKR,,DevLoader,0,tricord.vxd
```

```
[LogConfig_XYZ1234]
ConfigPriority=NORMAL
IOConfig=4@180-1B3%fff0(3::)
IRQConfig=4,5,9,10,11

[Strings]
String0="XYZ Corp"
String1="Tricorder Model 1234"
```

When viewed as a text file, an INF file seems disjointed and unstructured. But an INF file has an implicit hierarchical structure, with a root section that refers to branch sections, each which refer to other branch sections. The INF file makes more sense when viewed as a hierarchy, which is why many developers create and modify INF files with the INFEDIT tool in the DDK. The following pseudocode depicts the hierarchical structure of TRICORD.INF.

```
[Manufacturer]
"XYZ Corp" --> [SECTION 0]
             "Tricorder Model 1234" --> [XYZ1234.Install]
                                  CopyFiles ---> [CopyFiles XYZ1234]
                                                 tricord.vxd

                                  AddReg    - > [AddReg XYZ1234]
                                                 HKR,,DevLoader,0,tricord.vxd

                                  LogConfig ---> [LogConfig XYZ1234]
                                                 ConfigPriority=NORMAL
                                                 IOConfig=20@200-3ff%3c0(3ff::)
                                                 IRQConfig=5,7,10,15
```

The INFEDIT view makes it clear that the TRICORD.VXD describes only a single device ("Tricorder Model 1234") from a single vendor ("XYZ Corp"). This Tricorder device requires three steps to install (three items in XYZ1234.Install). A single file (tricord.vxd) must be copied. A single registry entry must be added (DevLoader=tricord.vxd) to the device's hardware key under HKLM\Enum. And, the device supports a single logical configuration consisting of a range of 20h I/O ports (anywhere between 200h and 3ffh) and an IRQ of 5, 7, 10, or 15.

You can avoid worrying about the unusual syntax on items like AddReg and IOConfig by using the INFEDIT tool to create and modify your INF file. For more details on the exact syntax of any INF file section, see the DDK

Code Details

Like the earlier examples, the TRICORD source consists of two files. An ASM file [TRICORD.ASM (Listing 10.2, page 210)] contains the DDB and Device Control Procedure. A C file [PNP.C (Listing 10.1, page 204)] contains the message handler and callback functions.

TRICORD follows the basic procedures outlined earlier in this chapter. Its Device Control Procedure handles only three messages: Sys_Dynamic_Device_Init, Sys_-Dynamic_Device_Exit, and PNP_New_DevNode. The PNP_New_DevNode handler registers a Configuration Handler with the Configuration Manager. This Configuration Handler processes CONFIG_START, CONFIG_STOP, CONFIG_REMOVE, and CONFIG_TEST notifications.

By including the DYNAMIC keyword in the VxD DEF file and processing the Sys_Dynamic_Device_Init and Sys_Dynamic_Device_Exit messages, TRICORD becomes a dynamically loadable VxD. However, neither message handler does any real processing. Both OnSysDynamicDeviceInit and SysDynamicDeviceExit simply return TRUE, indicating success.

```
CONFIGRET OnPNPNewDevnode(DEVNODE DevNode, DWORD LoadType)
{
    CONFIGRET rc;
    switch (LoadType)
    {
    case DLVXD_LOAD_DEVLOADER:
        pDeviceContext = (DEVICE_CONTEXT *)_HeapAllocate(sizeof(DEVICE_CONTEXT),
                                                  HEAPZEROINIT );

        if (!pDeviceContext)
            return CR_FAILURE;
        rc = CM_Register_Device_Driver(DevNode, ConfigHandler, pDeviceContext,0);
        if (rc != CR_SUCCESS)
            return rc;
        return CR_SUCCESS;
    default:
        return(CR_DEFAULT);
    }
}
```

OnPNPNewDevnode does some simple processing. If the LoadType parameter is anything other than DLVXD_LOAD_DEVLOADER, the handler returns the CR_DEFAULT value defined by the Configuration Manager. If DevType is DLVXD_LOAD_DEVLOADER, the handler first allocates a DEVICE_CONTEXT structure for instance data (data about this particular devnode) and then registers as the device driver for the devnode by calling CM_Register_Device_Driver. As a device driver for the devnode, TRICORD will receive configuration notifications from the Configuration Manager through the ConfigHandler callback function, which was passed as a parameter to CM_Register_Device_Driver. It may seem backwards to register as a device driver during DLVXD_LOAD_DEVLOADER processing, and yet ignore the DLVXD_LOAD_DRIVER messages, but, as discussed earlier in this chapter, this is indeed proper behavior for a VxD that acts as both Plug and Play Device Loader and Device Driver.

```
CONFIGRET CM_HANDLER ConfigHandler(CONFIGFUNC cfFuncName,
                                   SUBCONFIGFUNC scfSubFuncName,
                                   DEVNODE dnToDevNode,
                                   DWORD dwRefData, ULONG ulFlags)
{
   CMCONFIG Config;
   DWORD  rc;
   DEVICE_CONTEXT *dev = (DEVICE_CONTEXT *)dwRefData;

   switch (cfFuncName)
   {
   case CONFIG_START:
      return ProcessConfigStart(dnToDevNode, dev );

   case CONFIG_TEST:
      return CR_SUCCESS;

   case CONFIG_STOP:
      return ProcessConfigStop(dnToDevNode, dev );

   case CONFIG_REMOVE:
      return ProcessConfigStop(dnToDevNode, dev );

   default:
      return CR_DEFAULT;
   }
}
```

The real work in TRICORD.VXD is done by ConfigHandler, the registered callback function. The Configuration Manager passes ConfigHandler a reason code parameter, cfFuncName, which tells ConfigHandler the reason for the callback. There are well over a dozen reason codes, but like most driver VxDs, TRICORD processes only a handful. Another parameter, dwRefData, is used as "reference data". It's actually a pointer to the DEVICE_CONTEXT structure that TRICORD allocated earlier in its OnPNPNewDevnode handler. At that time, TRICORD passed this DEVICE_CONTEXT pointer to the Configuration Manager in a call to CM_Register_Device_Driver, and the Configuration Manager now passes it back as the dwRefData parameter to ConfigHandler.

It is important that ConfigHandler return CR_DEFAULT for any function code that wasn't specifically processed. The Microsoft DDK specifically recommends this behavior for compatibility with future versions of Windows. In fact, the debug version of Windows 95 tests every VxD's default response by calling the Configuration Handler function with a bogus value of 0x12345678. If a VxD doesn't respond to this message with CR_DEFAULT, Windows will output an error message on the debugger screen.

Of all the notifications actually processed by ConfigHandler, CONFIG_TEST results in the least processing: TRICORD returns a value of CR_SUCCESS, giving the Configuration Manager permission to either remove or stop using the device. The most interesting action in ConfigHandler occurs for CONFIG_START, CONFIG_STOP, and CONFIG_REMOVE notifications. For each of these, ConfigHandler calls a subroutine to do the real work.

```
CONFIGRET ProcessConfigStart( DEVNODE devnode, void *p )
{
    DEVICE_CONTEXT  *dev = (DEVICE_CONTEXT *)p;
    CONFIGRET       rc;
    CMCONFIG        Config;
    MEMREGS         *regs;
    WORD            reg;
    IRQHANDLE       hndIrq;

    rc = CM_Get_Alloc_Log_Conf(&Config, devnode,
CM_GET_ALLOC_LOG_CONF_ALLOC);
    if (rc != CR_SUCCESS)
    {
      DPRINTF1(dbuf, "CM_Get_Alloc_Log_Conf failed rc=%x\n", rc );
        return CR_FAILURE;
    }

    Print_Assigned_Resources(&Config);
    if (! ((Config.wNumIRQs == 1) &&
        (Config.wNumIOPorts == 1 || Config.wNumMemWindows == 1)) )
    {
      DPRINTF0(dbuf, "Expected resources not assigned" );
        return CR_FAILURE;
    }

    if (Config.wNumMemWindows)
    {
        dev->MemBase = Config.dMemBase[0];
        dev->MemSize = Config.dMemLength[0];
        dev->pMem = (MEMREGS *)MyMapPhysToLinear( dev->MemBase,
                                            Config.dMemLength[0] );
        if (!dev->pMem)
        {
         DPRINTF0(dbuf, "MyMapPhysToLinear failed" );
            return CR_FAILURE;
        }
        dev->pMem->Ctrl = CTRL_START_DEVICE;
    }
```

```
else
{
  dev->IoBase = Config.wIOPortBase[0];
  reg = dev->IoBase + REG_CTRL;
  _outpdw( reg, CTRL_START_DEVICE );
}
```

A CONFIG_START notification tells TRICORD that its device has been assigned resources, and that the VxD can now communicate with its device. ProcessConfigStart begins by retrieving the assigned resources with a call to CM_Get_Alloc_Log_Conf, using the value CM_GET_ALLOC_LOG_CONF_ALLOC for the flags parameter. This flag value specifies the allocated logical configuration, as opposed to the logical configuration used at boot. The allocated logical configuration is returned in the CMCONFIG buffer provided by ConfigHandler. ConfigHandler calls a utility function, Print_Assigned_Resources, to decode the CMCONFIG buffer and print out the assigned resources.

Print_Assigned_Resources has four blocks, one for each resource type (I/O port, memory range, IRQ, and DMA channel). Each block first tests to see if one or more resources of that type was actually assigned and, if so, prints the name of the resource. Then a for loop prints information about each assigned resource of that type. For example, this block processes the I/O port resource:

```
if (pConfig->wNumIOPorts)
{
   DPRINTF0(dbuf, "IO resources\r\n" );
   for (i=0; i < pConfig->wNumIOPorts; i++)
   {
      DPRINTF1(dbuf, "Range #%d: ", pConfig->wNumIOPorts );
      DPRINTF2(dbuf, "starts at %x len is %d\r\n",
              pConfig->wIOPortBase[i],pConfig->wIOPortLength[i] );
   }
}
```

The `Print_Assigned_Resources` function is included mainly to illustrate decoding of the `CMCONFIG` structure. A VxD usually has some expectation about the number and type of resources it will use, while remaining flexible about exactly *which* IRQ or I/O port is assigned. This is true of `ProcessConfigStart`, which expects a single IRQ assignment and either a memory range or an I/O range. If these expectations aren't met, `ProcessConfigStart`, and in turn `ConfigHandler`, returns with an error.

```
if (! ((Config.wNumIRQs == 1) &&
        (Config.wNumIOPorts == 1 || Config.wNumMemWindows == 1)) )
{
    DPRINTF0("Expected resources not assigned" );
    return CR_FAILURE;
}
```

After verifying that resources are assigned as expected, `ProcessConfigStart` determines whether the device has been configured as memory-mapped or I/O-mapped. If memory-mapped, the function maps the assigned physical base memory address to a linear address, using a utility function `MyMapPhysToLinear`. `ProcessConfigStart` then uses the linear address as a pointer, writing an initialization value to the device's control register.

If the device wasn't assigned a memory range, TRICORD uses the assigned I/O range instead. Once again TRICORD writes an initialization value to the device's control register, but this time it uses an `OUT` instruction instead of a pointer. `ProcessConfigStart` uses the `_outpdw` macro to perform the `OUT` since the device has 32-bit registers and the C run-time doesn't include a 32-bit form of `in` or `out`.

Finally, `ProcessConfigStart` installs an interrupt handler by filling in a `VPICD_IRQ_DESCRIPTOR` structure and passing it to the VPICD service `VPICD_-Virtualize_IRQ`. The structure's `VID_IRQ_Number` field is the device's assigned IRQ (from the `CMCONFIG` structure). `VID_Options` is set to `VPICD_OPT_REF_DATA`. This field works together with the `VID_Ref_Data` field, which is set to point to the `DEVICE_CONTEXT` (passed in as `dwRefData` and originally allocated by the `PNP_New_DevNode` handler). When the VPICD calls the registered interrupt handler, it will pass `VID_Ref_Data` (really a `DEVICE_CONTEXT` pointer) as a parameter.

This interrupt handler is specified by the `VPICD_IRQ_DESCRIPTOR`'s `Hw_Int_Proc` field. The registered handler is `HwIntProcThunk` (in TRICORD's assembly module), but this thunk merely grabs the reference data parameter from the `EDX` register and pushes it on the stack before calling the `HwIntProcHandler` function in the C module to do the real handling.

`ConfigHandler`'s processing for `CONFIG_STOP` and `CONFIG_REMOVE` is much simpler than for `CONFIG_START`. For both of these messages, `ConfigHandler` calls `ProcessConfigStop`.

```
CONFIGRET ProcessConfigStop( DEVNODE devnode, DEVICE_CONTEXT dev )
{
    if (dev->pMem)
    {
        *(pMem->Ctrl) = CTRL_STOP_DEVICE;
        UnMapPhysToLinear( dev->pMem );
    }
    else if (dev->IoBase)
    {
        _outpdw( IoBase + REG_CTRL, CTRL_STOP_DEVICE );
    }
    VPICD_Force_Default_Behavior( dev->hndIrq );
    HeapFree( dev, 0 );

    return CR_SUCCESS;
}
```

This subroutine undoes the actions taken by `ProcessConfigStart`. First TRI-CORD commands the device itself to stop, then undoes the linear-to-physical memory mapping if necessary, and finally frees the `DEVICE_CONTEXT` structure originally allocated by the `PNP_New_DevNode` message handler.

Summary

While Plug and Play's Configuration Manager/Enumerator/Arbitrator mechanism is definitely complex, the system-to-VxD Plug and Play interface is reasonably straightforward. At that system boundary, Plug and Play support only involves handling a few well-defined messages and constructing an appropriate INF file.

Even so, drivers that fully support the flexibility possible under Plug and Play will be considerably more complex than, for example, a legacy driver. It's the old generality vs simplicity trade-off: a board that can be dynamically reconfigured to use a wide variety of resources won't be as simple as one with fixed addresses; code that "binds" to its resources at run-time won't be as simple as code that manipulates fixed addresses.

All the same, most commercial drivers probably should include Plug and Play support. The benefits to end-users (and thus the difference in marketability) are significant.

Listing 10.1   PNP.C

```c
#define WANTVXDWRAPS

#include <basedef.h>
#include <vmm.h>
#include <debug.h>
#include "vxdcall.h"
#include <vxdwraps.h>
#include "intrinsi.h"
#include <configmg.h>
#include <vpicd.h>
#include "wrappers.h"

#ifdef DEBUG
#define DPRINTF0(buf, fmt )            _Sprintf(buf, fmt ); Out_Debug_String( buf )
#define DPRINTF1(buf, fmt, arg1)       _Sprintf(buf, fmt, arg1 ); Out_Debug_String( buf )
#define DPRINTF2(buf, fmt, arg1, arg2) _Sprintf(buf, fmt, arg1, arg2 ); Out_Debug_String( buf )
#else
#define DPRINTF0(buf, fmt)
#define DPRINTF1(buf, fmt, arg1)
#define DPRINTF2(buf, fmt, arg1, arg2)
#endif

#define _outpdw( port, val )    _asm mov dx, port \
                                _asm mov eax, val \
                                _asm out dx, eax

#define REG_CTRL    0
#define REG_STATUS  1

#define CTRL_START_DEVICE   0x01
#define CTRL_STOP_DEVICE    0x00

typedef struct
{
    DWORD    Ctrl;
    DWORD    Status;
} MEMREGS;

typedef struct
{
    DWORD               MemBase;
    DWORD               MemSize;
    MEMREGS             *pMem;
    WORD                IoBase;
    WORD                Irq;
    IRQHANDLE           hndIrq;
    VPICD_IRQ_DESCRIPTOR IrqDescr;
} DEVICE_CONTEXT;

BOOL OnSysDynamicDeviceInit(void);
BOOL OnSysDynamicDeviceExit(void);
CONFIGRET OnPNPNewDevnode(DEVNODE DevNode, DWORD LoadType);
CONFIGRET CM_HANDLER ConfigHandler(CONFIGFUNC cfFuncName, SUBCONFIGFUNC scfSubFuncName,
                                   DEVNODE dnToDevNode, DWORD dwRefData, ULONG ulFlags);
CONFIGRET ProcessConfigStart( DEVNODE devnode, DEVICE_CONTEXT *dev );
CONFIGRET ProcessConfigStop( DEVNODE devnode, DEVICE_CONTEXT *dev );
void Print_Assigned_Resources( CMCONFIG *pConfig );
DWORD MyMapPhysToLinear( DWORD phys, DWORD size );
BOOL UnMapPhysToLinear( DWORD lin, DWORD size );
```

Listing 10.1 (continued)   `PNP.C`

```c
char dbuf[80];
DEVICE_CONTEXT *pDeviceContext;

// functions in asm module
void HwIntProcThunk( void );

BOOL OnSysDynamicDeviceInit()
{
    return TRUE;
}

BOOL OnSysDynamicDeviceExit()
{
    return TRUE;
}

CONFIGRET OnPNPNewDevnode(DEVNODE DevNode, DWORD LoadType)
{
    CONFIGRET    rc;

    switch (LoadType)
    {

        case DLVXD_LOAD_DEVLOADER:
            pDeviceContext = (DEVICE_CONTEXT *)_HeapAllocate( sizeof(DEVICE_CONTEXT),
                                                             HEAPZEROINIT );
            if (!pDeviceContext)
                return CR_FAILURE;
                    rc = CM_Register_Device_Driver(DevNode, ConfigHandler,
                                                   pDeviceContext,
                                                   CM_REGISTER_DEVICE_DRIVER_REMOVEABLE |
                                                   CM_REGISTER_DEVICE_DRIVER_DISABLEABLE );
            if (rc != CR_SUCCESS)
                return rc;

            return CR_SUCCESS;

        default:
            return(CR_DEFAULT);
    }
}
```

Listing 10.1 (continued)   `PNP.C`

```c
#pragma VxD_PAGEABLE_DATA_SEG
#pragma VxD_PAGEABLE_CODE_SEG

CONFIGRET CM_HANDLER ConfigHandler(CONFIGFUNC cfFuncName, SUBCONFIGFUNC scfSubFuncName,
                                   DEVNODE dnToDevNode, DWORD dwRefData, ULONG ulFlags)
{
    CMCONFIG       Config;
    DWORD          rc;
    DEVICE_CONTEXT *dev = (DEVICE_CONTEXT *)dwRefData;

    switch (cfFuncName)
    {
    case CONFIG_START:
        return ProcessConfigStart(dnToDevNode, dev );

    case CONFIG_TEST:
        return CR_SUCCESS;

    case CONFIG_STOP:
        return ProcessConfigStop(dnToDevNode, dev );

    case CONFIG_REMOVE:
        return ProcessConfigStop(dnToDevNode, dev );

    default:
        return CR_DEFAULT;
    }
}

CONFIGRET ProcessConfigStart( DEVNODE devnode, void *p )
{
    DEVICE_CONTEXT *dev = (DEVICE_CONTEXT *)p;
    CONFIGRET      rc;
    CMCONFIG       Config;
    MEMREGS        *regs;
    WORD           reg;
    IRQHANDLE      hndIrq;

    rc = CM_Get_Alloc_Log_Conf(&Config, devnode, CM_GET_ALLOC_LOG_CONF_ALLOC);
    if (rc != CR_SUCCESS)
    {
      DPRINTF1(dbuf, "CM_Get_Alloc_Log_Conf failed rc=%x\n", rc );
        return CR_FAILURE;
    }

    Print_Assigned_Resources(&Config);
    if (! ((Config.wNumIRQs == 1) && (Config.wNumIOPorts == 1 || Config.wNumMemWindows == 1)) )
    {
      DPRINTF0(dbuf, "Expected resources not assigned" );
        return CR_FAILURE;
    }
```

___

Listing 10.1 (continued)   `PNP.C`

```c
    if (Config.wNumMemWindows)
    {
        dev->MemBase = Config.dMemBase[0];
        dev->MemSize = Config.dMemLength[0];
        dev->pMem = (MEMREGS *)MyMapPhysToLinear( dev->MemBase, Config.dMemLength[0] );
        if (!dev->pMem)
        {
         DPRINTF0(dbuf, "MyMapPhysToLinear failed" );
            return CR_FAILURE;
        }
        dev->pMem->Ctrl = CTRL_START_DEVICE;
    }
    else
    {
      dev->IoBase = Config.wIOPortBase[0];
      reg = dev->IoBase + REG_CTRL;
      _outpdw( reg, CTRL_START_DEVICE );
    }

    dev->IrqDescr.VID_IRQ_Number = Config.bIRQRegisters[0];
    dev->IrqDescr.VID_Options = VPICD_OPT_REF_DATA;
    dev->IrqDescr.VID_Hw_Int_Ref = dev;
    dev->IrqDescr.VID_Hw_Int_Proc = HwIntProcThunk;
    hndIrq = VPICD_Virtualize_IRQ( &dev->IrqDescr );
    if (!hndIrq)
    {
      DPRINTF0(dbuf, "VPICD_Virt failed" );
        return CR_FAILURE;
    }

    return CR_SUCCESS;
}

CONFIGRET ProcessConfigStop( DEVNODE devnode, void *p )
{
    DEVICE_CONTEXT *dev = (DEVICE_CONTEXT *)p;
    WORD  reg;

    if (dev->pMem)
    {
        dev->pMem->Ctrl = CTRL_STOP_DEVICE;
        UnMapPhysToLinear( (DWORD)dev->pMem, dev->MemSize );
    }
    else if (dev->IoBase)
    {
     reg = dev->IoBase + REG_CTRL;
       _outpdw( reg, CTRL_STOP_DEVICE );
    }
    VPICD_Force_Default_Behavior( dev->hndIrq );
    _HeapFree( dev, 0 );
    return CR_SUCCESS;
}
```

Listing 10.1 (continued)   `PNP.C`

```c
void Print_Assigned_Resources( CMCONFIG *pConfig )
{
    int i;

    if (pConfig->wNumMemWindows)
    {
        DPRINTF0(dbuf, "Mem resources\r\n" );
        for (i=0; i < pConfig->wNumMemWindows; i++)
        {
            DPRINTF1(dbuf, "Range #%d: ", pConfig->wNumMemWindows );
            DPRINTF2(dbuf, "starts at %x len is %d\r\n",
                    pConfig->dMemBase[i],pConfig->dMemLength[i] );
        }
    }

    if (pConfig->wNumIOPorts)
    {
        DPRINTF0(dbuf, "IO resources\r\n" );
        for (i=0; i < pConfig->wNumIOPorts; i++)
        {
            DPRINTF1(dbuf, "Range #%d: ", pConfig->wNumIOPorts );
            DPRINTF2(dbuf, "starts at %x len is %d\r\n",
                    pConfig->wIOPortBase[i],pConfig->wIOPortLength[i] );
        }
    }

    if (pConfig->wNumIRQs)
    {
        DPRINTF0(dbuf, "IRQs: " );
        for (i=0; i < pConfig->wNumIRQs; i++)
        {
            DPRINTF1(dbuf, "%d ", pConfig->bIRQRegisters[i]);
        }
        DPRINTF0(dbuf, "\r\n");
    }

    if (pConfig->wNumDMAs)
    {
        DPRINTF0(dbuf, "DMA channels:" );
        for (i=0; i < pConfig->wNumDMAs; i++)
        {
            DPRINTF1(dbuf, "%d ", pConfig->bDMALst[i]);
        }
        DPRINTF0(dbuf, "\r\n");
    }

}
```

Listing 10.1 (continued)   PNP.C

```c
DWORD MyMapPhysToLinear( DWORD phys, DWORD size )
{
    DWORD lin;
    DWORD nPages = size / 4096;

    lin = _PageReserve( PR_SYSTEM, nPages, 0 );
    if (lin == -1)
        return 0;
    if (!_PageCommitPhys( lin >> 12, nPages, PC_INCR | PC_WRITEABLE, 0 ))
     return 0;
    if (!_LinPageLock( lin >> 12, nPages, 0 ))
        return 0;
    return lin;
}


BOOL UnMapPhysToLinear( DWORD lin, DWORD size )
{
    DWORD nPages = size / 4096;

    if (!_LinPageUnlock( lin, nPages, 0 ))
        return 0;
    if (!_PageDecommit( lin >> 12, nPages, 0))
     return 0;
    if (!_PageFree( (void *)lin >> 12, 0 ))
        return 0;
    return 1;
}


BOOL __stdcall HwIntProcHandler(VMHANDLE hVM, IRQHANDLE hIRQ, void *pRefData)
{
    DEVICE_CONTEXT *dev = (DEVICE_CONTEXT *)pRefData;

    return TRUE;
}
```

Listing 10.2   TRICORD.ASM

```
.386p

;*****************************************************************************
;                 I N C L U D E S
;*****************************************************************************

    include vmm.inc
    include debug.inc

;===========================================================================
;        V I R T U A L   D E V I C E   D E C L A R A T I O N
;===========================================================================

DECLARE_VIRTUAL_DEVICE     TRICORD, 1, 0, ControlProc, UNDEFINED_DEVICE_ID, \
                           UNDEFINED_INIT_ORDER

VxD_LOCKED_CODE_SEG

;===========================================================================
;
;    PROCEDURE: ControlProc
;
;    DESCRIPTION:
;     Device control procedure for the SKELETON VxD
;
;    ENTRY:
;     EAX = Control call ID
;
;    EXIT:
;     If carry clear then
;         Successful
;     else
;         Control call failed
;
;    USES:
;     EAX, EBX, ECX, EDX, ESI, EDI, Flags
;
;===========================================================================

BeginProc ControlProc
    Control_Dispatch SYS_DYNAMIC_DEVICE_INIT, _OnSysDynamicDeviceInit, cCall, <ebx>
    Control_Dispatch SYS_DYNAMIC_DEVICE_EXIT, _OnSysDynamicDeviceExit, cCall, <ebx>

    clc
    ret

EndProc ControlProc
```

Listing 10.2 (continued)   TRICORD.ASM

```
PUBLIC _HwIntProcThunk
_HwIntProcThunk PROC NEAR ; called from C, needs underscore

    sCall HwIntProcHandler, <eax, ebx, edx>
    or    ax, ax
    jz    clearc
    stc
    ret

clearc:
    clc
    ret

 _HwIntProcThunk ENDP

VxD_LOCKED_CODE_ENDS

    END
```

Listing 10.3   TRICORD.MAK

```
CFLAGS     = -DWIN32 -DCON -Di386 -D_X86_ -D_NTWIN -W3 -Gs -D_DEBUG -Zi
CVXDFLAGS  = -Zdp -Gs -c -DIS_32 -Z1 -DDEBLEVEL=1 -DDEBUG
LFLAGS     = -machine:i386 -debug:notmapped,full -debugtype:cv
             -subsystem:console kernel32.lib
AFLAGS     = -coff -DBLD_COFF -DIS_32 -W2 -Zd -c -Cx -DMASM6 -DDEBLEVEL=1 -DDEBUG

all: tricord.vxd

pnp.obj: pnp.c
        cl $(CVXDFLAGS) -Fo$@ %s

tricord.obj: tricord.asm
        ml $(AFLAGS) -Fo$@ %s

tricord.vxd: tricord.obj pnp.obj ..\wrappers\vxdcall.obj tricord.def
        echo >NUL @<<tricord.crf
-MACHINE:i386 -DEBUG -DEBUGTYPE:MAP -PDB:NONE
-DEF:tricord.def -OUT:tricord.vxd -MAP:tricord.map
-VXD vxdwraps.clb wrappers.clb vxdcall.obj tricord.obj pnp.obj
<<KEEP
        link @tricord.crf
        mapsym tricord
```

Listing 10.4 `TRICORD.DEF`

```
VXD TRICORD DYNAMIC
SEGMENTS
    _LTEXT      CLASS 'LCODE'    PRELOAD NONDISCARDABLE
    _LDATA      CLASS 'LCODE'    PRELOAD NONDISCARDABLE
    _TEXT       CLASS 'LCODE'    PRELOAD NONDISCARDABLE
    _DATA       CLASS 'LCODE'    PRELOAD NONDISCARDABLE
    _LPTEXT     CLASS 'LCODE'    PRELOAD NONDISCARDABLE
    _CONST      CLASS 'LCODE'    PRELOAD NONDISCARDABLE
    _BSS        CLASS 'LCODE'    PRELOAD NONDISCARDABLE
    _TLS        CLASS 'LCODE'    PRELOAD NONDISCARDABLE
    _ITEXT      CLASS 'ICODE'    DISCARDABLE
    _IDATA      CLASS 'ICODE'    DISCARDABLE
    _PTEXT      CLASS 'PCODE'    NONDISCARDABLE
    _PDATA      CLASS 'PCODE'    NONDISCARDABLE
    _STEXT      CLASS 'SCODE'    RESIDENT
    _SDATA      CLASS 'SCODE'    RESIDENT
    _MSGTABLE   CLASS 'MCODE'    PRELOAD NONDISCARDABLE IOPL
    _MSGDATA    CLASS 'MCODE'    PRELOAD NONDISCARDABLE IOPL
    _IMSGTABLE  CLASS 'MCODE'    PRELOAD DISCARDABLE IOPL
    _IMSGDATA   CLASS 'MCODE'    PRELOAD DISCARDABLE IOPL
    _DBOSTART   CLASS 'DBOCODE'  PRELOAD NONDISCARDABLE CONFORMING
    _DBOCODE    CLASS 'DBOCODE'  PRELOAD NONDISCARDABLE CONFORMING
    _DBODATA    CLASS 'DBOCODE'  PRELOAD NONDISCARDABLE CONFORMING
    _16ICODE    CLASS '16ICODE'  PRELOAD DISCARDABLE
    _RCODE      CLASS 'RCODE'
EXPORTS
    TRICORD_DDB @1
```

Listing 10.5   `TRICORD.INF`

```
[Version]
Signature=$CHICAGO$
Class=Unknown
Provider=%String0%
LayoutFile=<Layout File>

[DestinationDirs]
DefaultDestDir=10

[Manufacturer]
%String1%=SECTION_0

[SECTION_0]
%String2%=1234_Install,XYZ1234

[1234_Install]
CopyFiles=1234_NewFiles
AddReg=1234_AddReg
LogConfig=1234_LogConfig

[1234_NewFiles]
TRICORD.VXD

[1234_AddReg]
HKR,,DevLoader,0,TRICORD.VXD

[1234_LogConfig]
ConfigPriority=NORMAL
IOConfig=20@200-3ff%ffc0(3ff::)
IRQConfig=5,7,10,15

[ControlFlags]

[SourceDisksNames]
1=XYZ1234 Driver Disk,,0000-0000

[SourceDisksFiles]

[Strings]
String0="XYZ Corp."
String1="XYZ Corp."
String2="Tricorder Model 1234"
```

Chapter 11

Communication from Applications to VxDs

VxDs do much more than just "handle" hardware. In most cases, VxDs also offer an interface to applications, so an application can actually *do* something with the hardware. Both Windows 3.x and Windows 95 have mechanisms which allow VxDs and applications to communicate in both directions: application-to-VxD and VxD-to-application. This chapter will cover communication in the application-to-VxD direction. The next chapter will cover VxD-to-application interaction.

Instead of organizing this chapter around whether the VxD is running under Windows 3.x or Windows 95, I've divided the chapter into sections that address either the Win16 application interface or the Win32 application interface. That's because the interface between a Win16 application and VxD is the same for both Windows 3.x and Windows 95. The distinguishing feature is the bitness of the application, not the version of Windows the VxD runs on.

Win16 Application to VxD:
View from VxD Side

To provide an interface for a Win16 application, a VxD exports what is known as an "API procedure". More correctly, a VxD exports a PM API procedure and/or a V86 API procedure. The PM API procedure is used by 16-bit protected mode applications, which includes Win16 applications as well as any DOS-extended (DPMI) applications. The V86 mode API is, of course, used by DOS applications.

A VxD exports these procedures by naming them in the DDB. VxDs typically use the Declare_Virtual_Device macro to declare the DDB. In this case the API procedure names go in the V86_Proc and PM_Proc fields. When assembled or compiled, these function names become addresses in the DDB, which the VMM uses to call the VxD on behalf of the application.

A VxD must also declare a unique Device ID in the DDB in order to export an API procedure. (This field is referred to as the Device_Num in the Declare_Virtual_Device macro). Developers commonly use the value UNDEFINED_DEVICE_ID for VxDs, but that's not good enough for a VxD that exports an API procedure or a service. Microsoft reserves the values 0–1FFh, so you're free to choose any value above that as long as it's unique. You can ensure that it's unique by registering with Microsoft for your very own Device ID.

When the VMM calls a VxD API procedure on behalf of an application, it puts the VM handle in EBX and a pointer to the Client Register Structure in EBP. The VxD must examine the Client Register Structure for the parameters passed in by the application (including the reason for the call). The VxD developer has total control over all other aspects of the interface design. The developer decides what functions to support, what registers to use, and what parameters to pass in registers. A common convention is for AX to specify the function code, and to use AX=0 for "Get Version".

Except for pointer parameters, the VxD can examine and use parameter values directly. For example, if the convention was for CX to contain a buffer size, the VxD would use the construct [EBX].Client_CX, or from C, crs->Client_CX. Pointer parameters can't be used "as is", because pointers have a different representation in the application's 16-bit segmented environment than in the VxD's 32-bit flat-model environment.

In the segmented world of Win16 and DOS applications, pointers are 16:16 values: either selector:offset for PM applications, or segment:offset for V86 applications. In the flat world of VxDs, pointers are 32-bit linear addresses. When an application passes a pointer to a VxD, say in DS:DX, the VxD must transform the segmented representation into a linear address. The VMM service Map_Flat performs this translation. To use this service, you specify a segment/selector and an offset, where each component is a field in the Client_Regs_Struc. The VMM returns a Ring 0 linear address. When using the DDK and the WRAPPERS library, this service is accessed via the MAPFLAT macro, as in the following code fragment.

```
pBuf = MAPFLAT(CLIENT_DS, CLIENT_DX);
```

You don't even need to tell the VMM whether the pointer is from a PM application (selector:offset) or a V86 application (segment:offset). The VMM figures that out for itself, based on the execution mode of the currently executing VM. (The current VM is the appropriate context, because it will always be the VM that called the VxD's API.)

What about returning a pointer from the VxD to the application? Clearly a flat model linear address must be transformed into a selector:offset or segment:offset, but either party to the transaction could be responsible for the conversion. There are two ways to approach this. The VxD could perform the conversion and give the application a 16:16 pointer. Alternatively, the VxD could return a linear address to the application and leave the conversion to the application.

Having the VxD do the conversion might seem to be the more natural solution, but is actually more work, mainly because only low-level selector functions are available to a VxD. To perform this conversion, a VxD must first obtain a selector via Allocate_LDT_Selector, then fill in the associated descriptor with BuildDescriptorDWORDS. Using Allocate_LDT_Selector isn't too bad, but BuildDescriptorDWORDS is. Your VxD must deal with details such as DPL, granularity, and big/default, all requiring intimate knowledge of 80x86 descriptors.

An application, on the other hand, has a more useful set of high-level selector functions (AllocSelector, SetSelectorBase, and SetSelectorLimit) which it can use to transform a linear address into a usable pointer.

Win16 Application to VxD: View from Application Side

To call into a VxD, a Win16 application uses the Windows Get Device Entry Point function , accessed through INT 2Fh. The application puts the numeric Device ID in BX, the function code in AX (1684), and calls the VMM via the software interrupt 2Fh. On return, ES:DI is a function pointer the application uses to call the VxD's API procedure. This technique works for both protected mode (Win16 or DOS-extended) applications and for V86-mode applications (plain old DOS).

The following function, GetVxDApi, encapsulates the INT 2Fh call. Pass in a VxD ID, and it returns the function pointer used to call the VxD.

```
typedef void (far *PVOIDFN)(void);
PVOIDFN GetVxDApi(WORD vxdid)
{
    PVOIDFN pfApi;

    _asm {
    push di
    push es
    xor di, di
    mov es, di
    mov ax, 1684h
    mov bx, vxdid
    int 2fh
    WORD PTR pfApi+2, es
    WORD PTR pfApi, di
    pop es
    pop di
    }
    return( pfApi );
}
```

Notice that I said "function pointer the application *uses to call* the VxD's API procedure" and not "function pointer *to* the VxD's API procedure". That's because Ring 3 code can't call Ring 0 code directly. If you dump the code pointed to by ES:DI, you'll see INT 30h followed by another value. The INT 30h is the VMM's way of transferring control from a Ring 3 application to a Ring 0 VxD. Executing a software interrupt from Ring 3 causes the processor to switch to Ring 0. The INT 30h handler is really the VMM's "call VxD from application" procedure. The VMM uses the bytes after the INT 30h instruction to determine which VxD the application wants to call, gets that VxD's API procedure from the VxD's DDB, sets up EBX to point to the client register structure, and, finally, calls the VxD. (However, see the sidebar for information on an alternative that a Windows 95 application can use to call a VxD.)

The above magic is all transparent from the application's point of view. The application sees only a far call to the address returned by the call to Get Device Entry Point. The application passes parameters to the VxD in registers, which means you must use assembly (or at least embedded assembly) to fill in the parameters. As explained earlier, the registers used for the parameters are determined by the VxD developer.

Win16 Application to VxD: Example Code

This section details a simple Win16 application and VxD combination that illustrates the above techniques. In this example, the application requests the VxD to allocate a system DMA buffer on its behalf (something an application can't do itself). The application initializes a structure that describes the buffer required and gives the VxD a pointer to this structure. The VxD allocates a DMA buffer and fills in the application's structure with information about the buffer. The application then translates one of the structure members into a usable pointer.

The Example Application

The application, contained completely in the file WIN16APP.C (Listing 11.1, page 233), is one of the world's simplest Win16 applications. It doesn't even have a message loop, only a WinMain. In WinMain, it calls the VxD to allocate a DMA buffer, displays information about the allocated buffer, calls the VxD to free the buffer, and then exits.

```
pfDmaBufApi = GetVxdApiEntry( DMABUF_ID );
if (!pfDmaBufApi)
{
    printf("Error! Couldn't get DMABUF Api\n");
    exit(1);
}
```

Under Windows 95, An Application Can Call a VxD by Name Instead of by Device ID

When a Win16 application knows that it's running under Windows 95 and not Windows 3.x, the application can use the VxD's 8-byte name instead of its Device ID to find its entry point. The 8-byte name is the one the VxD declares in its DDB, which is usually space padded, and usually does not contain a NUL character at the end. This method also uses INT 2F AX=1684h and returns the same far function pointer. However, in this case BX must be set to 0 and ES:DI is used as an input parameter, pointing to the name.

Because Windows 95 supports this new VxD calling method, it is no longer strictly necessary to obtain a VxD ID in order to provide an API to 16-bit applications running under Windows 95. However, VxD developers that supply a 16-bit API will continue to require a VxD ID as long as they support customers running under Windows 3.x — and even after that if Win16 applications that use the old "call by ID" method are already in the field.

The application first calls a helper function, GetVxdApiEntry, to obtain a function pointer to the VxD entry point. The application then fills in the Size field of the DMA_BUFFER_DESCRIPTOR, telling the VxD what size DMA buffer is required. The VxD will fill in the other two fields, PhysAddr and LinAddr, with the physical address and linear address of the allocated buffer.

```
_asm
{
   mov ax, DMABUF_FUNC_ALLOCBUFFER
   lea si, dmadesc      ; small model, don't need to load DS
   call DWORD PTR pfDmaBufApi
  mov err, ax
}
```

The VxD expects DS:SI to point to the DMA_BUFFER_DESCRIPTOR, so the application uses embedded assembly to load the two registers with the address of the DMA_BUFFER_DESCRIPTOR structure and the AX register with the function code DMABUF_FUNC_ALLOCUFFER. With the registers initialized as expected by the VxD, the application calls the VxD entry point through the function pointer pfDmaBufApi.

If the VxD was unable to allocate the buffer, it returns with a non-zero value in AX. The application tests for this result, producing an error message and exiting. Otherwise, the VxD has allocated the buffer and filled in the PhysAddr and LinAddr fields. An application that was really doing DMA would use the PhysAddr to program the DMA controller; this example merely prints out the field's value.

```
_asm mov myds, ds
usSel = AllocSelector( myds );
SetSelectorBase( usSel, dmadesc.LinAddr );
SetSelectorLimit( usSel, dmadesc.Size );
DmaBufPtr = MAKELP( usSel, 0 );
```

The example application does use the LinAddr field to obtain a usable pointer to the allocated buffer. First, the application obtains a selector via AllocSelector. Next, it calls SetSelectorBase, passing the newly allocated selector and the linear address returned by the VxD. After that, the application uses SetSelectorLimit to set the size of the newly allocated selector. The example also limits the selector to the size of the requested buffer. With this restriction, overwriting the allocated buffer will cause a GP fault and the register will terminate the application immediately. The application completes the conversion by using the MAKELP macro to turn the selector into a pointer. The application now has a usable pointer that maps to the linear address returned by the VxD.

When this conversion is complete, the application displays a message box showing the DMA buffer's physical linear and logical (pointer) address. Finally, the application prepares for termination. It frees the selector that it just allocated, then calls the VxD again, this time using the function code DMABUF_FUNC_FREEBUFFER, to free the allocated buffer.

The Example VxD

The DMABUF VxD called by the WIN16APP application is also very simple. To support the Win16 application, the VxD needs only to handle the Init_Complete message and support a PM API with only two functions, AllocBuffer and FreeBuffer (Listing 11.5, page 236).

The only reason that DMABUF handles the Init_Complete message is that under Windows 3.x, physically contiguous pages must be allocated during initialization, and a system DMA buffer must consist of physically contiguous pages. In Windows 95, contiguous pages may be allocated at any time. To accommodate the difference, DMABUF's OnInitComplete function checks what version of Windows is running. If it is running under Windows 3.x, DMABUF preallocates a DMA buffer of a fixed size (64 Kb). The driver saves the buffer's linear and physical addresses in global variables, where they can be retrieved when an application calls the VxD. For more details on DMA buffer requirements and PageAllocate, see Chapter 6. The following code shows the OnInitComplete handler.

```
BOOL OnInitComplete(VMHANDLE hVM, PCHAR CommandTail)
{
    DWORD    ver;

    Get_VMM_Version();

    if (HIWORD(ver) <= 3)
    {
        // Win3.x, not 95
        bWin3x = TRUE;
        // must alloc phys contig pages now
        LinAddr = _PageAllocate(nPages, PG_SYS, 0, 0x0F, 0, 0x1000, &PhysAddr,
                                PAGEFIXED | PAGEUSEALIGN | PAGECONTIG );
    }
    return TRUE;
}
```

PM_Api_Handler (shown in the following paragraph of code) is the entry point for calls from Win16 applications. Since the application should specify a function code in the AX register (found in the Client_AX field of the CLIENT_STRUCT parameter), PM_Api_Handler switches on this value.

```
void __cdecl PM_Api_Handler(VMHANDLE hVM, CLIENT_STRUCT *pcrs)
{
    DMA_BUFFER_DESCRIPTOR *pBufDesc;

    switch( pcrs->CWRS.Client_AX )
    {
     case DMABUF_FUNC_ALLOCBUFFER:
        pBufDesc = MAPFLAT(Client_DS, Client_SI);
        pcrs->CWRS.Client_AX = AllocBuffer( pBufDesc );
        break;

     case DMABUF_FUNC_FREEBUFFER:
        pBufDesc = MAPFLAT(Client_DS, Client_SI);
        pcrs->CWRS.Client_AX = FreeBuffer( pBufDesc );
        break;

    default:
      pcrs->CWRS.Client_AX = DMABUF_INVALID_FUNC;
      break;
    }
}
```

The DMABUF API consists only of two functions, AllocBuffer and FreeBuffer. In both cases, the buffer in question is described by a DMA_BUFFER_DESCRIPTOR structure passed by the application in DS:SI. To access this buffer, the VxD must translate the application's selector:offset pointer into a usable flat pointer. PM_Api_Handler uses the VMM service Map_Flat, accessed via the macro MAPFLAT, to accomplish this conversion. Finally, PM_Api_Handler calls the appropriate subroutine, either AllocBuffer or FreeBuffer, passing in the flat pointer to the DMA_BUFFER_DESCRIPTOR. The AllocBuffer function is shown in the following code.

```
DWORD AllocBuffer( DMA_BUFFER_DESCRIPTOR *pBufDesc )
{
    DWORD rc;

    if (bOwned)
    {
       rc = DMABUF_ALREADY_ALLOCED;
    }
    else
    {
       bOwned = TRUE;
       if (bWin3x)
       {
          if (pBufDesc->Size > 16 * 4 * 1024)
             rc = DMABUF_SIZE_TOO_BIG;
          else
          {
             pBufDesc->PhysAddr = PhysAddr;
             pBufDesc->LinAddr = LinAddr;
          }
       }
```

```
    else
    {
        LinAddr = pBufDesc->LinAddr
                = PageAllocate(pBufDesc->Size / 4096,
                    PG_SYS, 0, 0x0F, 0, 0x1000,
                    &pBufDesc->PhysAddr,
                    PAGEFIXED | PAGEUSEALIGN | PAGECONTIG );
        if (!pBufDesc->LinAddr)
            rc = DMABUF_BUF_NOT_AVAIL;
    }
}
    return rc;
}
```

For the sake of simplicity, the DMABUF VxD allows only one application to allocate a DMA buffer at a time. To enforce this policy, AllocBuffer checks the global variable bOwned. If this boolean is set, AllocBuffer fails the call and returns with the error code DMABUF_ALREADY_ALLOCED.

If no other application has already claimed the buffer, AllocBuffer checks the bWin3x variable set by OnInitComplete. If this variable is set, then the VxD is running under Windows 3.x and the DMA buffer was preallocated during initialization. If the caller requested a larger buffer size than was allocated, the call fails with a return value of DMABUF_SIZE_TOO_BIG. If the buffer size is acceptable, the VxD copies the physical and linear addresses returned earlier by _PageAllocate into pBufDesc->PhysAddr and pBufDesc->LinAddr.

If the VxD is running under Windows 95, the buffer was not preallocated during initialization, and so must be allocated now, using the size requested by the caller. If _PageAllocate fails for any reason, AllocBuffer returns with DMABUF_NOT_AVAIL. If _PageAllocate succeeds, AllocBuffer returns to the caller, with pBufDesc->LinAddr and pBufDesc->PhysAddr values provided by _PageAllocate. Notice that DMABUF also stores the linear address in the global variable LinAddr — I'll explain why in a moment.

FreeBuffer first checks that bOwned was set by AllocBuffer. If not, the function returns immediately with DMABUF_NOT_ALLOCED. Next, the function verifies that the linear address specified by the caller is the same as the one in pBufDesc->LinAddr, which was returned by _PageAllocate. If the addresses don't match, FreeBuffer returns with DMABUF_NOT_ALLOCED. This precaution prevents the VxD from freeing an invalid address passed in by a buggy application. Finally, FreeBuffer may indeed free the buffer, but only if running under Windows 95. If under Windows 3.x, the buffer allocated during initialization must stay around for future use. The FreeBuffer function is shown in the following code.

```
DWORD FreeBuffer( DMA_BUFFER_DESCRIPTOR *pBufDesc )
{
   DWORD rc;

   if (bOwned)
   {
      bOwned = FALSE;
      if (pBufDesc->LinAddr == LinAddr)
      {
         if (!bWin3x)
         {
            PageFree( pBufDesc->LinAddr, 0 );
         }
      }
      else
      {
         rc = DMABUF_NOT_ALLOCED;
      }
   }
   else
   {
      rc = DMABUF_NOT_ALLOCED;
   }
   return rc;
}
```

Win32 Application to VxD:
View from VxD side

The interface from a Win32 application to a VxD is much different, both viewed from the VxD side and from the application side. As before, I'll first explain the VxD side, then the application side.

A VxD doesn't need to export a special procedure in order to support Win32 applications. Instead, its control procedure must handle a special message, called W32_DEVICEIOCONTROL. The VMM sends this message to the VxD on behalf of the calling application.

Parameters are passed, not through registers, but all bundled up into a DIOCPARAMETERS structure. The VMM puts a pointer to this structure in ESI. Here's the structure:

```
typedef struct DIOCParams    {
    DWORD     Internal1;
    DWORD     VMHandle;
    DWORD     Internal2;
    DWORD     dwIoControlCode;
    DWORD     lpvInBuffer;
    DWORD     cbInBuffer;
    DWORD     lpvOutBuffer;
    DWORD     cbOutBuffer;
    DWORD     lpcbBytesReturned;
    DWORD     lpoOverlapped;
    DWORD     hDevice;
    DWORD     tagProcess;
} DIOCPARAMETERS;
```

The DIOCPARAMETERS structure is defined in VWIN32.H, not VMM.H. Also note that VToolsD uses a different structure name (IOCTLPARAMS) and different field names.

The dwIoControlCode field tells the VxD which function to perform. The lpvInBuffer and cbInBuffer are pointers to a generic input buffer and the size of the input buffer, and lpvOutBuffer and cbOutBuffer are the same for the generic output buffer. Note that these pointer parameters don't need translation, but can be used directly by the VxD. Both the function code in dwIoControlField and the meaning of the buffer contents are defined by the VxD. This interface is generic on purpose, so that you can do more with a device than just read from and write to it. In most cases, both the application and the VxD will treat the generic buffer as a specific structure, casting the buffer pointer to and from a pointer to DIOCPARAMETERS as necessary.

The VMM will test your VxD to determine if it supports the Win32 DeviceIoControl interface by sending a W32_DEVICEIOCONTROL message with a dwIoControlCode of DIOC_GET_VERSION. If your VxD doesn't respond as expected, the VMM will not pass on further W32_DEVICEIOCONTROL messages. The response the VMM is expecting is a return value of zero from the message handler. Your VxD may return whatever version information it wishes (or none at all) in the lpvOutBuffer; all the VMM cares about is the return value.

Win32 Application to VxD: View from the Application Side

A Win32 application calls into a VxD by using the DeviceIoControl function. One of the parameters to this function is a device handle obtained via a call to CreateFile. That's right, the same call that creates or opens a file can also open a "channel" to a VxD. To open a VxD, rather than a normal file, with CreateFile, you use a special form in place of the filename:

"\\.\name"

When using this format in your C code, don't forget that backslash represents an escape sequence, so use two consecutive backslashes for each.

This strange format tells Windows that you don't really want to open a normal file; instead, you want Windows to find and load the VxD with that name, and give you a special handle to it. Your application can then use this handle with calls to DeviceIoControl. Windows turns this call into a W32_DEVICEIOCONTROL message, with all of the application's parameters neatly bundled up into a single DIOCPARAMETERS structure.

If the filename contains an extension, Windows looks in the standard search path for the VxD: current directory, Windows directory, then path environment variable. Specifying an extension is the usual method, and the extension is usually VXD. If there is no extension, Windows looks in the registry for the KnownVxDs key under HKLM\SYSTEM\CURRENTCONTROLSET\CONTROL\SESSIONMANAGER. If this key has an associated value, Windows treats the value as the VxD's full pathname. If Windows can't find the VxD there either, it treats the filename as a VxD module name, and searches its internal VxD list for an already-loaded VxD with that name.

When "opening" a VxD, the VxD name is considered case sensitive. To be safe, use all uppercase in both your application and VxD DDB declaration.

If CreateFile returns INVALID_HANDLE_VALUE, you should call GetLastError to get error information. There are two possible errors when opening a VxD. ERROR_FILE_NOT_FOUND indicates that all the methods described above have failed to find the specified VxD. ERROR_NOT_SUPPORTED indicates that the VxD was found but that it doesn't "support" the DeviceIoControl interface — which in many cases means the VxD wanted to support DeviceIoControl but didn't properly handle the DIOC_GET_VERSION, as described in the previous section.

Special care is needed in handling ERROR_NOT_SUPPORTED. The problem is that the VxD did load successfully (the actual error was in the VxD's response to the W32_DEVICEIOCONTROL) but CreateFile returned no handle that the application could use to close the VxD and thus unload it. To force the VxD to be unloaded, the application must call DeleteFile, using the VxD's module name in the DDB, not the filename. A VxD should choose a module name equal to the filename minus the extension, although the choice of module name is completely up to the VxD

If the VxD referenced in the CreateFile is dynamically loadable, the call to CreateFile may do more than open a "channel" to a VxD for future DeviceIoControl communication. If the VxD is dynamically loadable and isn't yet loaded, Windows will automatically load the VxD on behalf of the application and send it a Sys_Dynamic_Device_Init message. Windows maintains a reference, or usage, count for the VxD, so if it's already loaded, a call to CreateFile doesn't load another copy of the VxD. Applications should generally use the value FILE_FLAG_DELETE_ON_CLOSE for the fdwAttrsAndFlags parameter when calling CreateFile. This tells Windows to unload the VxD when the reference count goes to zero. (A zero reference count means that every application that had opened the VxD has now closed it.)

If the VxD returns with success to the `Sys_Dynamic_Device_Init` message, the VMM immediately sends the `W32_DEVICEIOCONTROL` message with the `dwIoControlCode` parameter set to `DIOC_GETVERSION`. A dynamic VxD does any per-application initialization here. As explained earlier, a VxD must return success for this message, otherwise the application sees an `ERROR_NOT_SUPPORTED` return code. If the VxD returns success, the VMM increments its internal reference count for the VxD. If another call to `CreateFile` is made before a `CloseHandle`, the VxD receives another message with a `dwIoControlCode` of `DIOC_GETVERSION` — but not another `Sys_Dynamic_Device_Init` message since the VxD is already loaded.

After getting a device handle with `CreateFile`, your application calls `DeviceIoControl`. The prototype for this function is:

```
BOOL DeviceIoControl(
    HANDLE        hDevice,
    DWORD         dwIoControlCode,
    LPVOID        lpInBuffer,
    DWORD         nInBufferSize,
    LPVOID        lpOutBuffer,
    DWORD         nOutBufferSize,
    LPDWORD       lpBytesReturned,
    LPOVERLAPPED  lpOverlapped
);
```

The first parameter is the handle returned by `CreateFile`, and the next four parameters should look familiar: the VxD receives those exact same parameters in its `W32_DEVICEIOCONTROL` message, though for the VxD they're all bundled up into a single `DIOCPARAMETERS` structure. The `lpBytesReturned` parameter is filled in by the VxD, telling the application how many bytes the VxD has copied to the output buffer.

When your application has finished communicating with the VxD, it closes the "channel" by calling `CloseHandle`, using the same device handle. If the VxD was dynamically loaded, this call to `CloseHandle` results in a `W32_DEVICEIOCONTROL` message with `dwIoControlCode` of `DIOC_CLOSEHANDLE`. When the final `CloseHandle` causes the reference count to go to zero, the VMM sends a final `Sys_Dynamic_Device_Exit` message and the VMM then unloads the VxD.

Don't forget to add the `DYNAMIC` keyword to the VxD statement in your `.DEF` file following the VxD's module name (Listing 11.8, page 242).

Win32 Application to VxD: Example Code

To illustrate how a Win32 application talks to a VxD, I've extended the same DMABUF VxD introduced earlier in this chapter and written a simple Win32 application that uses DeviceIoControl to talk to the VxD (Listing 11.9, page 243). Once again, the application is very simple (nothing but a main), and because it is a Win32 console application, we can simply use printf — no message boxes.

This Win32 application is similar in structure to its Win16 counterpart. The application "opens" the VxD, initializes a DMA_BUFFER_DESCRIPTOR and then calls the VxD to allocate a DMA buffer. When the VxD returns to the application, the VxD will have written the allocated buffer's physical and linear addresses into the DMA_BUFFER_DESCRIPTOR. Because this is a Win32 application, a linear address is a pointer, and no selector magic is needed.

```
const PCHAR VxDName = "\\\\.\\DMABUF.VXD";
hDevice = CreateFile(VxDName, 0,0,0,
                     CREATE_NEW, FILE_FLAG_DELETE_ON_CLOSE, 0);

if (hDevice == INVALID_HANDLE_VALUE)
{
    err = GetLastError();
    printf("Cannot load VxD, error=%08lx\n", err );
    if (err == ERROR_NOT_SUPPORTED)
    {
        DeleteFile("\\\\.\\DMABUF");
    }
    exit(1);
}
```

To "open" a channel to the VxD, the application calls CreateFile with the filename \\.\DMABUF.VXD. If the call fails, the application uses GetLastError to obtain the actual VxD return (error) code, and if the return was ERROR_NOT_SUPPORTED, the application calls DeleteFile to unload the VxD.

```
dmadesc.Size = 32 * 1024;
if (err = DeviceIoControl(hDevice, DMABUF_FUNC_ALLOCBUFFER,
    &dmadesc, sizeof(DMA_BUFFER_DESCRIPTOR),
    NULL, 0, &cbBytesReturned, NULL))
    printf("DeviceIoControl failed, error=%x\n", err);
```

If the open succeeded, the application initializes the DMA_BUFFER_DESCRIPTOR structure with the size of the requested buffer, then calls DeviceIoControl, using a dwIoControlCode of DMABUF_FUNC_ALLOCBUFFER. In this example, no output buffer is used. Instead, the VxD modifies the caller's input buffer (lpvInBuffer). Furthermore, because the VxD doesn't copy any bytes to the output buffer, it never fills in the application's cbBytesReturned variable. Bending the rules like this is perfectly acceptable under Windows 95, and by defining the interface in this way, I was able to re-use the exact same VxD code already written for the PM API portion of the VxD.

```
else
{
    printf("Physical=%081X\nLinear=%081X\n", dmadesc.PhysAddr,
        dmadesc.LinAddr );
    if (err = DeviceIoControl(hDevice, DMABUF_FUNC_FREEBUFFER,
        &dmadesc, sizeof(DMA_BUFFER_DESCRIPTOR),
        NULL, 0, &cbBytesReturned, NULL))
        printf("DeviceIoControl failed, error=%x\n", err);
}

CloseHandle( hDevice );
```

If the ALLOCBUFFER DeviceIoControl fails (non-zero return value), the application prints the error code and exits. Otherwise, the application prints the physical and linear addresses of the allocated buffer, and immediately frees the buffer with another DeviceIoControl call, but this time with a function code of DMABUF_FUNC_FREEBUFFER. Finally, the application closes the channel to the VxD with a call to CloseHandle. If no other application is using the VxD and the VxD is dynamically loadable, this close also unloads the VxD from memory.

To implement the VxD side, I merely added a W32_DEVICEIOCONTROL message handler (shown in the following paragraph of code) to the same DMABUF VxD developed for the Win16 application. This message handler is even simpler than the PM API function, because no translation of pointer parameters is necessary. Because both Win32 applications and VxDs use linear addresses, all pointers contained in the DIOCPARAMETERS structure are directly usable by the VxD.

```
DWORD OnW32Deviceiocontrol(PDIOCPARAMETERS p)
{
    DWORD rc;

    switch (p->dwIoControlCode)
    {
    case DIOC_OPEN:
    case DIOC_CLOSEHANDLE:
        return 0;

    case DMABUF_FUNC_ALLOCBUFFER:
        if (!_Assert_Range( p->lpvInBuffer,
                            sizeof( DMA_BUFFER_DESCRIPTOR ),
                            0, 0, ASSERT_RANGE_NULL_BAD))
            return DMABUF_INVALID_PARAMETER;
        else
            return( AllocBuffer( p->lpvInBuffer ) );

    case DMABUF_FUNC_FREEBUFFER:
        if (!_Assert_Range( p->lpvInBuffer,
                            sizeof( DMA_BUFFER_DESCRIPTOR ),
                            0, 0, ASSERT_RANGE_NULL_BAD))
            return DMABUF_INVALID_PARAMETER;
        else
            return( FreeBuffer( p->lpvInBuffer ) );

    default:
        return -1;
    }
}
```

The message handler specifically checks for dwIoControlCode values of DIOC_GETVERSION and DIOC_CLOSEHANDLE, returning 0 for each. Failure to do so will result in failure when the application calls CreateFile and CloseHandle, respectively. The VxD also returns an error code of -1 for unexpected control codes.

The two expected codes are DMABUF_FUNC_ALLOCBUFFER and DMABUF_FUNC_FREEBUFFER. In both cases, the VxD is expecting the caller's input buffer to be a pointer to a DMA_BUFFER_DESCRIPTOR, but before using the pointer as such, the VxD validates it. The cbInBuffer parameter, though ostensibly for this exact purpose, cannot be used to validate the buffer size. cbInBuffer isn't necessarily the size of the input buffer; it only reflects the caller's claims about the input buffer size. The VxD guards against both a null lpvInBuffer value and a buffer that's too small with a single call to the VMM service _AssertRange.

The Calling Interface for _Assert_Range

```
BOOL __cdecl _Assert_Range(DWORD pStruc, DWORD ulSize,
                           DWORD signature, DWORD lSignatureOffset,
                           DWORD ulFlags);
```

_Assert_Range verifies that the buffer pointed to by pStruc is at least ulSize in length. In addition, it can check for a signature value at the offset lSignatureOffset. However, DMABUF doesn't use this feature, passing in 0 for signature to disable it. DMABUF does use the value ASSERT_RANGE_NULL_BAD for the ulFlags parameter, however, so that a NULL value for pStruc will cause _Assert_Range to fail. If _Assert_Range fails, DMABUF returns to the application with a DMABUF_INVALID_PARAMETER error.

After this validation, the VxD simply casts the caller's input buffer, p->lpvInBuffer, to a pointer to a DMA_BUFFER_DESCRIPTOR, then passes that pointer directly to either AllocBuffer or FreeBuffer, depending on the value of p->dwIoControlCode. The return value from the helper function is passed directly back to the caller as the return from DeviceIoControl. Note these two helper functions are unchanged from the original DMABUF VxD, which contained only Win16 API support.

Summary

If you structure your code right, supporting both Win16 and Win32 applications in your VxD isn't much more trouble than supporting just one or the other. The message here is that you should put the real work of the API in subroutines that can be called from either your PM API procedure or from your W32_DEVICEIOCONTROL message handler. If you follow this practice, then all your PM API procedure will do is extract its caller's parameters from the CLIENT_STRUCT structure. Similarly, the W32_DEVICEIOCONTROL handler should merely extract its caller's parameters from the DIOCPARAMETERS structure. Both interface procedures then call the same helper subroutines.

The application interfaces described in this chapter support communications initiated by the application: when the application calls the VxD. The next chapter will cover the reverse direction: when a VxD calls into an application.

Listing 11.1   WIN16APP.C

```c
#include <string.h>
#include <windows.h>

#include "dmabuf.h"

typedef void (far * PVOIDFN)( void );

static char MsgBoxBuf[ 1024 ] = { 0 };
PVOIDFN  pfDmaBufApi;
DMA_BUFFER_DESCRIPTOR dmadesc;

PVOIDFN GetVxdApiEntry( int VxdId )
{
    PVOIDFN pfApi;

    _asm
    {
        xor di, di
        mov es, di
        mov bx, VxdId
        mov ax, 1684h
        int 2fh
        mov WORD PTR pfApi+2, es
        mov WORD PTR pfApi, di
    }

    return( pfApi );
}

int PASCAL WinMain( HANDLE hInstance, HANDLE hPrevInstance,
                    LPSTR lpCmdLine, int nCmdShow )
{
    char far *DmaBufPtr;
    unsigned short usSel, myds;
    WORD  err;

    pfDmaBufApi = GetVxdApiEntry( DMABUF_ID );
    if (!pfDmaBufApi)
    {
        MessageBox( NULL, "Error, couldn't get VxD API", "USEAPI", MB_OK );
    }
```

Listing 11.1 (continued)   `WIN16APP.C`

```c
else
    {
        dmadesc.Size = 32L * 1024L;
        _asm
        {
            mov ax, DMABUF_FUNC_ALLOCBUFFER
            lea si, dmadesc     ; small model, don't need to load DS
            call DWORD PTR pfDmaBufApi
            mov err, ax
        }
        if (err)
        {
            MessageBox( NULL, "Error calling AllocBuffer",
                        "USEAPI", MB_OK );
        }
        else
        {
            _asm mov myds, ds
            usSel = AllocSelector( myds );
            SetSelectorBase( usSel, dmadesc.LinAddr );
            SetSelectorLimit( usSel, dmadesc.Size );
            DmaBufPtr = MAKELP( usSel, 0 );
            wsprintf( MsgBoxBuf,
                        "Physical=%08lX\nLinear=%08lXSelector=%X\n",
                        dmadesc.PhysAddr, dmadesc.LinAddr, usSel );
            MessageBox( NULL, MsgBoxBuf, "USEAPI", MB_OK );

            FreeSelector( usSel );
            _asm
            {
                mov ax, DMABUF_FUNC_FREEBUFFER
                call DWORD PTR pfDmaBufApi
            }
        }
    }

    return 0;

}
```

Listing 11.2   *WIN16APP.MAK*

```
all: win16app.exe

win16app.obj: win16app.c
    cl -W3 -c -AS -Gsw2 -I..\vxd win16app.c

win16app.exe: win16app.def win16app.obj
    link win16app.obj,win16app.exe,win16app.map
        /MAP /CO,slibcew libw /nod,win16app.def
```

Listing 11.3   *WIN16APP.DEF*

```
NAME      WIN16APP
EXETYPE   WINDOWS
CODE      PRELOAD MOVEABLE DISCARDABLE
DATA      PRELOAD MOVEABLE
HEAPSIZE    4096
STACKSIZE   8192
```

Listing 11.4   DMABUF.H

```
// DMABUF.h - include file for VxD DMABUF
#define DMABUF_ID                 0xDB0
#define DMABUF_FUNC_ALLOCBUFFER    0x1000
#define DMABUF_FUNC_FREEBUFFER     0x1001


#define DMABUF_ALREADY_ALLOCED     0x0001
#define DMABUF_SIZE_TOO_BIG        0x0002
#define DMABUF_BUF_NOT_AVAIL       0x0003
#define DMABUF_BUF_NOT_ALLOCED     0x0004
#define DMABUF_INVALID_PARAMETER   0x0005
#define DMABUF_INVALID_FUNC        0x0006

typedef struct
{
    DWORD Size;
    DWORD PhysAddr;
    DWORD LinAddr;
} DMA_BUFFER_DESCRIPTOR;
```

Listing 11.5   DMABUF.C

```
#define WANTVXDWRAPS

#include <basedef.h>
#include <vmm.h>
#include <debug.h>
#include "vxdcall.h"
#include <vxdwraps.h>
#include <wrappers.h>
#include <vwin32.h>
#include "dmabuf.h"

#ifdef DEBUG
#define DPRINTF0(buf, fmt)            _Sprintf(buf, fmt ); Out_Debug_String( buf )
#define DPRINTF1(buf, fmt, arg1)      _Sprintf(buf, fmt, arg1 );
                                      Out_Debug_String( buf )
#define DPRINTF2(buf, fmt, arg1, arg2) _Sprintf(buf, fmt, arg1, arg2 );
                                      Out_Debug_String( buf )
```

Listing 11.5 (continued)   DMABUF.C

```
#else
#define DPRINTF0(buf, fmt)
#define DPRINTF1(buf, fmt, arg1)
#define DPRINTF2(buf, fmt, arg1, arg2)
#endif

BOOL bOwned = FALSE;
DWORD nPages = 16;      // 64K = 16 * 4K
void *LinAddr;
DWORD PhysAddr;
BOOL  bWin3x = FALSE;
char  dbgbuf[80];

DWORD AllocBuffer( DMA_BUFFER_DESCRIPTOR *pBufDesc );
DWORD FreeBuffer( DMA_BUFFER_DESCRIPTOR *pBufDesc );

BOOL OnSysDynamicDeviceInit()
{
    DPRINTF0(dbgbuf,"Loading\r\n");
    return TRUE;
}

BOOL OnSysDynamicDeviceExit()
{
    DPRINTF0(dbgbuf,"Unloading\r\n");
    return TRUE;
}

DWORD OnW32Deviceiocontrol(PDIOCPARAMETERS p)
{
    DPRINTF1(dbgbuf,"W32DevIoControl code=%x\n", p->dwIoControlCode );

    switch (p->dwIoControlCode)
    {
    case DIOC_GETVERSION:
    case DIOC_CLOSEHANDLE:      // file closed
        return 0;

    case DMABUF_FUNC_ALLOCBUFFER:
        if (!_Assert_Range( p->lpvInBuffer, sizeof( DMA_BUFFER_DESCRIPTOR ), 0, 0,
            ASSERT_RANGE_NULL_BAD))
            return DMABUF_INVALID_PARAMETER;
        else
            return( AllocBuffer( (DMA_BUFFER_DESCRIPTOR *)p->lpvInBuffer ) );

    case DMABUF_FUNC_FREEBUFFER:
        if (!_Assert_Range( p->lpvInBuffer, sizeof( DMA_BUFFER_DESCRIPTOR ), 0, 0,
                         ASSERT_RANGE_NULL_BAD))
            return DMABUF_INVALID_PARAMETER;
        else
            return( FreeBuffer( (DMA_BUFFER_DESCRIPTOR *)p->lpvInBuffer ) );

    default:
        return -1;
    }
}
```

Listing 11.5 (continued)   `DMABUF.C`

```c
void __cdecl PM_Api_Handler(VMHANDLE hVM, CLIENT_STRUCT *pcrs)
{
    DMA_BUFFER_DESCRIPTOR *pBufDesc;

    switch( pcrs->CWRS.Client_AX )
    {
     case DMABUF_FUNC_ALLOCBUFFER:
        pBufDesc = MAPFLAT(Client_DS, Client_SI);
        pcrs->CWRS.Client_AX = AllocBuffer( pBufDesc );
        break;

     case DMABUF_FUNC_FREEBUFFER:
        pBufDesc = MAPFLAT(Client_DS, Client_SI);
        pcrs->CWRS.Client_AX = FreeBuffer( pBufDesc );
        break;

    default:
        pcrs->CWRS.Client_AX = DMABUF_INVALID_FUNC;
        break;
    }
}

DWORD AllocBuffer( DMA_BUFFER_DESCRIPTOR *pBufDesc )
{
    DWORD rc = 0;

    if (bOwned)
    {
        rc = DMABUF_ALREADY_ALLOCED;
    }
    else
    {
        bOwned = TRUE;
        if (bWin3x)
        {
            if (pBufDesc->Size > 16 * 4 * 1024)
                rc = DMABUF_SIZE_TOO_BIG;
            else
            {
                pBufDesc->PhysAddr = PhysAddr;
                pBufDesc->LinAddr = LinAddr;
            }
        }
        else
        {
            // Win95, can alloc phys contig pages at any time
            pBufDesc->LinAddr = LinAddr = _PageAllocate(pBufDesc->Size >> 12,
                                            PG_SYS, 0, 0x0F, 0, 0x1000,
                                            &pBufDesc->PhysAddr,
                                            PAGEFIXED | PAGEUSEALIGN | \
                                            PAGECONTIG );

            if (!LinAddr)
                rc = DMABUF_BUF_NOT_AVAIL;
        }
    }
    return rc;
}
```

Listing 11.5 (continued) *DMABUF.C*

```c
DWORD FreeBuffer( DMA_BUFFER_DESCRIPTOR *pBufDesc )
{
    DWORD    rc = 0;

    if (bOwned)
    {
        bOwned = FALSE;
        // free buffer only if Win95
        // and don't free buffer unless it's the same one we allocated
        if (pBufDesc->LinAddr == LinAddr)
        {
            if (!bWin3x)
            {
                _PageFree( pBufDesc->LinAddr, 0 );
            }
        }
        else
        {
            rc = DMABUF_BUF NOT_ALLOCED;
        }
    }
    else
    {
        rc = DMABUF_BUF_NOT_ALLOCED;
    }
    return rc;
}

BOOL OnInitComplete(VMHANDLE hVM, PCHAR CommandTail)
{
    DWORD    ver;

    Get_VMM_Version();

    if (HIWORD(ver) <= 3)
    {
        // Win3.x, not 95
        bWin3x = TRUE;
        // must alloc phys contig pages now
        LinAddr = _PageAllocate(nPages, PG_SYS, 0, 0x0F, 0, 0x1000, &PhysAddr,
                            PAGEFIXED | PAGEUSEALIGN | PAGECONTIG );
    }
    return TRUE;
}
```

Listing 11.6   DMADDB.ASM

```
.386p

;*******************************************************************************
;                    I N C L U D E S
;*******************************************************************************

    include vmm.inc
    include debug.inc

;=============================================================================
;        V I R T U A L   D E V I C E   D E C L A R A T I O N
;=============================================================================
DMABUF_ID    EQU    0DB0H    ; must match ID in DMABUF.H

DECLARE_VIRTUAL_DEVICE        DMABUF, 1, 0, ControlProc, DMABUF_ID, \
                             UNDEFINED_INIT_ORDER, 0, PM_API

;extrn _PM_Api_Handler:near
;extrn _V86_Api_Handler:near

VxD_LOCKED_CODE_SEG

;=============================================================================
;
;    PROCEDURE: ControlProc
;
;    DESCRIPTION:
;     Device control procedure for the SKELETON VxD
;
;    ENTRY:
;     EAX = Control call ID
;
;    EXIT:
;     If carry clear then
;         Successful
;     else
;         Control call failed
;
;    USES:
;     EAX, EBX, ECX, EDX, ESI, EDI, Flags
;
;=============================================================================

BeginProc ControlProc
    Control_Dispatch INIT_COMPLETE, _OnInitComplete, cCall, <ebx>
    Control_Dispatch SYS_DYNAMIC_DEVICE_INIT, _OnSysDynamicDeviceInit, cCall, <ebx>
    Control_Dispatch SYS_DYNAMIC_DEVICE_EXIT, _OnSysDynamicDeviceExit, cCall, <ebx>
    Control_Dispatch W32_DEVICEIOCONTROL, _OnW32Deviceiocontrol, cCall, <esi>
    clc
    ret

EndProc ControlProc
```

Listing 11.6 (continued)   `DMADDB.ASM`

```
BeginProc PM_API

    cCall _PM_Api_Handler, <ebx, ebp>
    mov   [ebp].Client_EAX, eax
    ret

EndProc PM_API

VxD_LOCKED_CODE_ENDS

    END
```

Listing 11.7   `DMABUF.MAK`

```
CFLAGS      = -DWIN32 -DCON -Di386 -D_X86_ -D_NTWIN -W3 -Gs -D_DEBUG -Zi
CVXDFLAGS   = -Zdp -Gs -c -DIS_32  Z1  DDEBLEVEL=1 -DDEBUG
LFLAGS      = -machine:i386 -debug:notmapped,full -debugtype:cv
              -subsystem:console kernel32.lib
AFLAGS      = -coff -DBLD_COFF -DIS_32 -W2 -Zd -c -Cx -DMASM6 -DDEBLEVEL=1 -DDEBUG

all: dmabuf.vxd

dmabuf.obj: dmabuf.c
        cl $(CVXDFLAGS) -Fo$@ %s

dmaddb.obj: dmaddb.asm
        ml $(AFLAGS) -Fo$@ %s

dmabuf.vxd: dmaddb.obj dmabuf.obj ..\..\wrappers\vxdcall.obj
                                  ..\..\wrappers\wrappers.clb dmabuf.def
        echo >NUL @<<dmabuf.crf
-MACHINE:i386 -DEBUG -DEBUGTYPE:MAP -PDB:NONE
-DEF:dmabuf.def -OUT:dmabuf.vxd -MAP:dmabuf.map
-VXD vxdwraps.clb wrappers.clb vxdcall.obj dmaddb.obj dmabuf.obj
<<
        link @dmabuf.crf
        mapsym dmabuf
```

Listing 11.8   DMABUF.DEF

```
VXD DMABUF DYNAMIC
SEGMENTS
    _LTEXT     CLASS 'LCODE'   PRELOAD NONDISCARDABLE
    _LDATA     CLASS 'LCODE'   PRELOAD NONDISCARDABLE
    _TEXT      CLASS 'LCODE'   PRELOAD NONDISCARDABLE
    _DATA      CLASS 'LCODE'   PRELOAD NONDISCARDABLE
    _LPTEXT    CLASS 'LCODE'   PRELOAD NONDISCARDABLE
    _CONST     CLASS 'LCODE'   PRELOAD NONDISCARDABLE
    _BSS       CLASS 'LCODE'   PRELOAD NONDISCARDABLE
    _TLS       CLASS 'LCODE'   PRELOAD NONDISCARDABLE
    _ITEXT     CLASS 'ICODE'   DISCARDABLE
    _IDATA     CLASS 'ICODE'   DISCARDABLE
    _PTEXT     CLASS 'PCODE'   NONDISCARDABLE
    _PDATA     CLASS 'PCODE'   NONDISCARDABLE
    _STEXT     CLASS 'SCODE'   RESIDENT
    _SDATA     CLASS 'SCODE'   RESIDENT
    _MSGTABLE  CLASS 'MCODE'   PRELOAD NONDISCARDABLE IOPL
    _MSGDATA   CLASS 'MCODE'   PRELOAD NONDISCARDABLE IOPL
    _IMSGTABLE CLASS 'MCODE'   PRELOAD DISCARDABLE IOPL
    _IMSGDATA  CLASS 'MCODE'   PRELOAD DISCARDABLE IOPL
    _DBOSTART  CLASS 'DBOCODE' PRELOAD NONDISCARDABLE CONFORMING
    _DBOCODE   CLASS 'DBOCODE' PRELOAD NONDISCARDABLE CONFORMING
    _DBODATA   CLASS 'DBOCODE' PRELOAD NONDISCARDABLE CONFORMING
    _16ICODE   CLASS '16ICODE' PRELOAD DISCARDABLE
    _RCODE     CLASS 'RCODE'
EXPORTS
    DMABUF_DDB @1
```

Listing 11.9   `WIN32APP.C`

```c
#include <stdio.h>
#include <stdlib.h>
#include <conio.h>
#include <windows.h>
#include "dmabuf.h"

HANDLE         hDevice;
DMA_BUFFER_DESCRIPTOR dmadesc;

void main(int ac, char* av[])
{
    DWORD     cbBytesReturned;
    DWORD     err;

    const PCHAR VxDName = "\\\\.\\DMABUF.VXD";
    hDevice = CreateFile(VxDName, 0,0,0, CREATE_NEW, FILE_FLAG_DELETE_ON_CLOSE, 0);

    if (hDevice == INVALID_HANDLE_VALUE)
    {
        err = GetLastError();
            fprintf(stderr, "Cannot load VxD, error=%08lx\n", err );
        if (err == ERROR_NOT_SUPPORTED)
        {
            DeleteFile("\\\\.\\DMABUF");
        }
         exit(1);
    }

    dmadesc.Size = 32 * 1024;
    if (!DeviceIoControl(hDevice, DMABUF_FUNC_ALLOCBUFFER,
                        &dmadesc, sizeof(DMA_BUFFER_DESCRIPTOR), NULL, 0,
                        &cbBytesReturned, NULL))
    {
        printf("DeviceIoControl failed, error=%d\n", GetLastError() );
    }
    else
    {
        printf( "Physical=%08lX\nLinear=%08lX\n", dmadesc.PhysAddr, dmadesc.LinAddr );
        if (!DeviceIoControl(hDevice, DMABUF_FUNC_FREEBUFFER,
                        &dmadesc, sizeof(DMA_BUFFER_DESCRIPTOR), NULL, 0,
                        &cbBytesReturned, NULL) )
        {
            printf("DeviceIoControl failed, error=%d\n", GetLastError() );
        }
    }

    CloseHandle( hDevice );
}
```

Listing 11.10 WIN32APP.MAK

```
win32app.exe: win32app.obj
   link @<<
kernel32.lib user32.lib gdi32.lib winspool.lib comdlg32.lib advapi32.lib
shell32.lib ole32.lib oleaut32.lib uuid.lib /NOLOGO /SUBSYSTEM:console
/INCREMENTAL:no /PDB:none /MACHINE:I386 /OUT:win32app.exe win32app.obj
<<

win32app.obj: win32app.c
   cl /c /ML /GX /YX /Od /D "WIN32" /D "NDEBUG" /D "_CONSOLE" -I..\vxd win32app.c
```

Chapter 12

Communication from VxDs to Applications

While sometimes it's enough for an application to call into a VxD and get the information or services it needs immediately, other times an application needs to be notified by a VxD asynchronously, that is, when a particular event occurs. Both Windows 3.x and Windows 95 support mechanisms for communication in this direction (VxD to application), but the interface is more complicated compared to the application-to-VxD methods examined in the last chapter.

The last chapter was divided into two sections, Win16 and Win32. This chapter will use a different division: VxDs using `PostMessage`, VxDs using appy-time, and VxDs using Win32-specific techniques. VxDs under both Windows 3.x and Windows 95 can use `PostMessage` to communicate with Windows applications, 16- and 32-bit. Windows 95 VxDs have other options as well — using "appy-time" services to communicate with Win16 applications and two different techniques to communicate with Win32 applications.

Difficulties with Calling from a VxD to a Win16 Application

Assume that a Win16 application has used the INT 2Fh API to pass a VxD the address of a callback function inside the application. This VxD must overcome several obstacles before it can use the application's callback. A VxD executes outside the context of any VM, whereas the Ring 3 callback must execute in the proper VM context — the SystemVM that registered the callback. So a VxD must first schedule a VM event, and be called back in the context of the System VM, that is, when that VM is current. From inside this event handler, the VxD can use VMM nested execution services to execute the application callback in the System VM.

If the VxD uses only this simple mechanism, the application callback code is very limited in what it can do. In particular, the only Windows API function the callback is allowed to use is PostMessage. When called from a VxD via nested execution, an application callback function executes much like an ISR and is subject to the same kind of constraints. Like an ISR, the callback "interrupts" the VM's execution at some unpredictable point — perhaps even in the middle of performing a Windows system call. Because Windows isn't re-entrant, it isn't safe for the callback to execute any Windows API calls except PostMessage.

VxDs running under Windows 3.x were stuck with this unhappy state of affairs. Windows 3.x VxDs could schedule a VM event and then use nested execution to call back into a VM, but the application callback was limited to PostMessage. For this reason, it was common practice for the application to pass the VxD the address of the Windows PostMessage function along with a window handle, and have the VxD use nested execution to call PostMessage directly on behalf of the application.

Windows 95 offers two improvements for VxDs calling into Win16 applications. One is the service SHELL_PostMessage, which takes care of the details of nested execution on behalf of the calling VxD. The other is a set of "appy-time" (application time) services that allow a VxD to schedule an event to run when the system is in a "safe state". From the appy-time event, the VxD can use other VMM services (new for Windows 95) to call any function in a Win16 DLL, and the Win16 function can itself call any Windows function — because it is "safe".

The POSTVXD example in this chapter illustrates both the PostMessage and the appy-time technique. POSTVXD determines at run-time which version of Windows it's running under (3.x or 95) and uses the appropriate technique, so that it works correctly on both versions.

VxD *PostMessage* **under Windows 3.x**

To call into a Win16 application under Windows 3.x, a VxD must first schedule a VM event for the System VM, and then use nested execution services from the event handler to call into the application. Events were introduced in Chapter 7 when hardware interrupts were discussed. For a hardware interrupt handler, scheduling an event provides a convenient way to defer processing. The interrupt handler example used a global event, "global" meaning the VxD didn't care what VM context the event handler ran in. The VxD callback will use a VM event instead, that is, an event called in the context of a particular VM. In the present situation, the VxD should use a VM event instead of a global event because it needs to call PostMessage, which lives in the System VM.

The POSTVXD example (Listing 12.2, page 269) uses the techniques discussed above when running under Windows 3.x. POSTVXD supports a PM API that lets Win16 applications register with the VxD. Using this API, an application passes in a window handle and the address of the Windows PostMessage function. The VxD then posts a message to this window whenever a VM is created or destroyed. Before terminating, the application should also use the VxD API to deregister the window handle, so that the VxD stops posting messages to it.

To interface to a Win16 application, POSTVXD needs only a PM API procedure, two message handlers (OnVmInit and OnVmTerminate), and an event callback. The source code for the PM API handler follows.

```
VOID  cdecl PM_Api_Handler(VMHANDLE hVM, CLIENT_STRUCT *pcrs)
{
    switch (pcrs->CWRS.Client_AX)
    {
    case POSTVXD_REGISTER:
        PostMsghWnd = (HANDLE)pcrs->CWRS.Client_BX;
        PostMsgSelector = pcrs->CWRS.Client_CX;
        PostMsgOffset = pcrs->CWRS.Client_DX;
        bClientRegistered = TRUE;
        pcrs->CWRS.Client_AX = 0;
        break;
    case POSTVXD_DEREGISTER:
        bClientRegistered = FALSE;
        pcrs->CWRS.Client_AX = 0;
        break;

    default:
        pcrs->CWRS.Client_AX = 0xffff;
    }
}
```

The PM API procedure handles two function codes, POSTVXD_REGISTER and POSTVXD_DEREGISTER, which are defined in the VxD's header file, POSTVXD.H (Listing 12.1, page 269). The code that handles POSTVXD_REGISTER copies the caller's input parameters to the global VxD variables PostMsghWnd, PostMsgSelector, and PostMsgOffset. The application provides the PostMessage address in two separate pieces, selector and offset. This 16:16 form is the natural form of a pointer for a Win16 application, although VxDs generally deal with 32-bit flat addresses, in this case a 16:16 address is better, because the VxD isn't going to use the PostMessage address itself. Instead, POSTVXD will pass this address to the VMM Simulate_Far_Call service, which wants the address in 16:16 form.

The PM API also sets a global boolean, bClientRegistered, when POSTVXD_REGISTER is called, and clears it when POSTVXD_DEREGISTER is called. The create and destroy message handlers look at this variable, and only take steps to post a message if bClientRegistered has already been set. The code for the POSTVXD message handlers follows.

```
BOOL OnVmInit(VMHANDLE hVM)
{
    VMINFO    *pInfo;

    if (bClientRegistered)
    {
        if (bWin3x)
        {
            pInfo = (VMINFO *)_HeapAllocate( sizeof( VMINFO ), 0 );
            if (pInfo)
            {
                pInfo->hVM = hVM;
                pInfo->bVmCreated = TRUE;
                Call_Priority_VM_Event(LOW_PRI_DEVICE_BOOST, Get_Sys_VM_Handle(),
                                PEF_WAIT_FOR_STI+PEF_WAIT_NOT_CRIT,
                                pInfo, PriorityEventThunk, 0 );
            }
        }
        else
        {
            SHELL_PostMessage( PostMsghWnd, WM_USER_POSTVXD, 1, (DWORD)hVM,
                            PostMessageHandler, NULL );
        }
    }
    return TRUE;
}
```

```
VOID OnVmTerminate(VMHANDLE hVM)
{
   VMINFO    *pInfo;

   if (bClientRegistered)
   {
      if (bWin3x)
      {
         pInfo = (VMINFO *)_HeapAllocate( sizeof( VMINFO ), 0 );
         if (pInfo)
         {
            pInfo->hVM = hVM;
            pInfo->bVmCreated = TRUE;
            Call_Priority_VM_Event(LOW_PRI_DEVICE_BOOST, Get_Sys_VM_Handle(),
                             PEF_WAIT_FOR_STI+PEF_WAIT_NOT_CRIT,
                             pInfo, PriorityEventThunk, 0 );
         }
      }
      else
      {
         SHELL_PostMessage( PostMsghWnd, WM_USER_POSTVXD, 0, hVM,
                         PostMessageHandler, NULL );
      }
   }
}
```

The OnVmInit and OnVmTerminate message handlers are almost identical. After verifying that bClientRegistered is set, each handler then determines what version of Windows is running. In this section, we'll only discuss what happens if the version check indicates Windows 3.x — a later section will cover the code for the Windows 95 case. Each handler dynamically allocates a VMINFO structure (defined at the top of POSTVXD.C), initializes the structure, then schedules a VM event. The VMINFO structure contains the handle of the VM being created or destroyed and a boolean (which is set if the VM has been created or clear if destroyed). This data is encapsulated into a structure because an event handler gets only a single reference data parameter. By using a pointer to the VMINFO structure as reference data, the message handler can pass more than one piece of information to the event handler.

The message handlers schedule a VM event by calling Call_Priority_VM_Event. This service allows the VxD to specify not only the desired VM, but also additional restrictions on when the event handler can be called.

The Calling Interface for `Call_Priority_VM_Event`

```
EVENTHANDLE Call_Priority_VM_Event(DWORD PriorityBoost, VMHANDLE hVM,
                                DWORD Flags, CONST VOID * Refdata,
                                PEventHANDLER EventCallback,
                                DWORD Timeout );
PriorityBoost: while executing the event callback, increase VM priority
            by this amount; can be LOW_PRI_DEVICE_BOOST,
            HIGH_PRI_DEVICE_BOOST, CRITICAL_SECTION_BOOST,
            TIME_CRITICAL_BOOST
hVM: event callback will run in context of this VM
Flags: PEF_TIME_OUT - call event handler when Timeout occurs
       PEF_WAIT_FOR_STI - wait until VM has interrupts enabled
       PEF_WAIT_NOT_CRIT - wait until VM does not own critical section
Refdata: passed to event callback
EventCallback: pointer to event callback function
Timeout: timeout, in ms; ignored unless PEF_TIME_OUT is set
```

To schedule the event that will call PostMessage, POSTVXD specifies the System VM handle and the restricting flags PEF_WAIT_FOR_STI and PEF_WAIT_NOT_CRIT. These flags prevent the event from interrupting a VM that is executing with interrupts disabled, or one that is executing a critical section; presumably such a VM has something important and/or time-critical to do. Once the VM has re-enabled interrupts or has exited the critical section, then the event handler can run and call PostMessage.

Using Nested Execution Services

Once inside the event handler PriorityEventHandler (called via PriorityEventThunk in the VxD's assembly module), it's safe to call PostMessage using VMM's nested execution services. These services are the key to executing Ring 3 code from a VxD. In a nutshell, nested execution works like this:

- A VxD sets up a VM's registers and stack as desired, changes the VM's CS and IP to point to a Ring 3 address, and then tells the VMM "ok, let the VM execute now".
- The VMM executes the VM, and when the VM executes a RET, the VMM and then the VxD regain control.

After this series of "handoffs", the Ring 3 function has been executed, and the VxD has control again.

```
VOID __stdcall PriorityEventHandler(VMHANDLE hVM, PVOID Refdata,
                                    PCLIENT_STRUCT pRegs)
{
   CLIENT_STRUCT    saveRegs;
   VMINFO           *pInfo = Refdata;

   Save_Client_State(&saveRegs);
   Begin_Nest_Exec();
   Simulate_Push(PostMsghWnd);                      // hwnd
   Simulate_Push(WM_USER_POSTVXD);                  // message
   Simulate_Push(pInfo->bVmCreated);                // wParam
   Simulate_Push(((DWORD)pInfo->hVM >> 16) );       // lParam
   Simulate_Push(((DWORD)pInfo->hVM & 0xffff) );
   Simulate_Far_Call(PostMsgSelector, PostMsgOffset);
   Resume_Exec();
   End_Nest_Exec();
   Restore_Client_State(&saveRegs);
   _HeapFree( pInfo, 0 );
}
```

PriorityEventHandler first saves the current VM state with a call to the VMM service Save_Client_State. The VxD supplies the buffer storage, using a local CLIENT_STRUCT variable. POSTVXD then enters a "nested execution block" by calling Begin_Nest_Exec. This call tells the VMM to prepare to execute Ring 3 code. Inside this block, the VxD modifies the VM's environment, first its stack and then its registers.

Several calls to the VMM service Simulate_Push push onto the VM's stack (*not* the VxD's) the hWnd, message, and wParam and lParam parameters (both zero) expected by PostMessage. The VxD extracts these parameter values from the VMINFO structure passed as a reference parameter. Note that PriorityEventHandler splits the 32-bit VM Handle into two 16-bit WORDs and pushes each on the stack, instead of pushing a single 32-bit DWORD onto the stack; PostMessage is 16-bit code and expects 16-bit parameters.

Finally the VxD calls the VMM service Simulate_Far_Call, supplying the selector and offset of the target Ring 3 function (in this case stored in PostMsgSelector and PostMsgOffset). Simulate_Far_Call modifies both the VM's stack and its registers, pushing the VM's current CS and IP onto the stack (just as a real FAR CALL would) before setting the VM's CS and IP to the selector and offset given as parameters.

So far, the VM's execution environment has been modified (without its knowledge), but no VM code has been executed. The next call, to Resume_Exec, makes that happen. When a VxD calls Resume_Exec, the VMM temporarily stops executing Ring 0 code and lets the currently scheduled VM run. Because PriorityEventHandler has modified the System VM's environment, when the System VM runs, it executes the Windows function PostMessage, using the parameters supplied by the VxD. When the VM executes a FAR RET from PostMessage, the VMM traps the instruction, and the Resume_Exec service returns to POSTVXD.

Calling a Real Mode Interrupt Handler from a VxD

The nested execution services could also be used by a VxD to call a real mode interrupt handler from a VxD, for example DOS (INT 21h) or the video BIOS (INT 10h). Instead of using Simulate_Push to push parameters on the VM's stack, a VxD would fill in parameters in registers by modifying the Client_Reg structure. Then, instead of calling Simulate_Far_Call, a VxD would use Simulate_Int.

However, in most cases you do not want to use nested execution services. Instead, use Exec_VxD_Int, without a nested execution block. The VToolsD declaration for Exec_VxD_Int looks like

```
VOID Exec_VxD_Int(DWORD Intnum, ALLREGS* Registers)
```

To use it, your VxD fills in an ALLREGS structure with the register parameters to be passed to the real mode handler, then passes the service the number of the software interrupt to execute and a pointer to this register structure. Your VxD must not change the segment register fields of the ALLREGS structure. If the real mode handler expects a pointer to be passed in ES:BX, then your VxD loads a 32-bit flat pointer into the EBX field of ALLREGS, leaving the ES field alone. Similarly, if the real mode handler expects a pointer in DS:SI, load the flat pointer into the ESI field of ALLREGS.

Using Exec_VxD_Int in a VxD is rather simple, but underneath lies a good deal of complexity. Any flat pointer parameters must be translated into segmented pointers before the real mode hander can use them. Furthermore, the targeted buffer must be located below 1Mb in order for the real mode handler to access it. Yet any buffers owned by the VxD (either statically allocated in the VxD's data segment or dynamically allocated through _HeapAllocate/_PageAllocate) are located above 2Gb, so the buffers owned by the VxD must be copied down to a real mode addressable buffer and then the real mode service is given a (segmented) pointer to that translation buffer.

This raises an interesting question. How does the Exec_VxD_Int service even know which registers in ALLREGS contain pointers? In fact, it doesn't. Exec_VxD_Int blindly calls the VxD that has hooked the software interrupt in question. For example, if your VxD calls Exec_VxD_Int with an intnum parameter of 10h, this results in the BIOSXLAT VxD being called, because BIOSXLAT used VMM Set_PM_Vector to hook INT 10h during Sys_Critical_Init.

It's the VxD that hooked the software interrupt — in this example, BIOSXLAT — that translates pointers and copies the pointer data to a real mode addressable buffer. Only a VxD that knows about INT 10h would know what registers are supposed to contain pointers. The software interrupt hook VxD in turn relies on another VxD, the V86MMGR, for the most complex part of pointer translation. The V86MMGR VxD owns a real mode addressable translation buffer and provides services that other VxDs can use to borrow and copy from/to this buffer.

So Exec_VxD_Int really works only when no pointers are being passed to the real mode handler, or when pointers are being passed but another VxD has hooked the software interrupt to provide translation services. Fortunately, the standard VxDs provided with Windows do hook the most common software interrupts (INT 21h, INT 10h, INT 13h, etc.), so in most cases your VxDs can use Exec_VxD_Int.

If your VxD must pass pointers when the real mode interrupt is not hooked by another VxD (and thus does not have translation services provided), your VxD will have to do the translation using V86MMGR services. Then your VxD would use Simulate_Int inside a nested execution block to actually call the real mode handler.

Before exiting, PriorityEventHandler exits the nested execution block by calling End_Nest_Exec and restores the VM to its original state with a call to Restore_Client_State, passing a pointer to the same CLIENT_STRUCT that was used in the earlier call to Save_Client_State. The next time the VM is scheduled, it will continue executing from wherever it was interrupted, unaware that this flow of execution was temporarily interrupted to call PostMessage. Finally, the VxD frees the VMINFO structure. (It is safe to do so because PostMessage has been executed by the time Resume_Exec returns.)

As you can see, calling a Win16 application from a VxD under Windows 3.x is a lot of work. A VxD running under Windows 95 has an easier job. (See the sidebar "Calling a Real Mode Interrupt Handler from a VxD" on page 252 for information on how a VxD can also use nested execution services to call a real mode interrupt handler.)

VxD *PostMessage under Windows 95*

The new services provided by the SHELL VxD under Windows 95 make it much easier for a VxD to notify a Windows application through PostMessage. A single call to SHELL_PostMessage will do the trick.

The Calling Interface for SHELL PostMessage

```
BOOL SHELL_PostMessage(HANDLE hWnd, DWORD uMsg, WORD wParam, DWORD lParam,
                    PPostMessage_HANDLER pCallback, PVOID dwRefData);
```

The first four parameters correspond exactly to the real PostMessage parameters. The pCallback parameter is a pointer to a callback function that will be called when the PostMessage actually completes. The last parameter, dwRefData, is reference data to be passed to the callback function.

The SHELL_PostMessage function itself has a boolean return value, where FALSE indicates failure, usually caused by insufficient memory. Note this is *not* the return value from PostMessage, because the execution of PostMessage is asynchronous (hence the callback function). The return value of the actual call to PostMessage is passed to the callback function, along with a pointer to the same reference data passed in to SHELL_PostMessage.

The Calling Interface for SHELL_PostMessage *Callback*

```
void PostMessageHandler( DWORD dwPostMessageReturnCode, void *refdata);
```

So the two-part approach required under Windows 3.x — Call_Priority_VM_Event followed by nested execution services in the event handler — can be replaced by a single call to SHELL_PostMessage under Windows 95. If the version check indicates Windows 95, the OnVmInit and OnVmTerminate handlers in POSTVXD simply do:

```
SHELL_PostMessage( PostMsghWnd, WM_USER_POSTVXD,1,
                   (DWORD)hVM, PostMessageHandler, NULL );
```

Note that the dynamically allocated VMINFO structure is no longer required, because the message handler itself can pass the VM handle and the boolean directly to SHELL_PostMessage.

VxD to Win16 Application under Windows 95: Appy Time

Although it's nice to have SHELL_PostMessage available, a VxD running under Windows 95 isn't limited to calling PostMessage to communicate with Win16 code. Using the new "appy-time" functions (also provided by SHELL), a Windows 95 VxD can call any function in a Win16 DLL, and the Win16 callback itself is allowed to call any Windows API function.

To use the appy-time services, you first schedule an appy-time event by calling SHELL_CallAtAppyTime.

The Calling Interface for Scheduling an Appy-time Event

```
APPY_HANDLE SHELL_CallAtAppyTime(APPY_CALLBACK pfnAppyCallBack,
                                 void *dwRefData,
                                 DWORD dwFlags, DWORD dwTimeout);
pfnAppyCallback: pointer to function to be called back at appy time
dwRefData: passed as parameter to pfnAppyCallback
dwFlags: describe callback conditions
        if CAAFL_TIMEOUT is set, service will timeout and
        callback will be invoked if appy time isn't
        available within dwTimeout ms
dwTimeout: timeout used if CAAFL_TIMEOUT is set in Flags
```

As with other events, a VxD returns after scheduling an appy-time event. Later, when Windows 95 is in a "safe state", the SHELL VxD will call the event handler.

SHELL supplies two parameters to the event handler callback: the same reference data passed in to SHELL_CallAtAppyTime, and a flag that has CAAFL_TIMEOUT set if the timeout occurred. If CAAFL_TIMEOUT is set, then the event handler is not running during appy time and so can't call Win16 code.

The Calling Interface for SHELL_CallAtAppyTime **Callback**

```
void AppyTimeHandler( void *dwRefData, DWORD dwFlags );
```

If this flag is not set, the event handler can use another SHELL service, SHELL_CallDll, to call any function in any Win16 DLL. This service will take care of loading the DLL, thunking the parameters from 32-bit to 16-bit (see Chapter 18 for a full discussion of thunking), and unloading the DLL after the function returns.

The Calling Interface for SHELL_CallDll

```
DWORD SHELL_CallDll(PCHAR lpszDll, PCHAR lpszProcName,
                    DWORD cbArgs, void *lpvArgs);
lpszDLL: name of Win16 DLL
lpszProcName: name of function in DLL
cbArgs: number of bytes in arguments passed to function
lpvArgs: pointer to structure containing arguments
```

The first two parameters are self-explanatory. The other two parameters, cbArgs and lpvArgs, describe the arguments to be passed to the DLL function. This short piece of code taken directly from the DDK documentation illustrates their use.

```
/* PASCAL calling convention passes arguments backwards */
struct tagEXITWINDOWARGS {
    WORD  wReserved;
    DWORD dwReturnCode;
} ewa = { 0, EW_REBOOTWINDOWS };
SHELL_CallDll("USER", "EXITWINDOWS", sizeof(ewa), &ewa);
```

In this example, the VxD is calling the Windows API function ExitWindows, which is declared in WINDOWS.H as:

```
BOOL  _far _pascal ExitWindows(DWORD dwReturnCode, UINT wReserved);
```

The VxD declares a structure containing only these two parameters. The order of the parameters in the structure is "backward" compared to the function declaration because ExitWindows is declared with the _pascal keyword. If the DLL function was declared as _cdecl instead, the structure would contain parameters in the "normal" order.

Win32-Specific Techniques: Asynchronous Procedure Calls

To communicate with a Win32 application, a Windows 95 VxD may use a completely different approach, one that fits naturally with the multi-threaded support in the Win32 API. There are two slightly different techniques, though both rely on a VxD "waking up" a Win32 application thread.

The simplest mechanism for a VxD to communicate with a Win32 application is via an asynchronous procedure call, or APC. This method is relatively simple for both the application and the VxD. The application first opens the VxD (CreateFile) and uses DeviceIoControl to pass to the VxD the address of a callback function. The application then puts itself into an "asleep yet alertable" state using the Win32 call SleepEx. The application must use SleepEx, not plain old Sleep, because only SleepEx puts the thread into an "alertable" state. While the application's thread is asleep, the VxD can call the application's callback function using the QueueUserApc service provided by the VWIN32 VxD.

The APCVXD Example

The APCVXD example illustrates the techniques discussed above. Like the POSTVXD example, APCVXD notifies a registered application whenever a VM is created or destroyed. But where POSTVXD notified a Win16 application via PostMessage, APCVXD notifies a Win32 application via an Asynchronous Procedure Call.

APCVXD supports a W32_DEVICEIOCONTROL interface, which lets Win32 applications register a callback function with the VxD. The VxD later calls this application function whenever a VM is created or destroyed. The VxD passes to the callback the address of a VMINFO structure that contains the VM handle and a boolean value (TRUE if create, FALSE if destroy). Inside the callback, after the application has printed the contents of the VMINFO structure, it calls DeviceIoControl with the APCVXD_RELEASEMEM control code, telling the VM to free the VMINFO structure.

The application is a Win32 console application (Listing 12.11, page 279), which means it can use standard I/O functions like printf. It consists of nothing but a main and a callback function.

```
void main(int ac, char* av[])
{
    DWORD    cbBytesReturned;
    DWORD    err;

    const PCHAR VxDName = "\\\\.\\APCVXD.VXD";
    hDevice = CreateFile(VxDName, 0,0,0, CREATE_NEW, FILE_FLAG_DELETE_ON_CLOSE, 0);
```

```
if (hDevice == INVALID_HANDLE_VALUE)
{
    err = GetLastError();
    printf("Cannot load VxD, error=%08lx\n", err );
    if (err == ERROR_NOT_SUPPORTED)
    {
        DeleteFile("\\\\.\\APCVXD");
    }
    exit(1);
}

if (err = DeviceIoControl(hDevice, APCVXD_REGISTER, &CallbackFromVxD,
                          sizeof(CallbackFromVxD), NULL, 0, NULL, NULL))
{
    printf("DeviceIoControl failed, error=%x\n", err);
}
else
{
    while (TRUE)
        SleepEx(1000, TRUE);
}
CloseFile( hDevice );
}
```

The application's main function uses `CreateFile` to get a handle to the VxD, then `DeviceIoControl` to pass to the VxD the address of its callback function, `CallbackFromVxD`. Finally, the application puts itself into an alertable wait state, using the Win32 `SleepEx` function with a timeout parameter of one second, and `TRUE` for the `bAlertable` parameter. `SleepEx` will block until either the timeout has expired or the VxD has called the application's callback. When `SleepEx` returns, the thread checks for keyboard input. If input was detected, the program closes the VxD handle and exits. Otherwise, it immediately calls `SleepEx` again, waiting for another callback from the VxD or another timeout, whichever comes first.

Note that the timeout in `SleepEx` is only necessary because the application must intermittently test for user input. If the application handled user input in a separate thread, `SleepEx` would not require a timeout (-1 for timeout parameter) and would return only after the VxD called `CallbackFromVxD`.

```
DWORD WINAPI CallbackFromVxD(PVOID param)
{
    VMINFO *pVmInfo = param;

    printf("VM %08lx was %s\r\n", pVmInfo->hVM, pVmInfo->bCreated ? "created" : "destroyed" );
    DeviceIoControl(hDevice, APCVXD_RELEASEMEM, pVmInfo, sizeof(pVmInfo),0,0,0,0);
    return 0;
);
```

The callback function, `CallbackFromVxD`, first casts its reference data parameter to a pointer to a `VMINFO` structure. The `VMINFO` structure contains the handle of the VM that was created or destroyed and a boolean indicating creation or destruction. The callback prints these two items using `printf`, since the application is a console application. Finally, the callback uses `DeviceIoControl` to call back into the VxD with the control code `APCVXD_RELEASEMEM`. This code tells the VxD to free the `VMINFO` structure that was passed in as reference data.

The APCVXD code is equally simple (Listing 12.7, page 275). It consists only of three message handlers: `OnW32Deviceiocontrol`, `OnVmInit`, and `OnVmTerminate`.

```
DWORD OnW32Deviceiocontrol(PDIOCPARAMETERS p)
{
    DWORD rc;

    switch (p->dwIoControlCode)
    {
    case DIOC_OPEN:
        rc = 0;
        break;

    case DIOC_CLOSEHANDLE:
      bClientRegistered = FALSE;
        rc = 0;
        break;

    case APCVXD_REGISTER:
        VmEventApc = p->lpvInBuffer;
        appThread = Get_Cur_Thread_Handle();
      bClientRegistered = TRUE;
        rc = 0;      // return OK
        break;

    case APCVXD_RELEASEMEM:
        _HeapFree(p->lpvInBuffer, 0);
        rc = 0;
        break;

    default:
        rc = 0xffffffff;
    }

    return rc;
}
```

Note that `OnW32Deviceiocontrol` returns zero when the control code indicates either `DIOC_GETVERSION` or `DIOC_CLOSEHANDLE`. As mentioned in the last chapter, failure to do so will cause the application call to `CreateFile` or `CloseHandle` to fail. APCVXD also handles two other control codes [defined in `APCVXD.H` (Listing 12.6, page 275)]: `APCVXD_REGISTER` and `APCVXD_RELEASEMEM`.

To process `APVXD_REGISTER`, **APCVXD** grabs the callback function address from the `DIOCPARAMETERS` input buffer, then calls the VMM service `Get_Cur_Thread_Handle` to obtain the Ring 0 handle for the caller's thread. (This thread handle will be used later, during the callback process.) Both the callback address and the thread handle are saved in global variables. To process `APCVXD_RELEASEMEM`, the VxD frees the pointer passed in by the caller via the `DIOCPARAMETER` input buffer. The application should have loaded this pointer with the address of a structure that was allocated earlier by the VxD (during VM create or destroy) and passed to the application's callback.

The VM_Init and VM_Terminate handlers (see the following paragraph of code) look something like their counterparts in the earlier POSTVXD (VxD to Win16 application) example. Each verifies that the boolean bClientRegistered is already set and then allocates and initializes a VMINFO structure containing the VM handle and a boolean indicating VM creation or destruction. But where the handlers in POSTVXD scheduled a VM event, APCVXD uses the VWIN32 service _VWIN32_QueueUserApc to queue a call to the registered application callback.

```
BOOL OnVmInit(VMHANDLE hVM)
{
    VMINFO    *pVmInfo;

    if (bClientRegistered)
    {
        pVmInfo = _HeapAllocate( sizeof(VMINFO), 0 );
        if (pVmInfo)
        {
            pVmInfo->hVM = hVM;
            pVmInfo->bVmCreated = TRUE;
            _VWIN32_QueueUserApc(VmEventApc, (DWORD)pVmInfo, appThread);
        }
    }
    return TRUE;
}

VOID OnVmTerminate(VMHANDLE hVM)
{
    VMINFO    *pVmInfo;

    if (bClientRegistered)
    {
        pVmInfo = _HeapAllocate( sizeof(VMINFO), 0 );
        if (pVmInfo)
        {
            pVmInfo->hVM = hVM;
            pVmInfo->bVmCreated = FALSE;
            _VWIN32_QueueUserApc(VmEventApc, (DWORD)pVmInfo, appThread);
        }
    }
}
```

Although both Win32 applications and VxDs support the notion of "thread handles", a Ring 3 thread handle (obtained by calling the Win32 API function GetCurrentThread) is not the same as a Ring 0 thread handle. Because _VWIN32_QueueUserApc requires a Ring 0 thread handle, APCVXD calls the VMM service Get_Cur_Thread_Handle during W32_DEVICEIOCONTROL processing to obtain the Ring 0 handle of the caller's thread.

The Calling Interface for VWIN32_QueueUserApc

```
VOID _VWIN32_QueueUserApc(PVOID pR3Proc, DWORD Param, THREADHANDLE hThread);
pR3Proc: linear address of Ring 3 code to execute
Param: parameter to pass to Ring 3 code
hThread: Ring3 code runs in this thread context
        NOTE: this is a Ring 0 thread handle, not a Ring 3 thread handle
```

As the name of the VWIN32 service suggests, the callback is not executed immediately but is queued, to be executed at a later time (when the System VM is current, etc.) When _VWIN32_QueueUserApc returns, the APCVXD message handler also returns, having finished its processing.

Because APCVXD uses global variables to store both the callback address and the thread handle, only one Win32 application can use APCVXD at a time. In order to support usage by multiple Win32 applications at the same time, APCVXD would need to dynamically allocate a structure to store the callback address and the thread handle and then add the dynamically allocated structures to a linked list. The create and destroy VM handlers would then traverse the list, calling _VWIN32_QueueUserApc for each registered callback in the list.

Win32-Specific Techniques:
Win32 Events

Although using an APC is probably the easiest way to implement a VxD-to-application calling mechanism, there is a much more efficient method. If the Win32 application is multithreaded, the application can continue to do work in a main thread while a second thread is waiting on a wakeup from the VxD. For example, a main thread could monitor for user input while a second thread waits on a VxD that is buffering incoming data. When the buffered data reaches a threshold level, the VxD wakes up the waiting Win32 thread.

VxDs use thread events for interthread notification, much as multi-threaded Win32 applications do. In a multi-threaded Win32 application, Win32 events are often used to signal from one thread to another that an operation has been completed, for example that a buffer has been read from disk. One thread creates the event, starts the second thread, and then waits on the event (which will be signaled by the second thread). Assuming the waiting thread has nothing to do until the data is read, this structure is an efficient use of resources; the waiting thread is blocked and thus consumes minimal processor cycles.

The Win32 API contains the following event functions:

- `CreateEvent` to create the event and obtain an event handle
- `ResetEvent` to set the event to the unsignaled state
- `SetEvent` to set the event to the signaled state
- `PulseEvent` to set the event to the signaled state and then immediately set it to unsignaled
- `WaitForSingleObject` to block until the event is signaled
- `WaitForMultipleObjects` to block until any or all the events are signaled (depending on flag parameter)

The following paragraph of code presents a simple multithreaded Win32 application which illustrates the use of Win32 events. It consists of two threads, where the first thread signals the second whenever the users presses the 'S' key. The second thread prints a message whenever it is signaled.

```
DWORD WINAPI SecondThread( HANDLE hEvent )
{
   while (TRUE)
   {
      WaitForSingleObject(hEvent, INFINITE );
      printf("Second thread was signaled\n");
   }
   return 0;
}

void main(int ac, char *av[])
{
   BOOL   bExit = FALSE;
   HANDLE hEvent;
   char   c;
   DWORD  tid;

   hEvent = CreateEvent( 0, FALSE, FALSE, NULL );
   CreateThread( 0, 0x1000, SecondThread, hEvent, 0, &tid );

   printf("Press 'S' to signal second thread\n");
   printf("Press 'X' to exit\n");
```

```
while (!bExit)
{
   c = getch();
   switch( c )
   {
   case 'S':
   case 's':
      SetEvent( hEvent );
      break;

   case 'X':
   case 'x':
      bExit = TRUE;
      break;
   }
}
}
```

VxDs and Win32 Events

Under Windows 95, VxDs have access to the very same Win32 event API, through a set of services provided by the VWIN32 VxD. Using these services, a VxD can signal a waiting Win32 application thread, or wait to be signaled by a Win32 application thread. The VWIN32 event services are:

- _VWIN32_ResetWin32Event
- _VWIN32_SetWin32Event
- _VWIN32_PulseWin32Event
- _VWIN32_WaitSingleObject
- _VWIN32_WaitMultipleObjects

Unfortunately, a VxD can't obtain a Win32 event handle simply by calling the appropriate event service. (Note that a CreateEvent service is conspicuously missing in the above list.) Thus, obtaining an event handle that is usable to a VxD becomes a complicated process involving, among other things, an undocumented system call. The event is always created by the application, via the Win32 API CreateEvent. The application must then translate the event handle returned by CreateEvent into a VxD event handle, using the undocumented Win32 API function OpenVxDHandle. The application then passes the translated (Ring 0) event handle to the VxD via DeviceIoControl, and the VxD uses this handle as a parameter to the VWIN32 event functions.

The EVENTVXD example (Listing 12.15, page 282) uses a Win32 event to signal a Win32 application thread from a VxD. Like the POSTVXD and APCVXD examples introduced earlier in this chapter, EVENTVXD notifies a registered application whenever a VM is created or destroyed. But where APCVXD used an Asynchronous Procedure Call to notify a Win32 application, EVENTVXD uses a Win32 event.

EVENTVXD supports the W32_DEVICEIOCONTROL message, which lets a Win32 application register a Win32 event handle with the VxD. The Win32 thread that registered this event handle should then wait on the event, which the VxD will signal whenever a VM is created or destroyed. As part of the initial registration, the VxD returns to the application the address of a VMINFO structure. When the application thread is signaled, this structure will contain the handle of the VM that was created or destroyed and a boolean indicating creation or destruction.

Like the earlier APCVXD example, the code for EVENTVXD consists of only three message handlers: OnW32Deviceiocontrol, OnVmInit, and OnVmTerminate.

```
DWORD OnW32Deviceiocontrol(PDIOCPARAMETERS p)
{
    DWORD rc;

    switch (p->dwIoControlCode)
    {
    case DIOC_OPEN:
        rc = 0;
        break;

    case DIOC_CLOSEHANDLE:
      bClientRegistered = FALSE;
        rc = 0;
        break;

    case EVENTVXD_REGISTER:
        hWin32Event = p->lpvInBuffer;
        *((DWORD *)(p->lpvOutBuffer)) = (DWORD)&GlobalVMInfo;
        *((DWORD *)(p->lpcbBytesReturned)) = sizeof(DWORD);
        bClientRegistered = TRUE;
        rc = 0;
        break;

    default:
        rc = 0xffffffff;
    }

    return rc;
}
```

Like the other W32_DEVICEIOCONTROL message handlers we've seen, this one returns 0 when the control code indicates either DIOC_GETVERSION or DIOC_CLOSEHANDLE. If the control code is EVENTVXD_REGISTER, EVENTVXD copies the event handle from the DIOCPARAMETERS input buffer into the global variable hWin32Event.

```
BOOL OnVmInit(VMHANDLE hVM)
{
    if (bClientRegistered)
    {
        GlobalVMInfo.hVM = hVM;
        GlobalVMInfo.bVmCreated = TRUE;
        Call_Priority_VM_Event(LOW_PRI_DEVICE_BOOST, Get_Sys_VM_Handle(),
                        PEF_WAIT_FOR_STI+PEF_WAIT_NOT_CRIT,
                        hWin32Event, PriorityEventThunk, 0 );
    }
    return TRUE;
}

VOID OnVmTerminate(VMHANDLE hVM)
{
    if (bClientRegistered)
    {
        GlobalVMInfo.hVM = hVM;
        GlobalVMInfo.bVmCreated = FALSE;
        Call_Priority_VM_Event(LOW_PRI_DEVICE_BOOST, Get_Sys_VM_Handle(),
                        PEF_WAIT_FOR_STI+PEF_WAIT_NOT_CRIT,
                        hWin32Event, PriorityEventThunk, 0 );
    }
}
```

The VM_Init and VM_Terminate handlers (Listing 12.16, page 284) look more like their counterparts from the POSTVXD example than the ones from the APCVXD example. Like POSTVXD, EVENTVXD must postpone its real work (signaling the Win32 event) for a VM event handler, because the VWIN32 event functions may only be called when the System VM is current. Unlike POSTVXD, however, EVENTVXD does not dynamically allocate a VMINFO structure and pass the structure address to the event handler as reference data. Instead, EVENTVXD uses a global VMINFO structure, and passes the Win32 event handle as reference data to its event callback.

Where both APCVXD and POSTVXD pass a VMINFO pointer to the application (POSTVXD via the `lParam` of `PostMessage` and APCVXD as a reference data parameter), EVENTVXD has no way of passing reference data to the Win32 application. The VxD doesn't call a function in the Win32 application. The Win32 application simply wakes up from the event it has been waiting on.

Because the VxD can't pass reference data to the Win32 thread that it's unblocking, it must use a different method to pass data. The VxD tells the Win32 application ahead of time, through `DeviceIoControl`, the address of a VMINFO structure that will contain VM information. The VxD must then always use this same VMINFO structure, because that's the one the Win32 application knows about.

```
VOID __stdcall PriorityEventHandler(VMHANDLE hVM, PVOID Refdata, CRS *pRegs)
{
    HANDLE hWin32Event = Refdata;

    _VWIN32_SetWin32Event( hWin32Event );
}
```

The VMM calls the System VM event handler, `PriorityEventHandler`, once the System VM has been scheduled. At this time, `PriorityEventHandler` can safely call `_VWIN32_SetWin32Event`, using the reference data parameter as the Win32 event handle.

The accompanying Win32 application, which uses the EVENTVXD, is more complicated than the other Windows example applications, partly because it has two threads, but mostly because it must go to great lengths to obtain a usable event handle.

```
void main( int ac, char *av[] )
{
    hEventRing3 = CreateEvent( 0, FALSE, FALSE, NULL );
    if (!hEventRing3)
    {
        printf("Cannot create Ring3 event\n");
        exit(1);
    }

    hKernel32Dll = LoadLibrary("kernel32.dll");
    if (!hKernel32Dll)
    {
    printf("Cannot load KERNEL32.DLL\n");
    exit(1);
    }
```

```
pfOpenVxDHandle = (HANDLE (WINAPI *) (HANDLE))
GetProcAddress( Kernel32Dll, "OpenVxDHandle" );
if (!pfOpenVxDHandle)
{
   printf("Cannot get addr of OpenVxDHandle\n");
   exit(1);
}

hEventRing0 = (*pfOpenVxDHandle)(hEventRing3);
if (!hEventRing0)
{
  printf("Cannot create Ring0 event\n");
  exit(1);
}
```

The main thread must make four different Win32 API calls to create a Win32 event and then obtain a Ring 0 handle for this event usable by the VxD. Creating the event requires only a call to CreateEvent. The application uses FALSE for the bManualReset parameter to obtain an auto-reset event. Windows will automatically reset this type of event to the non-signaled state when it wakes up the waiting thread, saving the second thread from explicitly calling ResetEvent. The application also specifies FALSE as the bInitialValue parameter. Thus, initially the event will be in the non-signaled state, causing the second thread to block on the event immediately.

To translate the event handle returned by CreateEvent into a handle usable by the VxD, the application must call the OpenVxDHandle function in KERNEL32.DLL. This function is not documented and not in the Win32 import library, thus its address must be acquired via run-time dynamic linking. First the application uses LoadLibrary to load KERNEL32.DLL. Then it calls GetProcAddress, specifying both the name of the function ("OpenVxDHandle") and the instance handle returned by LoadLibrary.

GetProcAddress returns a function pointer, which the application uses to call the OpenVxDHandle function. This function takes as input a Ring 3 event handle, returned by CreateEvent and returns another handle for the event (one usable at Ring 0). The application stores this Ring 0 handle in hEventRing0, to be passed to the EVENTVXD via DeviceIoControl.

```
hDevice = CreateFile( VxDName, 0, 0, 0, CREATE_NEW,
                      FILE_FLAG_DELETE_ON_CLOSE, 0 );
if (!hDevice)
{
   printf("Cannot load VxD error=%x\n", GetLastError() );
   exit(1);
}
```

```
if (!DeviceIoControl( hDevice, EVENTVXD_REGISTER,
                      hEventRing0, sizeof(hEventRing0),
                      &pVMInfo, sizeof(pVMInfo),
                      &cbBytesReturned, 0 ))
{
   printf("DeviceIoControl REGISTER failed\n");
   exit(1);
}
```

The next part of main looks similar to the APC example application described earlier in this chapter. The application opens a channel to the VxD and uses DeviceIoControl to pass hEventRing0 to the VxD.

The function prototype for DeviceIoControl declares both the lpInBuffer and the lpOutBuffer parameters to be void pointers, but it is always up to the VxD to decide exactly how these pointers are used. EVENTVXD expects the input pointer for an EVENTVXD_REGISTER control code to be a Ring 0 event handle, not a pointer. EVENTVXD expects the output pointer to point to a DWORD, which it fills in with the address of a VMINFO structure.

After giving the event handle to the VxD, the main thread has nothing left to do but create the second thread (which will wait to be signaled by the VxD) and wait for user input. Because the main thread has nothing else to do but wait for input — it's the second thread that's doing the work — it uses the C library function getch, which blocks. When getch finally returns with a key, the main thread closes the channel to the VxD and returns.

```
CreateThread( 0, 0x1000, SecondThread, hEventRing3, 0, &tid );
printf("Press any key to exit...");
getch();
CloseHandle( hDevice );
```

You may notice that the main thread doesn't do anything to terminate the second thread. This may seem dangerous, and in fact, Windows 95 won't automatically kill off additional threads when the main thread ends. However, the C run-time exit code does terminate additional threads when main returns. If you want to be extra safe, you can explicitly terminate the secondary thread before exiting main by calling TerminateThread and passing in the (Ring 3) thread handle returned originally by CreateThread.

That wraps up the main thread of the application, which exists only to create a second thread which does the real work. The second thread, contained in the function SecondThread, is short and simple.

```
DWORD WINAPI SecondThread( PVOID hEventRing3 )
{
   while( TRUE )
   {
      WaitForSingleObject((HANDLE)hEventRing3, INFINITE );
      printf("VM %081x was %x", pVMInfo->hVM,
             pVMInfo->bCreated ? "created" : "destroyed" );
   }
   return 0;
}
```

The reference data parameter gives SecondThread the handle of a Win32 event to wait on. SecondThread then waits, with an infinite timeout, on this event. When the event is signaled, SecondThread uses the global variable pVMInfo to access a VMINFO structure that contains the VM handle and an indication of either creation or destruction. Then SecondThread waits again on the event. Note that SecondThread doesn't have to call ResetEvent because the event was created as an auto-reset event.

Summary

This chapter covered all the techniques used by VxDs to communicate with applications. All rely on an initial call to the VxD, initiated by the application, to pass information about a callback function or event handle which the VxD uses later to communicate back to the application. Under Windows 3.x, a VxD may not call arbitrary Win16 code but is essentially limited to calling PostMessage, using the window handle and PostMessage address passed in by the application. Under Windows 95, a VxD may still communicate with a Win16 application by calling PostMessage, but the VxD may also call any function in any Win16 DLL. A VxD has two different choices when communicating with a Win32 application: either the simple but not so elegant asynchronous procedure call (APC) or the more elegant use of Win32 events to signal a waiting Win32 application thread.

Listing 12.1   POSTVXD.H

```
#define POSTVXD_ID              0xBADD
#define POSTVXD_REGISTER        0x1000
#define POSTVXD_DEREGISTER      0x1001


// based on WM_USER in windows.h
#define WM_USER_POSTVXD         (0x0400+0x0100)
```

Listing 12.2   POSTVXD.C

```
#define WANTVXDWRAPS

#include <basedef.h>
#include <vmm.h>
#include <debug.h>
#include "vxdcall.h"
#include <vxdwraps.h>
#include <wrappers.h>
#include <vwin32.h>
#include "postvxd.h"

#ifdef DEBUG
#define DPRINTF0(buf, fmt) _Sprintf(buf, fmt ); Out_Debug_String( buf )
#define DPRINTF1(buf, fmt, arg1) _Sprintf(buf, fmt, arg1 ); Out_Debug_String( buf )
#define DPRINTF2(buf, fmt, arg1, arg2) _Sprintf(buf, fmt, arg1, arg2 );
                                        Out_Debug_String( buf )
#else
#define DPRINTF0(buf, fmt)
#define DPRINTF1(buf, fmt, arg1)
#define DPRINTF2(buf, fmt, arg1, arg2)
#endif

VOID _cdecl PostMessageHandler(DWORD dwPostMessageReturnCode, PVOID refdata);

// functions in asm module
void PriorityEventThunk( void );

BOOL bClientRegistered = FALSE;         // True when PM API called to register
WORD PostMsgOffset;
WORD PostMsgSelector;
HANDLE PostMsghWnd;
char dbgbuf[80];
BOOL  bWin3x;

typedef struct
{
    BOOL     bVmCreated;
    VMHANDLE hVM;
} VMINFO;
```

Listing 12.2 (continued)   POSTVXD.C

```
BOOL OnSysDynamicDeviceInit()
{
    DPRINTF0(dbgbuf,"Loading\r\n");
    return TRUE;
}

BOOL OnSysDynamicDeviceExit()
{
    DPRINTF0(dbgbuf,"Unloading\r\n");
    return TRUE;
}

BOOL OnInitComplete(VMHANDLE hVM)
{
    DWORD ver;

    ver = Get_VMM_Version();

    if (HIWORD(ver) <= 3)
    {
        // Win3.x, not 95
        bWin3x = TRUE;
    }
}

BOOL OnVmInit(VMHANDLE hVM)
{
    VMINFO   *pInfo;

    if (bClientRegistered)
    {
      if (bWin3x)
      {
        pInfo = (VMINFO *)_HeapAllocate( sizeof( VMINFO ), 0 );
        if (pInfo)
        {
            pInfo->hVM = hVM;
            pInfo->bVmCreated = TRUE;
            Call_Priority_VM_Event(LOW_PRI_DEVICE_BOOST, Get_Sys_VM_Handle(),
                            PEF_WAIT_FOR_STI+PEF_WAIT_NOT_CRIT,
                            pInfo, PriorityEventThunk, 0 );
        }
      }
      else
      {
          _SHELL_PostMessage( PostMsghWnd, WM_USER_POSTVXD, 1, (DWORD)hVM,
                            PostMessageHandler, NULL );
      }
    }
    return TRUE;
}
```

Listing 12.2 (continued)   *POSTVXD.C*

```
VOID OnVmTerminate(VMHANDLE hVM)
{
    VMINFO    *pInfo;

    if (bClientRegistered)
    {
      if (bWin3x)
      {
         pInfo = (VMINFO *)_HeapAllocate( sizeof( VMINFO ), 0 );
         if (pInfo)
         {
            pInfo->hVM = hVM;
            pInfo->bVmCreated = TRUE;
            Call_Priority_VM_Event(LOW_PRI_DEVICE_BOOST, Get_Sys_VM_Handle(),
                                   PEF_WAIT_FOR_STI+PEF_WAIT_NOT_CRIT,
                                   pInfo, PriorityEventThunk, 0 );
         }
      }
      else
      {
         _SHELL_PostMessage( PostMsghWnd, WM_USER_POSTVXD, 0, hVM,
                             PostMessageHandler, NULL );
      }
    }
}

VOID   stdcall PriorityEventHandler(VMHANDLE hVM, PVOID Refdata, CRS *pRegs)
{
    CLIENT_STRUCT   saveRegs;
    VMINFO          *pInfo = Refdata;

    Save_Client_State(&saveRegs);
    Begin_Nest_Exec();
    Simulate_Push(PostMsghWnd);                    // hwnd
    Simulate_Push(WM_USER_POSTVXD);                // message
    Simulate_Push(pInfo->bVmCreated);              // wParam
    Simulate_Push(((DWORD)pInfo->hVM >> 16) );     // lParam
    Simulate_Push(((DWORD)pInfo->hVM & 0xffff) );
    Simulate_Far_Call(PostMsgSelector, PostMsgOffset);
    Resume_Exec();
    End_Nest_Exec();
    Restore_Client_State(&saveRegs);
    _HeapFree( pInfo, 0 );
}

VOID _cdecl PostMessageHandler(DWORD dwPostMessageReturnCode, PVOID refdata)
{
    if (!dwPostMessageReturnCode)
        DPRINTF0(dbgbuf, "PostMessage failed!\r\n");
}
```

Listing 12.2 (continued)   *POSTVXD.C*

```
VOID __cdecl PM_Api_Handler(VMHANDLE hVM, CLIENT_STRUCT *pcrs)
{
    switch (pcrs->CWRS.Client_AX)
    {
    case POSTVXD_REGISTER:
      PostMsghWnd = (HANDLE)pcrs->CWRS.Client_BX;
      PostMsgSelector = pcrs->CWRS.Client_CX;
      PostMsgOffset = pcrs->CWRS.Client_DX;
        bClientRegistered = TRUE;
        pcrs->CWRS.Client_AX = 0;
        break;

    case POSTVXD_DEREGISTER:
        bClientRegistered = FALSE;
        pcrs->CWRS.Client_AX = 0;
        break;

    default:
        pcrs->CWRS.Client_AX = 0xffff;
    }
}
```

Listing 12.3   *POSTDDB.ASM*

```
.386p

;****************************************************************************
;                    I N C L U D E S
;****************************************************************************

    include vmm.inc
    include debug.inc

;============================================================================
;      V I R T U A L   D E V I C E   D E C L A R A T I O N
;============================================================================
POSTVXD_ID    EQU    0BADDh

DECLARE_VIRTUAL_DEVICE    POSTVXD, 1, 0, ControlProc, POSTVXD_ID, \
                         UNDEFINED_INIT_ORDER, 0, PM_API


VxD_LOCKED_CODE_SEG
```

Listing 12.3 (continued)   *POSTDDB.ASM*

```
;=============================================================
;
;   PROCEDURE: ControlProc
;
;   DESCRIPTION:
;    Device control procedure for the SKELETON VxD
;
;   ENTRY:
;    EAX = Control call ID
;
;   EXIT:
;    If carry clear then
;         Successful
;    else
;         Control call failed
;
;   USES:
;    EAX, EBX, ECX, EDX, ESI, EDI, Flags
;
;=============================================================

BeginProc ControlProc
    Control_Dispatch SYS_DYNAMIC_DEVICE_INIT, _OnSysDynamicDeviceInit, cCall, <ebx>
    Control_Dispatch SYS_DYNAMIC_DEVICE_EXIT, _OnSysDynamicDeviceExit, cCall, <ebx>
    Control_Dispatch INIT_COMPLETE, _OnInitComplete, cCall, <ebx>
    Control_Dispatch VM_INIT, _OnVmInit, cCall, <ebx>
    Control_Dispatch VM_TERMINATE, _OnVmTerminate, cCall, <ebx>
    clc
    ret

EndProc ControlProc

BeginProc PM_API

    cCall _PM_Api_Handler, <ebx, ebp>
    ret

EndProc PM_API

VxD_LOCKED_CODE_ENDS

VxD_CODE_SEG

BeginProc _PriorityEventThunk

    sCall PriorityEventHandler, <ebx,edx,ebp>
    ret

EndProc _PriorityEventThunk

VxD_CODE_ENDS

    END
```

Listing 12.4   POSTVXD.MAK

```
CFLAGS     = -DWIN32 -DCON -Di386 -D_X86_ -D_NTWIN -W3 -Gs -D_DEBUG -Zi
CVXDFLAGS  = -Zdp -Gs -c -DIS_32 -Zl -DDEBLEVEL=1 -DDEBUG
LFLAGS     = -machine:i386 -debug:notmapped,full -debugtype:cv
             -subsystem:console kernel32.lib
AFLAGS     = -coff -DBLD_COFF -DIS_32 -W2 -Zd -c -Cx -DMASM6 -DDEBLEVEL=1 -DDEBUG

all: postvxd.vxd

postvxd.obj: postvxd.c
        cl $(CVXDFLAGS) -Fo$@ %s

postddb.obj: postddb.asm
        ml $(AFLAGS) -Fo$@ %s

postvxd.vxd: postddb.obj postvxd.obj ..\..\wrappers\vxdcall.obj
                                     ..\..\wrappers\wrappers.clb postvxd.def
        echo >NUL @<<postvxd.crf
-MACHINE:i386 -DEBUG -DEBUGTYPE:MAP -PDB:NONE
-DEF:postvxd.def -OUT:postvxd.vxd -MAP:postvxd.map
-VXD vxdwraps.clb wrappers.clb vxdcall.obj postddb.obj postvxd.obj
<<
        link @postvxd.crf
        mapsym postvxd
```

Listing 12.5   POSTVXD.DEF

```
VXD POSTVXD DYNAMIC
SEGMENTS
    _LTEXT      CLASS 'LCODE'   PRELOAD NONDISCARDABLE
    _LDATA      CLASS 'LCODE'   PRELOAD NONDISCARDABLE
    _TEXT       CLASS 'LCODE'   PRELOAD NONDISCARDABLE
    _DATA       CLASS 'LCODE'   PRELOAD NONDISCARDABLE
    _LPTEXT     CLASS 'LCODE'   PRELOAD NONDISCARDABLE
    _CONST      CLASS 'LCODE'   PRELOAD NONDISCARDABLE
    _BSS        CLASS 'LCODE'   PRELOAD NONDISCARDABLE
    _TLS        CLASS 'LCODE'   PRELOAD NONDISCARDABLE
    _ITEXT      CLASS 'ICODE'   DISCARDABLE
    _IDATA      CLASS 'ICODE'   DISCARDABLE
    _PTEXT      CLASS 'PCODE'   NONDISCARDABLE
    _PDATA      CLASS 'PCODE'   NONDISCARDABLE
    _STEXT      CLASS 'SCODE'   RESIDENT
    _SDATA      CLASS 'SCODE'   RESIDENT
    _MSGTABLE   CLASS 'MCODE'   PRELOAD NONDISCARDABLE IOPL
    _MSGDATA    CLASS 'MCODE'   PRELOAD NONDISCARDABLE IOPL
    _IMSGTABLE  CLASS 'MCODE'   PRELOAD DISCARDABLE IOPL
    _IMSGDATA   CLASS 'MCODE'   PRELOAD DISCARDABLE IOPL
    _DBOSTART   CLASS 'DBOCODE' PRELOAD NONDISCARDABLE CONFORMING
    _DBOCODE    CLASS 'DBOCODE' PRELOAD NONDISCARDABLE CONFORMING
    _DBODATA    CLASS 'DBOCODE' PRELOAD NONDISCARDABLE CONFORMING
    _16ICODE    CLASS '16ICODE' PRELOAD DISCARDABLE
    _RCODE      CLASS 'RCODE'
EXPORTS
    POSTVXD_DDB @1
```

Listing 12.6   `APCVXD.H`

```
#define APCVXD_REGISTER        0x8100
#define APCVXD_RELEASEMEM      0x8101

typedef struct
{
    BOOL      bVmCreated;
    DWORD     hVM;
} VMINFO;
```

Listing 12.7   `APCVXD.C`

```
#define WANTVXDWRAPS

#include <basedef.h>
#include <vmm.h>
#include <debug.h>
#include "vxdcall.h"

#include <wrappers.h>
#include <vwin32.h>
#include "apcvxd.h"

#ifdef DEBUG
#define DPRINTF0(buf, fmt) _Sprintf(buf, fmt ); Out_Debug_String( buf )
#define DPRINTF1(buf, fmt, arg1) _Sprintf(buf, fmt, arg1 ); Out_Debug_String( buf )
#define DPRINTF2(buf, fmt, arg1, arg2) _Sprintf(buf, fmt, arg1, arg2 );
                                        Out_Debug_String( buf )
#else
#define DPRINTF0(buf, fmt)
#define DPRINTF1(buf, fmt, arg1)
#define DPRINTF2(buf, fmt, arg1, arg2)
#endif

typedef struct tcb_s *PTCB;
char dbgbuf[80];

BOOL     bClientRegistered = FALSE;
PVOID    VmEventApc = 0;
PTCB     appThread = 0;

BOOL OnVmInit(VMHANDLE hVM)
{
    VMINFO     *pVmInfo;

    if (bClientRegistered)
    {
        pVmInfo = _HeapAllocate( sizeof(VMINFO), 0 );
        if (pVmInfo)
        {
            pVmInfo->hVM = hVM;
            pVmInfo->bVmCreated = TRUE;
            _VWIN32_QueueUserApc(VmEventApc, (DWORD)pVmInfo, appThread);
        }
    }
    return TRUE;
}
```

Listing 12.7 (continued)   `APCVXD.C`

```
VOID OnVmTerminate(VMHANDLE hVM)
{
    VMINFO    *pVmInfo;

    if (bClientRegistered)
    {
        pVmInfo = _HeapAllocate( sizeof(VMINFO), 0 );
        if (pVmInfo)
        {
            pVmInfo->hVM = hVM;
            pVmInfo->bVmCreated = FALSE;
            _VWIN32_QueueUserApc(VmEventApc, (DWORD)pVmInfo, appThread);
        }
    }
}

BOOL OnSysDynamicDeviceInit()
{
    DPRINTF0( dbgbuf, "Loading\r\n");
    return TRUE;
}

BOOL OnSysDynamicDeviceExit()
{
    DPRINTF0( dbgbuf, "Unloading\r\n");
    return TRUE;
}

DWORD OnW32Deviceiocontrol(PDIOCPARAMETERS p)
{
    DWORD rc;

    switch (p->dwIoControlCode)
    {
    case DIOC_OPEN:
        rc = 0;
        break;

    case DIOC_CLOSEHANDLE:
      bClientRegistered = FALSE;
        rc = 0;
        break;

    case APCVXD_REGISTER:
        VmEventApc = p->lpvInBuffer;
        appThread = Get_Cur_Thread_Handle();
      bClientRegistered = TRUE;
        rc = 0;     // return OK
        break;

    case APCVXD_RELEASEMEM:
        _HeapFree(p->lpvInBuffer, 0);
        rc = 0;
        break;

    default:
        rc = 0xffffffff;
    }

    return rc;
}
```

Listing 12.8   APCDDB.ASM

```
.386p

;****************************************************************************
;                    I N C L U D E S
;****************************************************************************

    include vmm.inc
    include debug.inc

;===========================================================================
;          V I R T U A L   D E V I C E   D E C L A R A T I O N
;===========================================================================
DECLARE_VIRTUAL_DEVICE    APCVXD, 1, 0, ControlProc, UNDEFINED_DEVICE_ID, \
                          UNDEFINED_INIT_ORDER

VxD_LOCKED_CODE_SEG

;===========================================================================
;
;    PROCEDURE: ControlProc
;
;    DESCRIPTION:
;     Device control procedure for the SKELETON VxD
;
;    ENTRY:
;     EAX = Control call ID
;
;    EXIT:
;     If carry clear then
;         Successful
;     else
;         Control call failed
;
;    USES:
;     EAX, EBX, ECX, EDX, ESI, EDI, Flags
;
;===========================================================================

BeginProc ControlProc
    Control_Dispatch VM_INIT, _OnVmInit, cCall, <ebx>
    Control_Dispatch VM_TERMINATE, _OnVmTerminate, cCall, <ebx>
    Control_Dispatch SYS_DYNAMIC_DEVICE_INIT, _OnSysDynamicDeviceInit, cCall, <ebx>
    Control_Dispatch SYS_DYNAMIC_DEVICE_EXIT, _OnSysDynamicDeviceExit, cCall, <ebx>
    Control_Dispatch W32_DEVICEIOCONTROL, _OnW32Deviceiocontrol, cCall, <esi>
    clc
    ret

EndProc ControlProc

VxD_LOCKED_CODE_ENDS

    END
```

Listing 12.9   `APCVXD.MAK`

```
CFLAGS    = -DWIN32 -DCON -Di386 -D_X86_ -D_NTWIN -W3 -Gs -D_DEBUG -Zi
CVXDFLAGS = -Zdp -Gs -c -DIS_32 -Zl -DDEBLEVEL=1 -DDEBUG
LFLAGS    = -machine:i386 -debug:notmapped,full -debugtype:cv
            -subsystem:console kernel32.lib
AFLAGS    = -coff -DBLD_COFF -DIS_32 -W2 -Zd -c -Cx -DMASM6 -DDEBLEVEL=1 -DDEBUG

all: apcvxd.vxd

apcvxd.obj: apcvxd.c
        cl $(CVXDFLAGS) -Fo$@ %s

apcddb.obj: apcddb.asm
        ml $(AFLAGS) -Fo$@ %s

apcvxd.vxd: apcddb.obj apcvxd.obj ..\..\..\wrappers\vxdcall.obj
                                  ..\..\..\wrappers\wrappers.clb apcvxd.def
        echo >NUL @<<apcvxd.crf
-MACHINE:i386 -DEBUG -DEBUGTYPE:MAP -PDB:NONE
-DEF:apcvxd.def -OUT:apcvxd.vxd -MAP:apcvxd.map
-VXD vxdwraps.clb wrappers.clb vxdcall.obj apcddb.obj apcvxd.obj
<<
        link @apcvxd.crf
        mapsym apcvxd
```

Listing 12.10   `APCVXD.DEF`

```
VXD APCVXD DYNAMIC
SEGMENTS
    _LTEXT     CLASS 'LCODE'    PRELOAD NONDISCARDABLE
    _LDATA     CLASS 'LCODE'    PRELOAD NONDISCARDABLE
    _TEXT      CLASS 'LCODE'    PRELOAD NONDISCARDABLE
    _DATA      CLASS 'LCODE'    PRELOAD NONDISCARDABLE
    _LPTEXT    CLASS 'LCODE'    PRELOAD NONDISCARDABLE
    _CONST     CLASS 'LCODE'    PRELOAD NONDISCARDABLE
    _BSS       CLASS 'LCODE'    PRELOAD NONDISCARDABLE
    _TLS       CLASS 'LCODE'    PRELOAD NONDISCARDABLE
    _ITEXT     CLASS 'ICODE'    DISCARDABLE
    _IDATA     CLASS 'ICODE'    DISCARDABLE
    _PTEXT     CLASS 'PCODE'    NONDISCARDABLE
    _PDATA     CLASS 'PCODE'    NONDISCARDABLE
    _STEXT     CLASS 'SCODE'    RESIDENT
    _SDATA     CLASS 'SCODE'    RESIDENT
    _MSGTABLE  CLASS 'MCODE'    PRELOAD NONDISCARDABLE IOPL
    _MSGDATA   CLASS 'MCODE'    PRELOAD NONDISCARDABLE IOPL
    _IMSGTABLE CLASS 'MCODE'    PRELOAD DISCARDABLE IOPL
    _IMSGDATA  CLASS 'MCODE'    PRELOAD DISCARDABLE IOPL
    _DBOSTART  CLASS 'DBOCODE'  PRELOAD NONDISCARDABLE CONFORMING
    _DBOCODE   CLASS 'DBOCODE'  PRELOAD NONDISCARDABLE CONFORMING
    _DBODATA   CLASS 'DBOCODE'  PRELOAD NONDISCARDABLE CONFORMING
    _16ICODE   CLASS '16ICODE'  PRELOAD DISCARDABLE
    _RCODE     CLASS 'RCODE'
EXPORTS
    APCVXD_DDB @1
```

Listing 12.11   *APC/WIN32APP/WIN32APP.C*

```c
#include <stdio.h>
#include <conio.h>
#include <windows.h>
#include "apcvxd.h"

HANDLE    hDevice;
char      buf[80];
DWORD WINAPI CallbackFromVxD(PVOID param);


DWORD WINAPI CallbackFromVxD(PVOID param)
{
    VMINFO *pVmInfo = param;

    printf("VM %08lx was %s\r\n", pVmInfo->hVM,
            pVmInfo->bVmCreated ? "created" : "destroyed" );
    DeviceIoControl(hDevice, APCVXD_RELEASEMEM, pVmInfo, sizeof(pVmInfo),0,0,0,0);
    return 0;
}

void main(int ac, char* av[])
{
    DWORD err;
    const PCHAR VxDName = "\\\\.\\APCVXD.VXD";

    hDevice = CreateFile(VxDName, 0,0,0, CREATE_NEW, FILE_FLAG_DELETE_ON_CLOSE, 0);

    if (hDevice == INVALID_HANDLE_VALUE)
    {
        err = GetLastError();
        printf("Cannot load VxD, error=%08lx\n", err );
        if (err == ERROR_NOT_SUPPORTED)
        {
            DeleteFile("\\\\.\\APCVXD");
        }
        exit(1);
    }

    if ( !DeviceIoControl(hDevice, APCVXD_REGISTER, &CallbackFromVxD, sizeof(void *),
                    NULL, 0, NULL, NULL))
    {
            printf("DeviceIoControl failed, error=%d\n", GetLastError() );
    }
    else
    {
        printf("press ctrl-C to exit . . .\n");

        while (TRUE)
        {
            SleepEx(-1, TRUE);
        }
    }
}
```

Listing 12.12   *APC/WIN32APP/WIN32APP.MAK*

```
win32app.exe: win32app.obj
    link @<<
kernel32.lib user32.lib gdi32.lib winspool.lib comdlg32.lib advapi32.lib
shell32.lib ole32.lib oleaut32.lib uuid.lib /NOLOGO /SUBSYSTEM:console
/INCREMENTAL:no /PDB:none /MACHINE:I386 /OUT:win32app.exe win32app.obj
<<

win32app.obj: win32app.c
    cl /c /ML /GX /YX /Od /D "WIN32" /D "NDEBUG" /D "_CONSOLE" -I..\vxd win32app.c
```

Listing 12.13   *EVENTVXD.H*

```
typedef struct
{
    BOOL      bVmCreated;
    DWORD     hVM;
} VMINFO;

#define EVENTVXD_REGISTER      0x8100
#define EVENTVXD_RELEASEMEM    0x8101
```

Listing 12.14   *EVENTVXD.C*

```
#define WANTVXDWRAPS

#include <basedef.h>
#include <vmm.h>
#include <debug.h>
#include "vxdcall.h"

#include <wrappers.h>
#include <vwin32.h>
#include "eventvxd.h"

#ifdef DEBUG
#define DPRINTF0(buf, fmt) _Sprintf(buf, fmt ); Out_Debug_String( buf )
#define DPRINTF1(buf, fmt, arg1) _Sprintf(buf, fmt, arg1 ); Out_Debug_String( buf )
#define DPRINTF2(buf, fmt, arg1, arg2) _Sprintf(buf, fmt, arg1, arg2 );
                                    Out_Debug_String( buf )
#else
#define DPRINTF0(buf, fmt)
#define DPRINTF1(buf, fmt, arg1)
#define DPRINTF2(buf, fmt, arg1, arg2)
#endif
```

Listing 12.14 (continued)   `EVENTVXD.C`

```c
// functions in asm module
void PriorityEventThunk( void );

typedef VMINFO *PVMINFO;

VOID __stdcall PriorityEventHandler(VMHANDLE hVM, PVOID Refdata, CRS *pRegs);

BOOL      bClientRegistered = FALSE;
VMINFO    GlobalVMInfo;
HANDLE    hWin32Event;
char      dbgbuf[80];

BOOL OnVmInit(VMHANDLE hVM)
{
    if (bClientRegistered)
    {
        GlobalVMInfo.hVM = hVM;
        GlobalVMInfo.bVmCreated = TRUE;
        Call_Priority_VM_Event(LOW_PRI_DEVICE_BOOST, Get_Sys_VM_Handle(),
                        PEF_WAIT_FOR_STI+PEF_WAIT_NOT_CRIT,
                        hWin32Event, PriorityEventThunk, 0 );
    }
    return TRUE;
}

VOID OnVmTerminate(VMHANDLE hVM)
{
    if (bClientRegistered)
    {
        GlobalVMInfo.hVM = hVM;
        GlobalVMInfo.bVmCreated = FALSE;
        Call_Priority_VM_Event(LOW_PRI_DEVICE_BOOST, Get_Sys_VM_Handle(),
                        PEF_WAIT_FOR_STI+PEF_WAIT_NOT_CRIT,
                        hWin32Event, PriorityEventThunk, 0 );
    }
}

VOID __stdcall PriorityEventHandler(VMHANDLE hVM, PVOID Refdata, CRS *pRegs)
{
    HANDLE hWin32Event = Refdata;

    _VWIN32_SetWin32Event( hWin32Event );
}

BOOL OnSysDynamicDeviceInit()
{
    DPRINTF0( dbgbuf, "Loading\r\n");
    return TRUE;
}

BOOL OnSysDynamicDeviceExit()
{
    DPRINTF0( dbgbuf, "Unloading\r\n");
    return TRUE;
}
```

Listing 12.14 (continued)   *EVENTVXD.C*

```c
DWORD OnW32Deviceiocontrol(PDIOCPARAMETERS p)
{
    DWORD rc;

    switch (p->dwIoControlCode)
    {
    case DIOC_OPEN:
        rc = 0;
        break;

    case DIOC_CLOSEHANDLE:
      bClientRegistered = FALSE;
        rc = 0;
        break;

    case EVENTVXD_REGISTER:
        hWin32Event = p->lpvInBuffer;
        *((DWORD *)(p->lpvOutBuffer)) = (DWORD)&GlobalVMInfo;
        *((DWORD *)(p->lpcbBytesReturned)) = sizeof(DWORD);
        bClientRegistered = TRUE;
        rc = 0;
        break;

    default:
        rc = 0xffffffff;
    }

    return rc;
}
```

Listing 12.15   *EVENTDDB.ASM*

```asm
    .386p

;****************************************************************************
;                 I N C L U D E S
;****************************************************************************

    include vmm.inc
    include debug.inc

;==========================================================================
;     V I R T U A L   D E V I C E   D E C L A R A T I O N
;==========================================================================
DECLARE_VIRTUAL_DEVICE      EVENTVXD, 1, 0, ControlProc, UNDEFINED_DEVICE_ID, \
                            UNDEFINED_INIT_ORDER
```

Listing 12.15 *(continued)*   *EVENTDDB.ASM*

```
VxD_LOCKED_CODE_SEG

;============================================================
;
;   PROCEDURE: ControlProc
;
;   DESCRIPTION:
;    Device control procedure for the SKELETON VxD
;
;   ENTRY:
;    EAX = Control call ID
;
;   EXIT:
;    If carry clear then
;        Successful
;    else
;        Control call failed
;
;   USES:
;    EAX, EBX, ECX, EDX, ESI, EDI, Flags
;
;============================================================

BeginProc ControlProc
    Control_Dispatch VM_INIT, _OnVmInit, cCall, <ebx>
    Control_Dispatch VM_TERMINATE, _OnVmTerminate, cCall, <ebx>
    Control_Dispatch SYS_DYNAMIC_DEVICE_INIT, _OnSysDynamicDeviceInit, cCall, <ebx>
    Control_Dispatch SYS_DYNAMIC_DEVICE_EXIT, _OnSysDynamicDeviceExit, cCall, <ebx>
    Control_Dispatch W32_DEVICEIOCONTROL, _OnW32Deviceiocontrol, cCall, <esi>
    clc
    ret

EndProc ControlProc

VxD_LOCKED_CODE_ENDS

VxD_CODE_SEG

BeginProc _PriorityEventThunk

    sCall PriorityEventHandler, <ebx,edx,ebp>
    ret

EndProc _PriorityEventThunk

VxD_CODE_ENDS

    END
```

Listing 12.16 `EVENTVXD.MAK`

```
CFLAGS      = -DWIN32 -DCON -Di386 -D_X86_ -D_NTWIN -W3 -Gs -D_DEBUG -Zi
CVXDFLAGS   = -Zdp -Gs -c -DIS_32 -Zl -DDEBLEVEL=1 -DDEBUG
LFLAGS      = -machine:i386 -debug:notmapped,full -debugtype:cv
              -subsystem:console kernel32.lib
AFLAGS      = -coff -DBLD_COFF -DIS_32 -W2 -Zd -c -Cx -DMASM6 -DDEBLEVEL=1 -DDEBUG

all: eventvxd.vxd

eventvxd.obj: eventvxd.c
        cl $(CVXDFLAGS) -Fo$@ %s

eventddb.obj: eventddb.asm
        ml $(AFLAGS) -Fo$@ %s

eventvxd.vxd: eventddb.obj eventvxd.obj ..\..\..\wrappers\vxdcall.obj
                                        ..\..\..\wrappers\wrappers.clb eventvxd.def
        echo >NUL @<<eventvxd.crf
-MACHINE:i386 -DEBUG -DEBUGTYPE:MAP -PDB:NONE
-DEF:eventvxd.def -OUT:eventvxd.vxd -MAP:eventvxd.map
-VXD vxdwraps.clb wrappers.clb vxdcall.obj eventddb.obj eventvxd.obj
<<
        link @eventvxd.crf
        mapsym eventvxd
```

Listing 12.17 `EVENTVXD.DEF`

```
VXD EVENTVXD DYNAMIC
SEGMENTS
    _LTEXT     CLASS 'LCODE'   PRELOAD NONDISCARDABLE
    _LDATA     CLASS 'LCODE'   PRELOAD NONDISCARDABLE
    _TEXT      CLASS 'LCODE'   PRELOAD NONDISCARDABLE
    _DATA      CLASS 'LCODE'   PRELOAD NONDISCARDABLE
    _LPTEXT    CLASS 'LCODE'   PRELOAD NONDISCARDABLE
    _CONST     CLASS 'LCODE'   PRELOAD NONDISCARDABLE
    _BSS       CLASS 'LCODE'   PRELOAD NONDISCARDABLE
    _TLS       CLASS 'LCODE'   PRELOAD NONDISCARDABLE
    _ITEXT     CLASS 'ICODE'   DISCARDABLE
    _IDATA     CLASS 'ICODE'   DISCARDABLE
    _PTEXT     CLASS 'PCODE'   NONDISCARDABLE
    _PDATA     CLASS 'PCODE'   NONDISCARDABLE
    _STEXT     CLASS 'SCODE'   RESIDENT
    _SDATA     CLASS 'SCODE'   RESIDENT
    _MSGTABLE  CLASS 'MCODE'   PRELOAD NONDISCARDABLE IOPL
    _MSGDATA   CLASS 'MCODE'   PRELOAD NONDISCARDABLE IOPL
    _IMSGTABLE CLASS 'MCODE'   PRELOAD DISCARDABLE IOPL
    _IMSGDATA  CLASS 'MCODE'   PRELOAD DISCARDABLE IOPL
    _DBOSTART  CLASS 'DBOCODE' PRELOAD NONDISCARDABLE CONFORMING
    _DBOCODE   CLASS 'DBOCODE' PRELOAD NONDISCARDABLE CONFORMING
    _DBODATA   CLASS 'DBOCODE' PRELOAD NONDISCARDABLE CONFORMING
    _16ICODE   CLASS '16ICODE' PRELOAD DISCARDABLE
    _RCODE     CLASS 'RCODE'
EXPORTS
    EVENTVXD_DDB @1
```

Listing 12.18   *W32EVENT/WIN32APP/WIN32APP.C*

```
#include <stdio.h>
#include <stdlib.h>
#include <conio.h>
#include <windows.h>
#include "..\vxd\eventvxd.h"

HANDLE hDevice;
VMINFO *pVMInfo;

DWORD WINAPI SecondThread( PVOID hEventRing3 );

DWORD WINAPI SecondThread( PVOID hEventRing3 )
{
   while( TRUE )
   {
      WaitForSingleObject((HANDLE)hEventRing3, INFINITE );
      printf("VM %08lx was %s\n", pVMInfo->hVM,
             pVMInfo->bVmCreated ? "created" : "destroyed" );
   }
   return 0;
}

void main( int ac, char *av[] )
{
    HINSTANCE hKernel32Dll;
    HANDLE    hEventRing3, hEventRing0;
    DWORD     tid;
    HANDLE    (WINAPI *pfOpenVxDHandle)(HANDLE);
    DWORD     cbBytesReturned;

    const PCHAR VxDName = "\\\\.\\EVENTVXD.VXD";

    hEventRing3 = CreateEvent( 0, FALSE, FALSE, NULL );
    if (!hEventRing3)
    {
       printf("Cannot create Ring3 event\n");
       exit(1);
    }

    hKernel32Dll = LoadLibrary("kernel32.dll");
    if (!hKernel32Dll)
    {
       printf("Cannot load KERNEL32.DLL\n");
       exit(1);
    }

    pfOpenVxDHandle = (HANDLE (WINAPI *) (HANDLE))
                    GetProcAddress( hKernel32Dll, "OpenVxDHandle" );
    if (!pfOpenVxDHandle)
    {
       printf("Cannot get addr of OpenVxDHandle\n");
       exit(1);
    }
```

Listing 12.18 (continued)   `W32EVENT/WIN32APP/WIN32APP.C`

```c
   hEventRing0 = (*pfOpenVxDHandle)(hEventRing3);
   if (!hEventRing0)
   {
      printf("Cannot create Ring0 event\n");
      exit(1);
   }

   hDevice = CreateFile( VxDName, 0, 0, 0, CREATE_NEW, FILE_FLAG_DELETE_ON_CLOSE, 0 );
   if (!hDevice)
   {
      printf("Cannot load VxD error=%x\n", GetLastError() );
      exit(1);
   }

   if (!DeviceIoControl( hDevice, EVENTVXD_REGISTER, hEventRing0,
                         sizeof(hEventRing0), &pVMInfo, sizeof(pVMInfo),
                         &cbBytesReturned, 0 ))
   {
      printf("DeviceIoControl failed, error=%x\n", GetLastError() );
      exit(1);
   }
   CreateThread( 0, 0x1000, SecondThread, hEventRing3, 0, &tid );
   printf("Press any key to exit...");
   getch();
   CloseHandle( hDevice );
}
```

Listing 12.19   `W32EVENT/WIN32APP/WIN32APP.MAK`

```
win32app.exe: win32app.obj
    link @<<
kernel32.lib user32.lib gdi32.lib winspool.lib comdlg32.lib advapi32.lib
shell32.lib ole32.lib oleaut32.lib uuid.lib /NOLOGO /SUBSYSTEM:console
/INCREMENTAL:no /PDB:none /MACHINE:I386 /OUT:win32app.exe win32app.obj
<<

win32app.obj: win32app.c
    cl /c /ML /GX /YX /Od /D "WIN32" /D "NDEBUG" /D "_CONSOLE" -I..\vxd win32app.c
```

Part 2

DLL-based Drivers

## Chapter 13

Introduction to 16-bit Driver DLLs

Why Driver DLLs are Always 16-bit

Back in the days of Windows 3.x, Microsoft recommended that developers package all hardware drivers as VxDs, the "true" device drivers for Windows. However, many developers — including Microsoft itself — ignored this advice and instead put driver functions into DLLs. After all, the learning curve for VxDs was very steep, and a driver packaged as a DLL could do the job adequately. (Notable exceptions were drivers, like those for the serial port, that required very fast interrupt response times.)

Today, Microsoft recommends that developers for Windows 95 package hardware drivers as VxDs. This time, however, the recommendation is much more difficult to ignore, because Win32 DLLs are forbidden from performing most "driver" type operations. The list of prohibited operations includes

- accessing memory-mapped hardware,
- performing DMA transfers,
- handling hardware interrupts, and
- issuing software interrupts.

As you can see, the only type of driver DLL you could package as a 32-bit DLL without breaking these rules is the simplest type: a polled-mode driver (no interrupts) for an I/O-mapped device.

Many Windows 95 developers are therefore heeding Microsoft's advice and writing VxDs. Nonetheless, you can still write a complex driver as a DLL if you build it as a 16-bit DLL because 16-bit DLLs aren't governed by the same limitations as Win32 DLLs. In fact, many of the standard drivers provided by Microsoft (including the mouse driver and multimedia drivers) remain 16-bit. Using a 16-bit DLL under Windows 95, however, requires writing another DLL in addition to the driver DLL: a thunk DLL.

Interfacing 16-bit DLL to 32-Bit Application Requires a Thunk

In Windows 95, driver DLLs are always 16-bit, regardless of whether the application interfacing to the DLL is an old 16-bit Windows 3.x application or a new 32-bit Win32 application. If you want your 16-bit driver DLL to be used by Win32 applications, you must write a translation layer to convert between the 32-bit world and the 16-bit world. This translation layer is called a thunk DLL.

Thunk DLLs will be covered in detail in Chapter 18. For now, just note that choosing to implement a driver as a 16-bit DLL implies creating a thunk DLL if you'll be supporting Windows 95. Considering the extra work required for the thunk, it may make more sense for you to write the driver as a VxD instead.

The remainder of this chapter introduces the basics of 16-bit Windows DLLs, and introduces a skeleton driver DLL. The next two chapters cover how a driver DLL interfaces to hardware and handles hardware interrupts.

The material in this chapter applies specificaly to 16-bit DLLs, and much of it is not relevant to 32-bit DLLs.

Static versus Dynamic Libraries

Although a DLL can be linked to and executed by an application much as a static library can, the DLL is not really a part of any single application. Understanding how a DLL differs from a static library will help clarify why drivers are more useful if packaged as a DLL rather than as a static library.

A static library (such as the run-time library for the C compiler) is nothing more than a collection of one or more precompiled functions. The static library is packaged as a single piece from which the linker can extract necessary components. At link time, the linker searches your application for references to functions outside of the application and resolves these references by searching for them in the library. The library functions are then copied into your application's .EXE image. After linking, an application calls one of the library functions using the same mechanism it would use to call one of its own internal functions.

A DLL is also a set of precompiled functions. When these functions are packaged, two pieces are created: an import library (.LIB) and a DLL. The DLL contains the actual code and data. The import library contains only name and module information for the functions. An application that wants to use functions in a DLL links with the import library, not the DLL. The linker doesn't fully resolve the application's references to external functions in the DLL, i.e. the linker does not copy the functions into the application's .EXE image. Instead, the linker stores only the function name and module from the import library as a placeholder in the .EXE.

The real magic happens at run time, when the application loads. At that time, the loader also loads the DLL into memory, thus giving all the DLL functions an address. The loader then goes back to the .EXE and fills in the placeholders left at link time with the DLL function addresses.

Why Package Drivers in a DLL?

DLLs are a convenient way to package driver functions because drivers are often used by more than one application, and also because drivers often need to change independently of the application. With the driver functions in a separate file from the applications, the driver itself can be updated without disturbing the application that uses the driver. If a driver is used by multiple applications, a DLL saves memory because only one copy of the DLL is in memory, whereas building a driver as a statically linked library would force each application that used the driver to have its own copy of the driver functions.

The most important reason for packaging a Windows driver as a DLL is to make it possible to replace the DLL with a "true" device driver (VxD or NT kernel mode driver). By isolating hardware-dependent code in a DLL, you can replace the DLL with a true driver without completely rewriting the applications that use the DLL. Of course, the applications will still require some changes, because usually the interface presented to the application by your custom driver DLL won't be exactly the same as the interface presented by the true driver.

Applications versus DLLs

DLLs are different than applications in several fundamental respects. These differences have implications for how a DLL is coded and how it is built. The most obvious difference to a user is that a DLL can't be executed directly from the Windows shell. It has no life of its own but is loaded when an application that uses it is loaded and is unloaded when that application is terminated.

Here's a comparison of DLLs and applications, from a developer's point of view:

- An application has its own stack segment. A DLL does not. A DLL uses the stack of the calling application.

- If multiple instances of an application are running, all instances share a single copy of the application's code segment. Similarly, if more than one application uses a DLL, there is still only a single copy of the DLL's code. (That's one of the advantages of DLLs compared to static libraries.)

- If multiple instances of an application are running, each instance gets its own copy of the application's data segment. This is not true for DLLs. If more than one application uses a DLL there is still only a single copy of the DLL's data segment.

- Memory dynamically allocated by a DLL may belong to either the calling application or to the DLL itself, depending on the exact method of allocation.

These items have ramifications for the DLL developer during both the coding and the build process. The following sections address each of these issues in detail.

DLLs and Stack Segments

A DLL doesn't have its own stack segment. This leads to some subtle difficulties with passing pointers as parameters to other DLL functions. As the following example illustrates, pointers passed to a DLL can easily turn into subtle bugs.

Suppose an application has the following function:

```
void FAR Foo1( void )
{
    int x;
    Foo2( &x );
}
```

and a DLL contains this function:

```
void Foo2( int *x )
{
    int y;
    y = *x;
}
```

If this code is compiled as small model, the &x expression in the call to Foo2 is a near pointer. That means the code generated to push the address of x onto the stack will push only the offset of the variable x. When Foo2 gets this offset from the stack and dereferences it, Foo2 incorrectly assumes that since this is a near pointer, the offset is relative to DS (the DLL's data segment). Foo2 doesn't know the offset is really relative to SS (the caller's stack segment). The result is that the expression *x accesses the wrong location, and Y is assigned an incorrect value.

Why do DLLs have this problem and normal applications don't? Because when a normal application is running, DS has the same value as SS, so SS-relative is the same thing as DS-relative. In a DLL, DS != SS. Does this mean DLLs can't pass the address of a local variable as a parameter? No, it just means that you must always pass a far pointer to a DLL, not a model-dependent (no far/near attribute) pointer.

There is a compiler option (add w to the memory model option) that will generate a warning when the address of a local variable is passed as a model-dependent pointer. Use this option, heed the warning, and change those parameters to far pointers.

DLLs and Data Segments

Even when used simultaneously by different applications, a DLL has only one copy of its data segment. This means extra work for the developer if the DLL needs to maintain some information on a per-application basis. For example, the DLL could encapsulate all per-application information into an AppInstance structure and allocate a new structure for each application using the DLL. Then at each entry point the DLL would need to figure out which application was calling it and reference the appropriate AppInstance structure.

I won't cover this topic any further, because driver DLLs don't usually have this problem. Typically, a driver DLL serves to serialize access to a device by multiple applications. In other words, if Application 1 is using the device through the DLL, Application 2 isn't allowed to use the device. On the other hand, Application 1 might use Device 1 and while Application 2 uses Device 2 (where both devices are managed by the same DLL). But that situation can be managed as per-device instead of per-application data. I'll cover per-device data in more detail in the next chapter.

DLLs and Ownership of Dynamically Allocated Memory

A Windows DLL must be careful when dynamically allocating memory, for this memory may be owned either by the calling application or by the DLL itself, depending on exactly how the allocation was made. Dynamic allocations can be made using either the GlobalAlloc call to the Windows memory manager or via the C run-time malloc function.

When a DLL calls GlobalAlloc directly, the DLL specifies whether the memory is to be owned by the DLL or by the calling application. If the DLL uses the GMEM_SHARE flag when calling GlobalAlloc, the DLL owns the memory; if not, then the application owns it. (I'll explain the parameters used by GlobalAlloc in more detail later; for now the only relevant parameter is GMEM_SHARE.) For VC++ 1.x, the malloc routine in the C run-time library always uses the GMEM_SHARE flag when called by a DLL, so any malloc-allocated buffers are owned by the DLL.

Either an application or a DLL can explicitly free a buffer via a call to `GlobalFree` or `free`. The ownership issue becomes important if nobody explicitly frees a dynamically allocated buffer. If the DLL allocates the memory but the application owns it, then the memory is freed automatically by Windows when the application exits. If the DLL owns the memory itself, then the memory is freed by Windows only when the DLL unloads — which doesn't happen until all applications using the DLL have exited.

So who should own a driver-allocated buffer — application or DLL? If a driver is not interrupt-driven, it doesn't really matter. In this case, the DLL executes only as a result of calls from an application. If the application goes away, it won't call the DLL anymore, and that means the DLL won't use the buffer.

For a driver that does handle interrupts, any dynamically allocated buffers used at interrupt time should be owned by the driver DLL. An example will clarify the issues involved. Suppose a driver has an Open entry point and a Close entry point. During Open, a buffer is allocated and an interrupt handler installed. At Close, the buffer is freed and the handler removed. Now suppose that an application exited without calling Close, perhaps because it crashed. Windows itself frees the allocated buffer when the application exits. Then the driver's interrupt handler accesses the freed buffer and bad things happen — you can't reference memory after it's been freed. If, on the other hand, the driver owned the buffer, Windows would not have freed it, and the interrupt handler could continue to access the buffer safely.

DLL Initialization and Termination

A Windows DLL may contain a special initialization entry point called `LibMain` and a special termination entry point called `WEP` (for Windows Exit Procedure). If present, `LibMain` is called when the DLL is loaded. If the DLL contains a `WEP`, it is called when the DLL is unloaded.

For many DLLs, the initialization code in `LibMain` does things like registering window classes and initializing the local heap (which is in the DLL's data segment). However, driver DLLs don't register windows and generally use the global heap instead of the local heap (the local heap is too small). The driver DLLs in this book don't contain a `LibMain`. Instead, I prefer to do initialization in another entry point called explicitly by an application using the DLL, an Open routine. Similarly, most DLLs use the `WEP` entry point to un-register window classes. Driver DLLs don't have window classes, and the driver DLLs in this book do cleanup in a Close entry point.

DLL Function Requirements

Although a DLL can be compiled with any memory model, all functions that the DLL exports to an application must be declared `far`. The reason is simple: a DLL has its own code and data segments, which are separate from the calling application's code and data segments. If you compile your DLL as medium or large model, all functions are, by definition, `far`. On the other hand, if you compile your DLL as small model, you must explicitly declare the DLL entry points with the `far` keyword. I've chosen to compile the driver DLLs in this book as small model (for reasons I'll explain in the next chapter), so all the driver entry points in these DLLs are declared `far`.

The DLL entry points are also declared with the `_export` keyword. This tells the linker to generate a special export definition record for these functions, which the loader uses at run time to resolve references to a DLL.

An exported DLL function must contain a special section of code at the beginning and at the end of the function. The code at the beginning is called the prologue: its purpose is to fix up the DS register (meaning it must load DS with the DLL's data segment). The DS fix-up is necessary because on entry to the DLL, DS contains the calling application's data segment, which is different from the DLL's data segment. The code at the end is called the epilogue; it restores DS to the caller's original data segment.

A special compiler option tells the compiler which functions need prologue and epilogue code. In VC++ 1.x, the `/GD` flag tells the compiler to generate prologue/epilogue code for all functions declared as `_export`. The makefiles for the DLLs in this book use this `/GD` flag.

The Skeleton Driver

The first sample DLL driver is a skeleton or template driver (Listings 13.1–13.4, pages 300–302). It doesn't interface with any hardware, but it exports a set of functions that are general enough to apply to most types of drivers: `DeviceOpen`, `DeviceClose`, `DeviceWrite`,`DeviceRead`,`DeviceGetWriteStatus`,`DeviceGetReadStatus`,`Device-GetDriverParams`,`DeviceSetDriverParams`, and `DeviceGetDriverCapabilities`.

As the example drivers become more involved, I'll add functionality to the functions in this skeleton, piece by piece. Of course, your driver may need additional capabilities that aren't covered by these functions. In that case, you're free to add functions as needed because Windows doesn't dictate a driver interface for non-standard devices.

Each function in the skeleton driver does nothing more than output a trace message containing its function name. The driver outputs these messages through the Windows API function `OutputDebugString`. `OutputDebugString` uses only a simple string parameter, but you can also use the more powerful `DebugOutput` function. `DebugOutput` uses a format string and a variable number of parameters, like `sprintf`, providing more useful formatting. Windows redirects these strings to the AUX device (serial port), but you can also use the DBWIN utility to display them in a window (more on DBWIN later).

Building the Skeleton Driver

The steps involved in building the driver DLL are:

* compile the .C file,

* link into a .DLL,

* run the resource compiler, and

* create the import library.

To automate the steps required to properly build the driver, I use `nmake` and a makefile (Listing 13.3 on page 302). If you copy my makefile and source files to the current directory, you can build the skeleton driver from scratch, simply by typing `nmake -fskeleton.mak`.

Choosing the proper compiler options is critical to correctly building a Windows DLL. The options used by the skeleton DLL are listed in Table 13.1, along with a notation of whether the option is mandatory for all DLLs.

The link process for a Windows DLL is similar to building a DOS application, except that you must specify a .DEF file when linking. This .DEF file must include the statements `EXETYPE WINDOWS` and `LIBRARY`. These statements tell the linker to build a Windows DLL. The other mandatory DEF statements are `CODE` and `DATA`, which determine the attributes of the DLL's code and data segments. I'll discuss these attributes in Chapter 15.

Table 13.1   Options used by the skeleton DLL.

Option	Requirement	Description
C	mandatory	compile only (no link)
GD	mandatory	generate function prologue to fix up DS
AS	optional	small model
Gs	mandatory	disable stack probes
Aw	recommended	generate warnings whenever a DLL uses a near pointer to take the address of a local variable
W3	recommended	warning level of 3 (highest is 4)
G2	recommended	generate code for 286+ (speeds execution)
Zi	optional	generate CodeView debug information
Oi	optional	use intrinsics (faster inp/outp/strcpy/memcpy)
Fc	optional	generate assembly output

The linker command line for a DOS application doesn't usually specify a library. It's not necessary because the C compiler embeds information in the .OBJ file that tells the linker which library (small, medium, large) to use. Windows DLLs need a special version of the C library, ?dllcew.lib instead of ?libcew.lib, where ? is an abbreviation for the memory model. When using VC++ 1.x, you should use the /NOD option so that the linker does not bring in the C library named in the .OBJ file. You should also explicitly list the DLL version of the library (?dllcew.lib) as the library argument. In addition, you should specify LIBW as a second library. This is the import library containing the Windows API functions.

Last, an import library for the DLL is built using IMPLIB. IMPLIB uses the DLL's .DEF file as input and builds a .LIB file containing exported function names and modules. This .LIB file is then linked, as a library, to an application that uses the DLL. There are no option switches required for IMPLIB.

The last step in my makefile, copying the driver DLL to the Windows directory, isn't strictly required, but it's useful. At run time, Windows uses the same method to locate a DLL that it does to locate an .EXE file: search the current directory, the Windows directory, the Windows system directory, and the directories listed in the PATH. By copying the driver to the Windows directory, I can invoke the application (and thus the DLL) regardless of the current directory or PATH variable.

DLL Requires an Application

A Windows DLL can't execute on its own. It must be called by a Windows program. I've supplied a sample Windows application, TESTDRIV.EXE, which can exercise all the functions supported by the driver.

TESTDRIV is a very rudimentary Windows application. Its user interface contains a single menu with several submenu items, one for each exported driver function: DeviceOpen, DeviceClose, DeviceWrite, DeviceRead, DeviceGetWriteStatus, DeviceGetReadStatus, DeviceGetDriverParams, and DeviceGetDriverCapabilities. Select a menu item, and TESTDRIV calls that function in the driver (Figure 13.1). TESTDRIV uses hard-coded values for all driver parameters — you can't specify from the user interface which device to open or what data to write. You could easily extend TESTDRIV to support user input of driver parameters.

If the driver function returns with an error code, TESTDRIV will display a message box. If the function is one of the three with output parameters (`DeviceGetReadStatus`, `DeviceGetWriteStatus`, or `DeviceGetDriverCapabilities`), the output parameters are displayed in a message box.

The next two chapters present two more driver DLLs, each adding more functionality to the SKELETON DLL introduced in this chapter. When developing this series of DLLs, I was careful to ensure that the three DLLs export exactly the same set of functions. For this reason, you can use the same TESTDRIV application with all three driver DLLs. In fact, because the driver functions are dynamically, not statically, linked to the application, you don't even need to relink TESTDRIV when you change the driver DLL implementation.

Figure 13.1　　`TESTDRIV.EXE` — *a sample Windows application that makes calls to our driver DLL with* `DBWIN` *active in the right-hand window.*

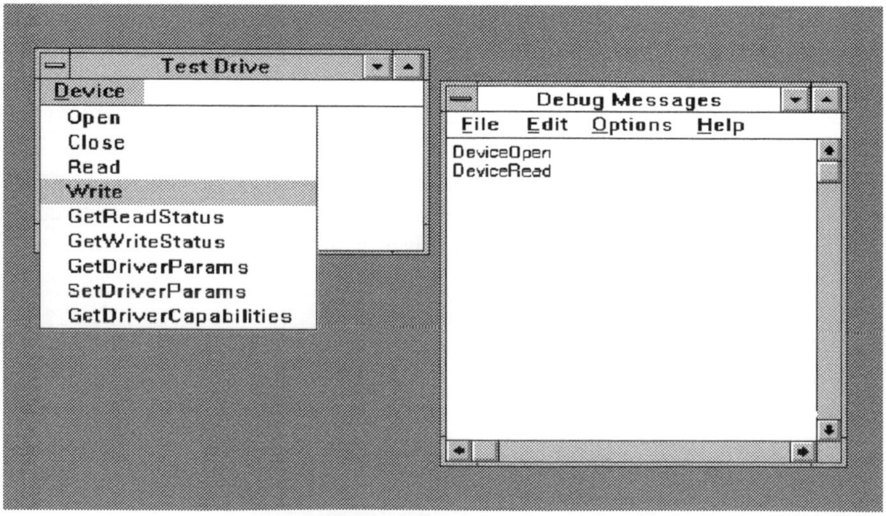

Debugging Tools for Driver DLLs

When using TESTDRIV to explore a driver, it also is useful to run the Windows DBWIN application, a utility included with VC++ 1.x. DBWIN captures all the strings that Windows applications and DLLs output via calls to `OutputDebugString` and `DebugOutput`. DBWIN redirects the strings to either its client window, a secondary monochrome monitor, or a serial port. By adding more of these `OutputDebugString` calls to your driver, you can trace its execution path and thus perform rudimentary debugging.

These trace statements, however, won't replace the need for a real debugger. An application-level debugger, such as the one provided in the VC++ Integrated Development Environment, can be used to debug some driver DLLs, unless the driver handles interrupts. A better choice would be a system-level debugger, either WDEB386 or SoftIce/Windows.

Summary

With the information in this chapter, you can set up a test environment for DLL driver development and confirm that you have your tools properly configured to create a DLL that links to an application. Although this chapter's skeleton driver doesn't really do anything, you can still exercise it to confirm that it communicates with an application. This sets the stage for producing a DLL that actually manipulates some hardware — the topic of the next chapter.

Listing 13.1   `SKELDLL\DRIVER.H`

```
typedef struct
{
    WORD        usDevNumber;
    LPBYTE      lpReadBuffer;
} DEVICECONTEXT, FAR *HDEVICE;

typedef struct
{
    WORD      usReadBufSize;
} DRIVERPARAMS, FAR * PDRIVERPARAMS;

typedef struct
{
    WORD      version;
} DRIVERCAPS, FAR * PDRIVERCAPS;
typedef PDRIVERCAPS FAR * PPDRIVERCAPS;

HDEVICE FAR PASCAL DeviceOpen( void );
int FAR PASCAL DeviceClose( HDEVICE );
int FAR PASCAL DeviceGetWriteStatus( HDEVICE, LPWORD pusStatus );
int FAR PASCAL DeviceGetReadStatus( HDEVICE, LPWORD pusStatus );
int FAR PASCAL DeviceWrite( HDEVICE, LPBYTE lpData, LPWORD pcBytes );
int FAR PASCAL DeviceRead( HDEVICE, LPBYTE lpData, LPWORD pcBytes );
int FAR PASCAL DeviceSetDriverParams( HDEVICE, PDRIVERPARAMS pParms );
int FAR PASCAL DeviceGetDriverParams( HDEVICE, PDRIVERPARAMS pParms );
int FAR PASCAL DeviceGetDriverCapabilities( HDEVICE, PPDRIVERCAPS ppDriverCaps );
```

Listing 13.2   `SKELDLL\SKELETON.C`

```
#include <windows.h>
#include "driver.h"

DEVICECONTEXT Device1 = { 0 };
DRIVERPARAMS DefaultParams = { 1024 };

HDEVICE FAR PASCAL _export DeviceOpen( void )
{
    OutputDebugString( "DeviceOpen\n" );

    return &Device1;
}

int FAR PASCAL _export DeviceClose( HDEVICE hDevice )
{
    OutputDebugString( "DeviceClose\n" );

    return 0;
}
```

Listing 13.2 (continued)   *SKELDLL\SKELETON.C*

```c
int FAR PASCAL _export DeviceGetWriteStatus( HDEVICE hDevice, LPWORD pusStatus )
{
    OutputDebugString( "DeviceGetWriteStatus\n");

    return 0;
}

int FAR PASCAL _export DeviceGetReadStatus( HDEVICE hDevice, LPWORD pusStatus )
{
    OutputDebugString( "DeviceGetReadStatus\n");

    return 0;
}

int FAR PASCAL _export DeviceWrite( HDEVICE hDevice, LPBYTE lpData, LPWORD pcBytes )
{
    OutputDebugString( "DeviceWrite\n");

    return 0;
}

int FAR PASCAL _export DeviceRead( HDEVICE hDevice, LPBYTE lpData, LPWORD pcBytes )
{
    OutputDebugString( "DeviceRead\n");

    return 0;
}

int FAR PASCAL _export DeviceSetDriverParams( HDEVICE hDevice, PDRIVERPARAMS pParms )
{
    OutputDebugString( "DeviceSetDriverParams\n");

    return 0;
}

int FAR PASCAL _export DeviceGetDriverParams( HDEVICE hDevice, PDRIVERPARAMS pParms )
{
    OutputDebugString( "DeviceGetDriverParams\n");

    return 0;
}

int FAR PASCAL _export DeviceGetDriverCapabilities( HDEVICE hDevice,
                                          PPDRIVERCAPS ppDriverCaps )
{
    OutputDebugString( "DeviceGetDriverCapabilities\n");

    return 0;
}
```

Listing 13.3   *SKELDLL\SKELETON.MAK*

```
all: skeleton.dll

# DRIVER DLL

skeleton.obj: skeleton.c driver.h
  cl -c -W3 -ASw -GD2s -Oi -Fc $*.c

skeleton.dll: skeleton.def skeleton.obj
  link skeleton,skeleton.dll,skeleton.map /MAP, sdllcew libw /nod/noe,skeleton.def
  implib skeleton.lib skeleton.dll
  copy skeleton.dll \windows\driver.dll
```

Listing 13.4   *SKELDLL\SKELETON.DEF*

```
LIBRARY    DRIVER
DESCRIPTION "Skeleton Driver"
EXETYPE  WINDOWS
DATA     PRELOAD MOVEABLE SINGLE
CODE     PRELOAD MOVEABLE DISCARDABLE
```

## Chapter 14

Driver DLLs:
Connecting to the Hardware

Unlike DOS, which allows programmers to directly manipulate any device at any time, Windows is somewhat protective of the physical machine resources. In a sophisticated, high-performance driver, the Windows protection mechanisms can significantly complicate device access. In simple, polled-mode drivers though, device access can still be quite straightforward.

This chapter shows how to convert the previous chapter's skeleton driver from an empty framework into a complete, yet very simple, polled-mode driver that actually manipulates a physical device. I'll first illustrate the more simple I/O-mapped case by giving a complete polled-mode serial port driver and then show how to modify the port-mapped version to access an imaginary memory-mapped device.

DLLs and Port-access

One of the big myths of Windows programming is that applications and DLLs are not allowed to use _inp or _outp. Here's the real story.

Under Windows 3.x, there is absolutely nothing wrong with using _inp or _outp from a DLL to access a non-standard I/O location: the access will go through to the hardware port, without being trapped by Windows. If you access one of the *standard* I/O ports — keyboard, timer, etc. — then a VxD will trap your access and your code may not work as expected. But standard devices require special system driver DLLs and VxDs with interfaces defined by Windows, so you shouldn't be doing this from a custom driver DLL anyway.

It is also perfectly acceptable for an application or DLL running under Windows 95 to use _inp or _outp to access a non-standard I/O location. This is true for both 16-bit and 32-bit DLLs. However, if you choose to do this in your DLL, your DLL is not truly Win32-compatible. The *correct* way to access hardware from a Win32 application or DLL (notice I said "Win32", not "Windows 95") is through a true device driver, which, under Windows 95, takes the form of a VxD and, under Windows NT, is a kernel-mode driver. Windows NT will terminate any Windows application or DLL that attempts to access a hardware device, either IO-mapped or memory-mapped. Windows 95 happens to be a more forgiving Win32 platform than Windows NT, but Microsoft warns that future versions of Windows 95 may be less lenient. To play it safe, put all hardware access in a "true" driver, that is, a VxD.

A Port-mapped Example

Listings 14.1–14.5 (pages 318–324) make up a simple polled-mode driver for a standard serial port. The serial port makes a good example because every system has one, located in the I/O space, and Windows doesn't insist on taking over this device at startup. So one can easily install a replacement for the serial port handler without causing any confusion for Windows: not true for other standard PC devices like the keyboard or timer.

This driver DLL exports the same public interface as the SKELETON DLL introduced in Chapter 13, but this version's routines are more than just stubs. This example is not meant to be a high-performance, commercial-quality driver. I've tried to keep the driver simple, without sacrificing generality. Thus, it doesn't achieve very high throughputs, but it can easily be adapted to support multiple ports or different Universal Asynchronous Receiver-Transmitters (UARTS). Because the point is to illustrate techniques and principles that apply to a wide range of drivers, I've also tried to avoid getting bogged down by the intricacies of the hardware and the details of serial communications. (Figure 14.1 outlines the essentials of the 8250 UART.) By stripping the handler to its bare essentials, I hope to make the core structure clear enough that you can easily see what is essential and apply that to your own hardware.

Although Windows doesn't insist on taking control of the serial port, that is its default behavior. When using the serial port driver DLLs under Windows 3.x, you should prevent the Virtual Comm Device (VCD) from interfering with all serial ports by commenting out the device=*vcd statement in the [386Enh] section of SYSTEM.INI. When using the serial port driver DLLs under Windows 95, you must modify the registry to prevent the VCOMM VxD from interfering with a particular serial port. Change the PortDriver entry under the port's software key

HKLM\SYSTEM\CURRENTCONTROLSET\SERVICES\PORTS\000X

to something other than serial.vxd, for example _serial.vxd. In both cases, remember to undo these changes when you're finisished with the serial port driver DLL.

Figure 14.1   Outline of 8250 UART registers.

Offset	Name	Access
6	Modem Status	R
5	Line Status (LSR)	R
4	Modem Control	R/W
3	Line Control	R/W
2	Interrupt Ident	R
1	Interrupt Enable	R/W
0	Receive Data	R
0	Transmit Data	W
1 when LCR bit 7 = 1	Baud Rate Divisor MSB	R/W
0 when LCR bit 7 = 1	Baud Rate Divisor LSB	R/W

Driver Design Conventions

All of the driver DLLs in this book share certain design elements. Each uses a device context structure to store all state and addressing information for a single instance of the device, specifically the COM1 serial port. Many developers use this technique because it makes it easier to adapt the driver to support multiple devices. The address of the context structure is used as a handle to the device. The handle is returned by a call to DeviceOpen and used as a parameter to all other calls into the driver.

As explained in Chapter 13, a DLL may be called by multiple applications. A driver DLL that allows applications to "share" a device would need to store all context information specific to one application in an instance data structure. The example Driver DLLs presented here do not use an application-specific instance data structure, because the driver interface is designed to allow only a single application to use the device at a time. With this restriction, the driver DLL needs only the device context structure to find all the relevant data.

Some build issues (specifically, the SS != DS problem described in Chapter 13) can be simplified by compiling a DLL as large model. However, the example drivers here are all compiled small model, not large model. Actually, both the skeleton driver in the last chapter and the polled-mode driver in this chapter would work fine if compiled as large model. But the interrupt-driven driver of the next chapter must be small model to work as designed. (Interrupt handlers must load their own data segments. Because a large model DLL has multiple data segments, compiling as large model would complicate accessing data in the interrupt handler. Although this data access issue can indeed be resolved, it is simpler to keep the driver as small model.) Very few drivers will bump up against the 64Kb code or 64Kb data limit of small model.

All example driver DLL statically allocate their device context, as well as most other important data structures. If your driver allocates any memory at run time, it is important that the memory be allocated with the GMEM_SHARE flag. As discussed in Chapter 13, memory dynamically allocated by a DLL is owned by the calling application, not the DLL, unless the DLL uses GMEM_SHARE

The malloc provided by VC++ 1.x uses GMEM_SHARE, so if you're using it you may safely use malloc for any dynamic allocations in a polled-mode driver. An interrupt-driven driver, which will be discussed in detail in the next chapter, can dynamically allocate also, but it must use GlobalAlloc instead of malloc.

The Polled-mode Driver

This chapter doesn't contain a detailed discussion of the polled-mode driver code. The driver is both small and simple. However, a brief discussion of the data structures and the parameter validation code used by all of the driver entry points is in order.

In the example, the capabilities word simply contains the driver's version number. More sophisticated drivers might probe the device to determine its specific capabilities. For example, a multi-model scanner driver might query the attached driver and store maximum resolution and color depth in a capabilities structure. This information could then control the behavior of other driver routines and also could be used by the calling application if appropriate.

The example driver doesn't use the DRIVERPARAMS structure. Again, a more complex driver might offer several configuration options. These options could be recorded in the parameters structure and then be referenced by all affected routines.

When asked to open a new device, the DeviceOpen routine initializes hDevice with the address of the static device context structure. The DeviceOpen routine then configures the UART, as shown in the following code.

```
// Configure UART.
outp( hDevice->usIoBase + UART_REG_IER, 0 );
outp( hDevice->usIoBase + UART_REG_LCR, UART_LCR_DLAB );
outp( hDevice->usIoBase + UART_REG_BAUDLO, BAUD_1200 );
outp( hDevice->usIoBase + UART_REG_BAUDHI, 0 );
outp( hDevice->usIoBase + UART_REG_LCR, UART_LCR_8N1 );
outp( hDevice->usIoBase + UART_REG_MCR, UART_MCR_LOOP );
inp( hDevice->usIoBase + UART_REG_LSR );
inp( hDevice->usIoBase + UART_REG_RDR );

SET( hDevice->bFlags, FLAGS_OPEN);
```

The DeviceOpen routine then sets the FLAGS_OPEN bit. The driver's other routines can then check the FLAGS_OPEN bit to verify that a requested service is appropriate for the specified device. This chapter's example uses only the FLAGS_OPEN bit in the status field, although your driver might record additional state information here.

To make the driver robust, each routine validates the hDevice pointer and the device's current state. For example, to prevent the driver from attempting to manipulate a nonexistent device structure, the driver entry points will validate the hDevice pointer with the test:

```
if (!ValidHandle( hDevice ))
    return -1;
```

To prevent the driver from being used on a port that hasn't yet been opened, the driver routine will test the bFlags field:

```
if ((hDevice->bFlags & FLAGS_OPEN) == 0)
    return -1;
```

The example driver statically allocates only one device control structure. Thus, even though the code is structured to support multiple devices, the example is limited to one device. To use this driver with more than one serial port, you would need to allocate additional device structures (either statically or dynamically when DeviceOpen is called) and modify ValidHandle to keep track of all such structures.

The DeviceRead and DeviceWrite routines have an interface similar to the Standard C Library read and write routines. The DeviceRead routine expects a handle to a device context, a pointer to a data buffer (lpData), and a pointer to the number of bytes to read (pcBytes). A polling loop then copies the data from the UART to the buffer, one byte at a time, until it has collected the requested number of bytes. The DeviceWrite routine works identically, but in the reverse direction. It copies the specified number of bytes from the buffer to the UART transmit register. The following code shows the main polling loop for each function.

```
int FAR PASCAL _export DeviceWrite( HDEVICE hDevice, LPBYTE lpData,
                                     LPWORD pcBytes )
{
    WORD    i;

    OutputDebugString( "DeviceWrite\n" );

    if (!lpData)
        return -1;

    if (!ValidHandle( hDevice ))
        return -1;

    if ((hDevice->bFlags & FLAGS_OPEN) == 0)
        return -1;

    for (i=0; i < *pcBytes; i++)
    {
        while ((inp( hDevice->usIoBase + UART_REG_LSR ) & UART_LSR_THRE) == 0)
            ;
        outp( hDevice->usIoBase + UART_REG_THR, lpData[ i ] );
    }

    return 0;
}
```

```
int FAR PASCAL _export DeviceRead( HDEVICE hDevice, LPBYTE lpData,
                                   LPWORD pcBytes )
{
    WORD        i;

    OutputDebugString( "DeviceRead\n" );

    if (!lpData)
        return -1;

    if (!ValidHandle( hDevice ))
        return -1;

    if ((hDevice->bFlags & FLAGS_OPEN) == 0)
        return -1;

    for (i=0; i < *pcBytes; i++)
    {
        while ((inp( hDevice->usIoBase + UART_REG_LSR ) & UART_LSR_RXRDY) == 0)
            ;
        lpData[i] = inp( hDevice->usIoBase + UART_REG_RDR );
    }

    return 0;
}
```

It may seem unnecessary to require the caller of `DeviceRead` and `DeviceWrite` to provide a pointer to the number of bytes requested. This interface is indeed overkill for a polled-mode driver, where the number of bytes requested is always the same as the number of bytes processed. But this feature will support the next chapter's interrupt-driven driver without any changes to the interface. Keeping the same exact interface means the TESTDRIV application introduced in Chapter 13 works with both the polled-mode and the interrupt-driven drivers without even recompiling TESTDRIV.

Note that each polling loop sits in a busy loop while waiting for the UART to finish processing the current byte. Thus, if the application tried to transmit a full buffer of data with a single write, it would lose all data that might be received during the time required to transmit the entire buffer. Also, if the application calls the read function when no data has been received, the driver will simply hang in a busy loop until it receives some data.

Even so, one can successfully use a driver of this form for low-speed, full-duplex communications by following these conventions:

- Transmit one byte per write or read.

- Never attempt a read unless a call to `DeviceGetReadStatus` indicates a byte is available.

Accessing Memory-mapped Devices

The designers of the original PC system purposely left a hole in the processor's physical address space between A0000h and F0000h. No RAM exists at these memory locations, leaving them free to be used by memory-mapped devices. To access a memory-mapped device under DOS, you form a pointer that addresses that location, then dereference the pointer. The basic idea is the same to access the device from a 16-bit Windows DLL. But the procedure is complicated by address translation issues. (See Chapter 3 and Appendix A for a review of these issues.)

If your device is mapped somewhere in the unused A0000h–F0000h range of physical address space, there is a very simple method to get a pointer to access the device. Windows provides pre-allocated selectors for physical locations A0000h, B0000h, C0000h, D0000h, E0000h, and F0000h. These selectors are actually variables exported from the Windows system DLLs and are named appropriately: _A000h, _B000h, etc. Windows has already set up both the selector's base address and the associated page table entries appropriately, so that selector _B000h really does map to physical address B0000h. Each selector has its limit set to 64Kb, so _A000h maps A0000–AFFFF, _B000h maps B0000–BFFFF, etc.

Win32 applications or DLLs may not use these prefabricated selectors because they are exported from the KERNEL16 module, not by the KERNEL32 module linked in by 32-bit code.

To form a pointer to a device, choose the appropriate selector and offset. For example, _D000h and an offset of 8000h combine to point to a device at D8000h. Converting this selector/offset combination into a usable pointer is a bit more complicated than just a simple MAKELP(_D000h, 8000h). The following code fragment illustrates the three steps necessary.

```
//IMPORTANT: double underscore in KERNEL.__D000h
//           single underscore in #define SelD000h( &_MyD000h)
// Access memory-mapped adapter at physical D0000h
// MUST import the selector in your .DEF file:
// IMPORTS
//    __MyD000h = KERNEL.__D000h
extern WORD _MyD000h;
#define SelD000h (&_MyD000h)

char far *lpAdapter = MAKELP( SelD000h, 0x8000 );
```

A Memory-mapped Version

Although you aren't likely to ever encounter a real memory-mapped serial port, if you did, you'd find it quite simple to adapt this chapter's example driver. Assuming a similar layout of registers, the changes consist primarily of some code in DeviceOpen that sets up a pointer to the base address of the device and of modifications throughout that substitute memory references for _inp and _outp calls.

Assuming that the port was mapped to physical location D8000h, then

```
// IMPORTS in .DEF file:
//   __MyD000h = KERNEL.__D000h

extern WORD _MyD000h;
#define SelD000h (&_MyD000h)

DEVICECONTEXT Device1 = { 0, MAKELP(SelD000h, 0x8000), 0, NULL };
```

would set up the base pointer, assuming that the DEVICECONTEXT structure had been modified so that the address field has type char far *.

The main read loop would then become

```
for (i=0; i  *pcBytes; i++)
{
    while (( *( hDevice->usIoBase + UART_REG_LSR) & UART_LSR_RXRDY) == 0)
    ;
    lpData[i] = *( hDevice->usIoBase + UART_REG_RDR );
}
```

If you are willing to forego some of the efficiencies available to memory-mapped I/O, you can handle both memory-mapped and port-mapped devices in the same source code by conditionally defining appropriate access and initialization macros.

Advanced Memory Issues

Many memory-mapped devices occupy fewer than 64Kb of space in the
A0000h–EFFFFh range. However, devices can be larger and/or located in high memory
(above 1Mb). If you need to manipulate a device that is larger than 64Kb or that is
located in high memory, you will not be able to use the pre-constructed selectors. For
such devices you will need to call a DOS Protected Mode Interface (DPMI) service to
build the appropriate selector.

DPMI is a set of services that are provided to applications by Windows but
accessed through INT 31h instead of through an API function call. DPMI provides
low-level services for manipulating selectors, manipulating the interrupt vector table,
switching between real mode and protected mode, and manipulating the page tables.
Windows 3.x and Windows 95 both support DPMI v0.9. (A later specification for
DPMI v1.0 exists, but is not supported by either. See also the sidebar "DPMI History"
on page 313.)

Bypassing the 64Kb Limit

By using DPMI, you can bypass the 64Kb segment size limit to create a single
selector that maps a device larger than 64Kb. Although the Windows API function
SetSelectorBase won't accept a limit greater than 64Kb, the DPMI service
SetSelectorLimit will. The tricky part is generating code that uses 32-bit offsets.

Under Windows 3.x, programs run in a 16-bit code segment, which means memory references use 16-bit offsets. It is possible to override the offset size and force the
processor to use a 32-bit offset by inserting a prefix byte before each instruction. This
will require coding in assembly. If you'll be moving a lot of bytes, the extra effort is
probably worth it.

Software Interrupts Are Not Allowed in Win32

DPMI services are accessed via a software interrupt, and thus are not available to Win32 applications or DLLs
because the software interrupt handler in the VMM makes assumptions about the "bitness" of its caller. Specifically, the handler assumes its caller is 16-bit, and saves only 16-bit registers on the stack. Attempting to
call any software interrupt from 32-bit code therefore results in a crash.

The code in the following paragraph (found in the file POLLBASI\MOVE32.ASM on the code disk) allocates a single selector that addresses a memory adapter larger than 64Kb, and then uses that selector to zero out the entire region. Even better, the locations are zeroed 4 bytes (a DWORD) at a time, using 32-bit instructions. This code is written in pure assembly, because that's the easiest way to generate 32-bit instructions under Windows 3.x.

```
.MODEL SMALL
  .CODE
  .386

zero32 PROC C PhysSize:DWORD, PhysBase: DWORD

    mov ax, 0            ; DPMI Alloc Selector
    mov cx, 1
    int 31h
```

DPMI History

Driver developers often use the DPMI services provided by Windows as a back door into Windows to do things that the 500-plus functions in the standard Windows API won't let them do: access devices in physical memory and communicate with DOS drivers and TSRs. But Windows really supports DPMI for a completely different reason.

When Windows 3.0 was under development, PC software vendors were already working on several products that would break the DOS 640Kb barrier. These products included:

- DOS extenders, like the one used in Lotus 1-2-3, which let a DOS program use up to 16Mb of memory;

- expanded memory managers, like Qualitas' 386MAX, which allow a DOS program to use memory above 640Kb, although only in 16Kb chunks;

- and DOS-based multitasking environments, like Quarterdeck's Desqview.

Microsoft worked with the vendors who made these products, among them Intel, Phar Lap, and Rational Systems, to design an interface that would allow Windows 3.0 to coexist peacefully with all these products. All of these types of products, Windows included, do their magic by using the 80386 processor's advanced features, such as paging. The interface that was designed, which became DPMI v0.9, put a single program, the DPMI server, in charge of the 80386 advanced features. Other programs then took advantage of the features by using services exported by the DPMI server.

```
        mov bx, ax              ; selector from Alloc
        mov cx, WORD PTR [PhysSize+2]
        mov dx, WORD PTR [PhysSize]
        mov ax, 08h             ; DPMI Set Selector Limit
        int 31h

        mov dx, bx              ; save selector
        mov bx, WORD PTR [PhysBase+2]
        mov cx, WORD PTR [PhysBase]
        mov si, WORD PTR [PhysSize+2]
        mov di, WORD PTR [PhysSize]
        mov ax, 0800h           ; DPMI Map Physical Address
        int 31h

        push dx                 ; save selector
        mov cx, bx              ; HI word of linear base
        mov dx, cx              ; LO word of linear base
        pop bx                  ; restore selector
        mov ax, 07h             ; DPMI Set Selector Base
        int 31h

        mov es, bx
        xor edi, edi
        mov ecx, PhysSize
        shr ecx, 2
        xor eax, eax
        rep stosd es:[edi]

zero32  ENDP

        END
```

Devices Mapped Above 1Mb Require DPMI Services

Although most memory-mapped devices are located between A0000h and FFFFFh, it is possible to locate a device above FFFFFh (1Mb). RAM is always mapped contiguously above the 1Mb boundary, so a device located above FFFFFh must be located beyond the last byte of physical memory. Otherwise, a hardware conflict occurs when both RAM and the device attempt to decode the same physical address, and the system won't function properly.

Forming a pointer to a memory-mapped device involves setting up both steps in the two-step (logical-linear, linear-physical) address translation process described in Chapter 3. The first step is setting up the selector's base address and limit. You can use the Windows API selector functions (`AllocSelector`, `SetSelectorBase`, and `SetSelectorLimit`) for this. The second step is setting up the page table entries so that the selector's base address maps to the desired physical address. There are no Windows API functions that manipulate page tables, but DPMI does provide a `MapPhysicalAddress` function which will do the job.

DPMI `MapPhysicalAddress` takes a physical address as input and returns the linear address that maps (through the page tables) to the physical address. To see how this DPMI call can be used, it's helpful to think of the two-step address translation process in a different way. Suppose you want a pointer to physical address A0000h. Because of the effect of paging, allocating a selector and setting its base address to A0000h doesn't guarantee that the selector translates to a physical address of A0000h. But notice that it doesn't really matter what the linear address is, as long as it maps to physical A0000h.

So, build the mapping backwards, starting with physical address A0000h. Give the physical address to DPMI `MapPhysicalAddress`; it will return a linear address, call it X. Now give that linear address to `SetSelectorBase`. The result is a selector that maps to linear address X, which maps to physical address A0000h.

There is one detail I haven't covered. The page tables work with 4Kb pages, so that a selector with a limit of more than 4Kb is composed of multiple pages. Each page can reside anywhere in physical memory — pages do not have to be physically contiguous. Devices, however, understand only physically contiguous memory. Thus, a useful selector strategy needs to guarantee not only that the first page maps to A0000h–A0FFFh, but also that the next page maps to A1000–A1FFFh, etc. In fact, DPMI `MapPhysicalAddress` does guarantee that the mapped pages are physically contiguous, although that's not obvious from the DPMI documentation.

The following code gives a function that uses Windows selector functions and the DPMI `MapPhysicalAddress` service to get a pointer to a memory-mapped device located above 1Mb. The code does nothing more than the four steps described above:

- allocates a selector,
- sets its limit,
- uses DPMI to get a linear address corresponding to a given physical address,
- then sets the selector base to that linear address.

```
void far *MapPhysToPtr( DWORD PhysBase, DWORD PhysSize )
{
   WORD myDs,sel;
   WORD HiBase, LoBase;

   _asm mov myDs, ds
   sel = AllocSelector( myDs );
   SetSelectorLimit( sel, PhysSize );

   _asm
   {
      mov cx, PhysBase
      mov bx, PhysBase+2
      mov di, PhysSize
      mov si, PhysSize+2
      mov ax, 0800h                      // DPMI Map Phys
      int 31h
      mov HiBase, bx
      mov LoBase, cx
   }
   // Set selector's linear address as given by DPMI Map Phys
   SetSelectorBase( sel, MAKELONG( LoBase, HiBase) );

   return( MAKELP( sel, 0 ) );
}
```

The only trick to this code is in the call to `AllocSelector`. This call takes one parameter, a template selector. Because the function is creating a selector to access data (not code), the code passes the current value of `DS`, a valid data selector, as the template selector parameter.

The function in the previous code fragment has a limitation: it works properly only for a size of less than 64Kb. You can easily adapt it to regions greater than 64Kb by replacing the Windows selector functions, which don't properly support limits greater than 64Kb, with analogous DPMI selector functions, which do support greater than 64Kb. Then you would access your device with assembly language code and 32-bit offsets as illustrated earlier in this chapter.

Summary

Certainly this example isn't a commercial-quality driver, but many of its weaknesses are deliberate simplifications that have nothing to do with the Windows environment. For example, a commercial-quality driver should test for receiver overrun and various framing errors. These tests can be added easily, without any concern for Windows-specific implementation issues.

The major shortcomings, though, are a direct consequence of trying to perform full-duplex operations by polling. The polled-mode design forces the application to handle the data one byte at a time — or risk missing significant amounts of data in the reverse direction. For a simple one-way device, like a dumb printer, such a polled-mode driver could perform quite satisfactorily. Thus, for certain devices, Windows device drivers can be just this simple. For a bi-directional device like the serial port to provide reliable, two-way communication without byte-wise supervision from the application, however, requires an interrupt-driven driver.

Windows does impose additional implementation constraints on interrupt-driven drivers, especially those that use memory buffers for communication between the application and the driver. The next chapter explains these issues and shows how to convert this chapter's polled-mode example into an interrupt-driven driver.

Listing 14.1   POLLED.H

```
typedef struct
{
    WORD    usReadBufSize;
} DRIVERPARAMS, FAR * PDRIVERPARAMS;

typedef struct
{
    WORD    version;
} DRIVERCAPS, FAR * PDRIVERCAPS;
typedef PDRIVERCAPS FAR * PPDRIVERCAPS;

typedef struct
{
    WORD          usDevNumber;
    WORD          usIoBase;
    BOOL          bFlags;
    LPBYTE        lpReadBuf;
    DRIVERPARAMS  params;
} DEVICECONTEXT, FAR *HDEVICE;

HDEVICE FAR PASCAL DeviceOpen( HWND hwnd );
int FAR PASCAL DeviceClose( HDEVICE );
int FAR PASCAL DeviceGetWriteStatus( HDEVICE, LPWORD pusStatus );
int FAR PASCAL DeviceGetReadStatus( HDEVICE, LPWORD pusStatus );
int FAR PASCAL DeviceWrite( HDEVICE, LPBYTE lpData, LPWORD pcBytes );
int FAR PASCAL DeviceRead( HDEVICE, LPBYTE lpData, LPWORD pcBytes );
int FAR PASCAL DeviceSetDriverParams( HDEVICE, PDRIVERPARAMS pParms );
int FAR PASCAL DeviceGetDriverParams( HDEVICE, PDRIVERPARAMS pParms );
int FAR PASCAL DeviceGetDriverCapabilities( HDEVICE, PPDRIVERCAPS ppDriverCaps );
```

Listing 14.2   UART.H

```
#define UART_REG_THR            0x00
#define UART_REG_RDR            0x00
#define UART_REG_IER            0x01
#define UART_REG_IIR            0x02
#define UART_REG_LCR            0x03
#define UART_REG_MCR            0x04
#define UART_REG_LSR            0x05
#define UART_REG_BAUDLO         0x00
#define UART_REG_BAUDHI         0x01

#define UART_IIR_NONE           0x01
#define UART_IIR_THRE           0x02
#define UART_IIR_RXRDY          0x04
#define UART_IER_THRE           0x02
#define UART_IER_RXRDY          0x01
#define UART_MCR_OUT2           0x08
#define UART_MCR_LOOP           0x10
#define UART_LSR_THRE           0x20
#define UART_LCR_DLAB           0x80
#define UART_LCR_8N1            0x03
#define UART_LSR_RXRDY          0x01
#define BAUD_1200               0x60
#define BAUD_110                0x0417L
```

Listing 14.3   POLLED.C

```c
#define _WINDLL

#include <windows.h>
#include <conio.h>
#include "polled.h"
#include "uart.h"

#define FLAGS_OPEN          0x04

#define SET( value, mask )    value |= mask
#define CLR( value, mask )    value &= (~mask)

DEVICECONTEXT Device1 = { 0, 0x3F8, 0, NULL };
DRIVERPARAMS DefaultParams = { 1024 };
DRIVERCAPS DriverCaps = { 0x0101 };

BOOL ValidHandle( HDEVICE hDevice );

HDEVICE FAR PASCAL _export DeviceOpen( HWND hwnd )
{
    HDEVICE     hDevice;

    OutputDebugString( "DeviceOpen\n" );

    hDevice = &Device1;

    if (hDevice->bFlags & FLAGS_OPEN)
        return (HDEVICE)-1;

    hDevice->params = DefaultParams;

    // Configure UART.
    outp( hDevice->usIoBase + UART_REG_IER, 0 );
    outp( hDevice->usIoBase + UART_REG_LCR, UART_LCR_DLAB );
    outp( hDevice->usIoBase + UART_REG_BAUDLO, BAUD_1200 );
    outp( hDevice->usIoBase + UART_REG_BAUDHI, 0 );
    outp( hDevice->usIoBase + UART_REG_LCR, UART_LCR_8N1 );
    outp( hDevice->usIoBase + UART_REG_MCR, UART_MCR_LOOP );
     inp( hDevice->usIoBase + UART_REG_LSR );
     inp( hDevice->usIoBase + UART_REG_RDR );

    SET( hDevice->bFlags, FLAGS_OPEN);

    return hDevice;
}
```

Listing 14.3 (continued)   POLLED.C

```
int FAR PASCAL _export DeviceClose( HDEVICE hDevice )
{
    OutputDebugString( "DeviceClose\n");

    if (!ValidHandle( hDevice ))
        return -1;

    if ((hDevice->bFlags & FLAGS_OPEN) == 0)
        return -1;

    CLR( hDevice->bFlags, FLAGS_OPEN );

    return 0;
}

int FAR PASCAL _export DeviceGetWriteStatus( HDEVICE hDevice, LPWORD pusStatus )
{
    OutputDebugString( "DeviceGetWriteStatus\n");

    if (!ValidHandle( hDevice ))
        return -1;

    if ((hDevice->bFlags & FLAGS_OPEN) == 0)
        return -1;

    if (inp( hDevice->usIoBase + UART_RFG_LSR ) & UART_LSR_THRE)
    {
        *pusStatus - 1;          // ready to transmit
    }
    else
    {
        *pusStatus = 0;          // not ready to transmit
    }

    return 0;
}

int FAR PASCAL _export DeviceGetReadStatus( HDEVICE hDevice, LPWORD pusStatus )
{
    OutputDebugString( "DeviceGetReadStatus\n");

    if (!ValidHandle( hDevice ))
        return -1;

    if ((hDevice->bFlags & FLAGS_OPEN) == 0)
        return -1;

    if (inp( hDevice->usIoBase + UART_REG_LSR ) & UART_LSR_RXRDY)
    {
        *pusStatus = 1;          // data ready
    }
    else
    {
        *pusStatus = 0;          // no data ready
    }

    return 0;
}
```

Listing 14.3 (continued) POLLED.C

```c
int FAR PASCAL _export DeviceWrite( HDEVICE hDevice, LPBYTE lpData, LPWORD pcBytes )
{
    WORD    i;

    OutputDebugString( "DeviceWrite\n" );

    if (!lpData)
        return -1;

    if (!ValidHandle( hDevice ))
        return -1;

    if ((hDevice->bFlags & FLAGS_OPEN) == 0)
        return -1;

    for (i=0; i < *pcBytes; i++)
    {
        while ((inp( hDevice->usIoBase + UART_REG_LSR ) & UART_LSR_THRE) == 0)
            ;
        outp( hDevice->usIoBase + UART_REG_THR, lpData[ i ] );
    }

    return 0;
}

int FAR PASCAL _export DeviceRead( HDEVICE hDevice, LPBYTE lpData, LPWORD pcBytes )
{
    WORD        i;

    OutputDebugString( "DeviceRead\n" );

    if (!lpData)
        return -1;

    if (!ValidHandle( hDevice ))
        return -1;

    if ((hDevice->bFlags & FLAGS_OPEN) == 0)
        return -1;

    for (i=0; i < *pcBytes; i++)
    {
        while ((inp( hDevice->usIoBase + UART_REG_LSR ) & UART_LSR_RXRDY) == 0)
            ;
        lpData[i] = inp( hDevice->usIoBase + UART_REG_RDR );
    }

    return 0;
}
```

Listing 14.3 (continued)   `POLLED.C`

```c
int FAR PASCAL _export DeviceSetDriverParams( HDEVICE hDevice, PDRIVERPARAMS pParams )
{
    OutputDebugString( "DeviceSetDriverParams\n");

    if (!pParams)
        return -1;

    if (!ValidHandle( hDevice ))
        return -1;

    if ((hDevice->bFlags & FLAGS_OPEN) == 0)
        return -1;

    hDevice->params = *pParams;

    return 0;
}

int FAR PASCAL _export DeviceGetDriverParams( HDEVICE hDevice, PDRIVERPARAMS pParams )
{
    OutputDebugString( "DeviceGetDriverParams\n");

    if (!pParams)
        return -1;

    if (!ValidHandle( hDevice ))
        return -1;

    if ((hDevice->bFlags & FLAGS_OPEN) == 0)
        return -1;

    *pParams = hDevice->params;

    return 0;
}

int FAR PASCAL _export DeviceGetDriverCapabilities( HDEVICE hDevice,
                                                    PPDRIVERCAPS ppDriverCaps )
{
    OutputDebugString( "DeviceGetDriverCapabilities\n");

    if (!ppDriverCaps)
        return -1;

    if (!ValidHandle( hDevice ))
        return -1;

    if ((hDevice->bFlags & FLAGS_OPEN) == 0)
        return -1;

    *ppDriverCaps = &DriverCaps;

    return 0;
}

BOOL ValidHandle( HDEVICE hDevice )
{
    return (hDevice == &Device1);
}
```

Listing 14.4   `POLLED.MAK`

```
all: polled.dll

# DRIVER DLL

polled.obj: polled.c polled.h
  cl -c -W3 -ASw -Gsw2 -Oi $*.c

polled.dll: polled.def polled.obj
  link polled,polled.dll,polled.map /CO /MAP, sdllcew libw /nod/noe,polled.def
  implib driver.lib polled.dll
  copy polled.dll \windows\driver.dll
```

Listing 14.5   `POLLED.DEF`

```
LIBRARY     DRIVER
DESCRIPTION "Polled Mode Driver"
EXETYPE     WINDOWS
DATA        PRELOAD MOVEABLE SINGLE
CODE        PRELOAD MOVEABLE DISCARDABLE
```

## Chapter 15

Driver DLL:
Interrupt Handling

This chapter will show how to build a 16-bit, interrupt-driven driver DLL. While a polled-mode driver DLL (like that of the last chapter) is certainly simple to build, a basic interrupt-driven version is only slightly more complex and offers significant advantages. Interrupt-driven drivers can usually offer improved throughput. Interrupt-driven drivers are also more "Windows polite" than polled-mode drivers, because the interrupt-driven driver doesn't tie up the processor while waiting for the device.

The basic structure of a Windows interrupt-driven driver is quite similar to the structure of a DOS driver: a real-time component (the Interrupt Service Routine, or ISR) services hardware events, and a higher-level component (which I'll just refer to as the driver) handles communication with the application or operating system. The driver and ISR typically communicate through a buffer that must be managed as a shared resource.

An interrupt-driven driver DLL is by definition a 16-bit DLL, because Win32 DLLs cannot install interrupt handlers. There is no Win32 API to install an interrupt handler — because that job should be done in a true driver — and the DOS `Set Vector` call used by Win16 DLLs is not available to Win32 DLLs.

Although a Windows driver has a familiar structure, it is complicated by Windows' tendency to virtualize services to protect the underlying hardware resources. In polled-mode drivers, these virtual services are nearly invisible to the programmer, but in an interrupt-driven environment some of these virtualizing mechanisms, in particular the virtual memory system, become more visible. To avoid breaking your application — or even breaking Windows — you need to understand something about how some of these virtual services work and about the conventions you must follow to write compatible interrupt-driven code. For the purposes of this chapter, that means:

- understanding how the Windows memory manager works so that you can create interrupt-safe code and data structures,
- knowing the conventions you must observe when accessing an interrupt-safe buffer or data structure,
- knowing the conventions you must observe when installing your interrupt handler.

The first half of this chapter is devoted to explaining these three topics. The second half of the chapter shows how to use this information to convert the polled-mode driver of the last chapter into a basic interrupt-driven driver.

Windows Memory Management Strategy Overview

When the Windows memory manager gets an allocation request for a larger block than is available, it takes one of the following three actions to free up memory to meet this new demand:

- Discard the current contents of an already-allocated block. Here, discard means reuse the same block without first saving its contents to disk — presumably because a valid copy is known to already exist on disk.
- Rearrange (move) the current contents of memory to create a larger block of contiguous memory.
- Swap the current contents of a block to disk.

Each of these three actions has potentially disastrous implications for an interrupt-driven device driver. In the next few sections, I'll explain:

- why swapped, discarded, or moved blocks create problems for an interrupt-driven handler, and
- how to write code and allocate data that won't be swapped, discarded, or moved.

(See also the sidebar "Layered Memory Managers" on page 327.)

What Is Discardable?

Win16 applications are organized into logical components called segments. There are three types of segments: code, data, and resource. Code segments contain a program's code, data segments contain a program's data (including stack and local heap), and resource segments contain user interface resources like menus, icons, bitmaps, etc.

All segments, whether allocated statically as part of an executable or dynamically by a running application, are allocated by calls to the Win16 memory manager API. When a program is first loaded into memory, the Windows loader allocates on behalf of the application, making calls to allocate segments to be used for the application's code, data, and resources. When an executing program needs additional memory, it calls the memory manager API directly.

Each segment, whether it be code, data, or resource, possesses a set of attributes that are tracked by the memory manager. These attributes determine what the memory manager may or may not do with that segment. The memory manager can only discard a segment if it is marked as discardable. Other attributes include: non-discardable, fixed, moveable, swappable, and non-swappable (also called pagelocked).

Attributes are specified in one of two ways: statically as part of the linking process or dynamically as part of the allocation request to the memory manager. The attributes of a program's static code, data, and resource segments are specified at link time, in the module definition (.DEF) file. The loader then allocates segments with these attributes on behalf of the application. The attributes of a segment dynamically allocated by a program, via a direct call to the memory manager, are specified as parameters to the function call. The program may later change a segment's attributes by another call to the memory manager.

Layered Memory Managers

In 16-bit Windows, the memory manager functionality is really provided by two different system components. One is KERNEL, which is a user-mode DLL. The other is the VMM, a Ring 0 VxD. Win16 applications use the memory management functions provided by KERNEL, like GlobalAlloc. KERNEL itself uses the services of the VMM (Virtual Machine Manager). So you can think of KERNEL as being layered on top of VMM.

The KERNEL memory manager deals with segments, which are mapped via the descriptor tables to linear address space. To satisfy allocation requests for segments, KERNEL allocates linear address blocks from the VMM. KERNEL performs two kinds of memory "management", discarding and moving, which we'll discuss in a later section. Both apply strictly to segments, not to the pages that actually make up segments. KERNEL has nothing to do with the third kind of "management", which is virtual memory, also known as paging.

Virtual memory is implemented by the VMM (Virtual Machine Manager, not Virtual Memory Manager). The VMM memory manager is responsible for managing physical memory: managing the paging tables, which map linear addresses to physical addresses, and swapping pages to and from disk. VMM only deals with pages, never segments.

The KERNEL module described here still exists in Windows 95 — it must, because Windows 95 supports Win16 applications — but is renamed to KERNEL16. Win32 applications use a different memory management API (e.g. VirtualAlloc instead of GlobalAlloc), which is provided by the KERNEL32 module. But KERNEL32 is nothing more than a thin wrapper around calls to the VMM, which is still the virtual memory manager in Windows 95 as well as in Windows 3.x.

When the Windows memory manager discards a segment, the segment is not written to disk but is literally discarded. When a program accesses a segment which has been discarded, the processor generates a Segment Not Present fault, and the fault handler will reread the segment from disk into memory. This behavior implies that every discarded segment must be read-only (never modified) and always available on disk. Code and resource segments are usually allocated as discardable. On the other hand, data segments should not be discardable, because they can't be recreated by reading the original segment from disk.

What is Moveable?

When Windows loads a Win16 program's resources, it loads them segment by segment, placing each individual segment into a contiguous block of linear memory. Thus, to load a segment, it isn't enough for Windows to have enough *free* memory, it must have enough *contiguous* free memory. (To be precise, it must have enough free linearly contiguous memory; the difference between linear and physical memory was discussed in Chapter 3). If the free memory is highly fragmented, then the memory manager may need to compact the active blocks to create larger free blocks (see the sidebar "Fragmentation").

Fragmentation

When a program requests an allocation from the memory manager, that request can be refused, even if free memory is available, if that free memory is scattered in several small pieces instead of a single larger one. This problem is known as fragmentation. It exists even under DOS, but multitasking makes the problem much worse. Much of the complexity of the Windows memory manager and the memory management API exists to combat this problem, so fragmentation merits a closer look.

Figure 15.1 illustrates the process of fragmentation. Initially, all available memory resides in a single large block, called the heap. The first application runs and allocates memory that carves off a block from this heap. Next, the same application allocates a second block. Now another application starts up and requests a block. Then the first application deallocates the first block it had allocated. Note that this leaves a hole in the heap, so that the heap is now composed of two blocks of free memory.

This condition of having holes in the heap is called fragmentation, and here's why it's a problem. Suppose that the second application now requests another allocation, but this time the size of the request is larger than either of the two blocks of free memory. The memory manager cannot satisfy the request, even though the size of the request is actually less than the total available free memory.

To handle this problem, the Windows memory manager moves blocks (copies the block's contents from one location to another) to coalesce scattered free blocks into a single large free block. Figure 15.2 (see page 330) is a picture of the fragmented heap in Figure 15.1, before moving blocks and after. Before the move, the largest available block was 192Kb. Afterward, it is 320Kb.

The memory manager can combat fragmentation effectively if all allocated memory is moveable. However, there are situations where programmers need to fix a block in place, preventing the memory manager from moving the block. Too many such fixed blocks create sandbars in the heap, as illustrated in Figure 15.3 (see page 330), and lead to excessive fragmentation. Thus, fixed blocks should be used only when absolutely necessary.

Windows relies on the processor's protected mode to efficiently implement move-able memory. In protected mode, a pointer is a logical address consisting of a selector and an offset. Because a selector doesn't directly specify a physical address — it directly specifies an index into a descriptor table — implementing moveable memory in protected mode is easy. To move a segment, Windows copies the segment's contents from one linear location to another, then updates the segment's base address in

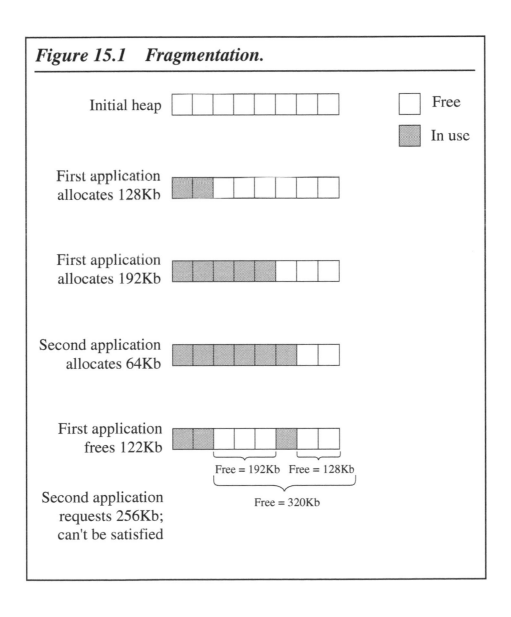

Figure 15.1   Fragmentation.

the descriptor table. The segment value itself doesn't change, so the index still points to the same entry in the descriptor table. Only the base address stored in the descriptor changes. This process is illustrated in Figure 15.4

This means that Windows can move segments around without the application's knowledge, because Windows can return a selector at the time of allocation. If the system later moves the allocated block, the application would be unaffected because the selector returned at allocation time still points (indirectly) to the block. As we'll see later, driver DLLs that handle interrupts *are* often affected by moveable segments.

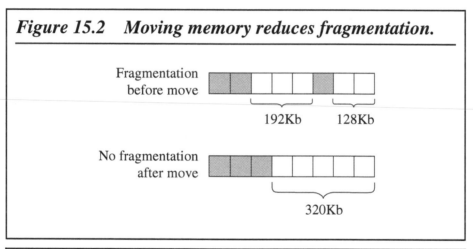

Figure 15.2   Moving memory reduces fragmentation.

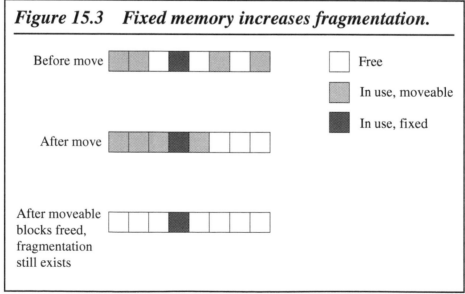

Figure 15.3   Fixed memory increases fragmentation.

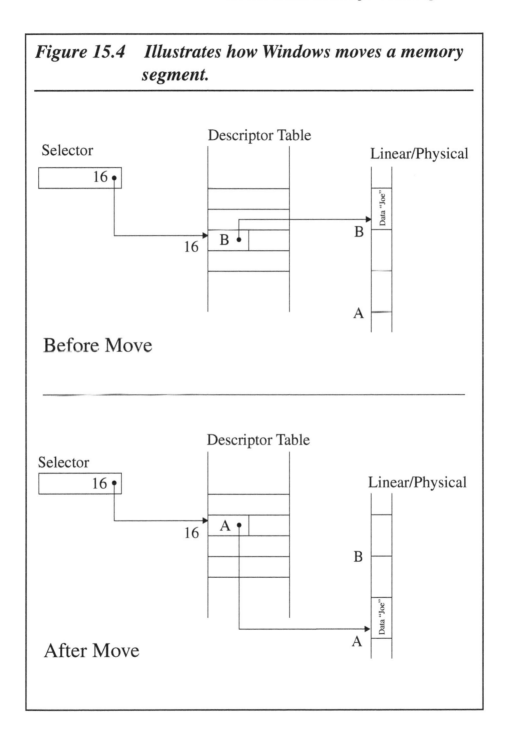

Figure 15.4   Illustrates how Windows moves a memory segment.

What is Swappable?

Moving and discarding memory blocks are both useful techniques for the Windows memory manager, but don't help when memory is really tight — when there are no free blocks remaining and no segments to be discarded. The remaining tool of the memory manager, swapping or paging, is the most powerful of all, allowing Windows to implement a technique known as virtual memory. Virtual memory is a neat trick which allows the memory manager to provide more memory than is physically available in the system.

Virtual memory requires some help from the processor hardware, specifically the paging feature. When memory gets low, the memory manager writes (swaps) a page to disk and marks the page as not present in the page tables. If a process later tries to access a location in that page, the not-present flag will cause the processor to generate a page fault. This fault (a processor exception) will suspend the current process and transfer control and the number of the missing page to the page fault handler. This handler, which is part of the memory manager, uses the page number to locate the page on disk and then reads it into memory. This entire paging process is handled dynamically by the operating system in a way that is completely transparent to the process that made the memory access.

If a page fault occurs when there are no free blocks in physical memory, then the memory manager must create free space by swapping a currently present page to disk.

Memory Requirements for an Interrupt-safe Driver

A driver that handles hardware interrupts has strict requirements on the type of memory it allocates. All code and data used at interrupt time must be fixed (non-moveable), pagelocked, and non-discardable. This includes the code for the interrupt handler itself, any data in the driver's data segment used at interrupt time, any dynamically allocated buffers used at interrupt time, and application-allocated buffers passed to the driver and used at interrupt time.

In the following few paragraphs, I'll explain why an interrupt handler has each of these three requirements. These requirements aren't unique to interrupt handlers running under Windows; they also are shared by handlers running under other 80x86 environments, such as UNIX or OS/2. So in the next few paragraphs when I use the terms fixed, pagelocked, and non-discardable, these are all generic memory attributes offered by many operating systems.

After this generic discussion, I'll talk about how Windows implements these same three attributes. When I'm referring specifically to these attributes as implemented by Windows, I'll use a slightly different nomenclature: FIXED, PAGELOCKED, and NONDISCARDABLE. This distinction is useful because, for example, FIXED doesn't always mean fixed, and PAGELOCKED always means both pagelocked and fixed.

Reason for Fixed

To understand why any data used by a hardware interrupt handler must be in a fixed segment, consider the following scenario. Suppose the memory manager moves the data segment used by an interrupt handler because the segment is marked moveable. After a few bytes are copied, a hardware interrupt occurs and the interrupt handler executes. The handler updates a variable that resides in the first byte of its data segment (the one that's being moved). Now when the handler finishes executing, the memory manager continues with the rest of the copy but the variable just updated by the handler is now incorrect. The memory manager copied the old value to the new segment and has no way of knowing that the handler later modified the value. Clearly the data segment used by a handler must be fixed to prevent this problem.

Reason for Non-discardable

The code segment containing a hardware interrupt handler has a different restriction: it must be non-discardable. (Note that the data segment must be non-discardable as well, but this is true of all data segments, while code segments are usually discardable.) Suppose that a hardware interrupt occurred, but the memory manager had discarded the handler's code segment. A Segment Not Present fault would occur, and the Windows fault handler would attempt to reload the segment from disk. If the interrupt occurred while the system was already in DOS (for some other reason), the result would be an attempt to re-enter DOS. DOS, however, is not reentrant code. Thus, the code segment must be non-discardable.

Reason for Pagelocked

Both code and data of a hardware interrupt handler must be pagelocked. The reasons are similar to those that force the segment to also be non-discardable. An interrupt handler accessing a swappable buffer would result in a page fault if that buffer had been swapped to disk. The interrupt handler code itself could have been swapped to disk, resulting in a page fault during execution. In either case, the page fault could cause DOS to be re-entered.

Static Interrupt-safe Code and Data: The Easy Way

The easiest way to insure that driver code and data segments are interrupt-safe is to mark code and data segments as FIXED and NONDISCARDABLE in the driver's module definition file. Note that this technique relies on two well-known Windows behaviors. First, Windows ignores the FIXED attribute when used by applications but respects it when used by DLLs — thus the driver must be a DLL. Second, when segments are marked as FIXED in the module definition file, Windows pagelocks the memory in addition to fixing it in linear memory.

When running under Windows 95, using the FIXED keyword in the module definition file is an easy way for a developer to make his static code and data interrupt-safe. The very same technique, when used under Windows 3.x, is easy for the developer, but has terrible side effects for the user. This allocation method can easily result in a situation where Windows is unable to start new Windows applications, and the user gets an "Insufficient memory to start the application" error message. This can happen even when there is plenty of free memory and free system resources. How can this be?

Use the Right Way under Windows 3.x

Each time a Windows application runs, the Windows loader allocates a 512-byte block for a data structure called the Program Segment Prefix (PSP). The PSP is used by DOS as well as Windows, so it must be located below 1Mb. If there is no memory available for the PSP, Windows can't run the application. This behavior is true under Windows 3.x, and is still true under Windows 95.

The problem with Windows 3.x is the strategy used by the 3.x memory manager: FIXED blocks are allocated from as low in the heap as possible, DISCARDABLE blocks come from high in the heap, and MOVEABLE from in between. This strategy helps to reduce fragmentation but often results in FIXED allocations using up precious low DOS memory, even when the users of FIXED memory don't need the memory to come from below 1Mb.

The Windows 95 memory manager uses a slightly different heap strategy, so that FIXED allocations do *not* use low DOS memory. Therefore, if your 16-bit driver DLL will run under Windows 95 only, you're safe to take the easy way out and use FIXED in your DEF file.

Here's the right way, which avoids using up precious low memory. First mark your driver's code and data segments as MOVEABLE — not FIXED — in the module definition file. At run time, before any interrupts occur, you explicitly fix and pagelock the segments. But you must be careful with this second step. The Windows API function GlobalPageLock will both fix and pagelock a segment but GlobalPageLock will also move the segment down to low memory (because FIXED blocks should be low in the heap), exactly what we want to avoid.

What's needed is a way to prevent the memory manager from moving the segment before pagelocking it. This can be done by first allocating all the memory below 1Mb, calling GlobalPageLock, and then freeing all the low memory. I've provided a function — called SafePageLock — which does just this, and I'll examine it in more detail later. First I'll examine a related issue: how to dynamically allocate interrupt-safe buffers. As with the driver's static segments, there is an easy way and a right way to do this under Windows 3.x.

Dynamically Allocating Interrupt-safe Buffers: The Easy Way

The easy way to dynamically allocate an interrupt-safe buffer is to call GlobalAlloc and specify that you want a buffer that is both fixed and pagelocked. GlobalAlloc takes two parameters, a bit-mapped value, representing the attributes of the segment to be allocated, and the size of the segment. Allowable values for the flags parameter include: GMEM_FIXED, GMEM_MOVEABLE, GMEM_DISCARDABLE, GMEM_NODISCARD, and GMEM_SHARE.

The GMEM_SHARE flag was introduced in Chapter 13. Although there is no flag to specify an attribute of pagelocked, when used by a DLL the GMEM_FIXED flag always has the side effect of pagelocking memory. So an allocation for an interrupt-safe buffer would use the flags GMEM_FIXED, GMEM_NODISCARD, and GMEM_SHARE.

Although the size parameter to GlobalAlloc is a 32-bit value, the largest allocation permitted is 16Mb–64Kb, much less than 2^{32}.

A return value of NULL from GlobalAlloc means the segment could not be allocated, usually because a free block of that size wasn't available. A non-NULL return value is the handle of the memory object. (More about handles and how to turn them into useable pointers in the next section.)

Dynamically Allocating Interrupt-safe Buffers: The Right Way

This easy method results in exactly the same problem discussed above with Windows 3.x and fixed driver code and data segments: the buffer is fixed, pagelocked, and non-discardable — but is also usually located below 1Mb. The right way is to first allocate from GlobalAlloc using GMEM_MOVEABLE instead of GMEM_FIXED and to fix and pagelock the buffer later with the SafePageLock function. This function is not part of the Windows API, but a function I will present in a later section. The safe function is necessary to prevent the memory manager from moving the buffer to low memory during the pagelock operation, as it would with a simple call to GlobalPageLock.

Before examining the code for SafePageLock, I will need to cover one more topic relevant to 16-bit drivers under both Windows 3.x and 95: the relationship between handles, selectors, and pointers. GlobalAlloc returns a handle. Functions such as GlobalPageLock and GlobalFix expect a selector, and accessing a dynamically allocated buffer requires a pointer.

Using the Buffer: Handles, Selectors, and Pointers

The handle returned by GlobalAlloc is not a pointer, it's just a value with special meaning to the memory manager. To access the associated memory object, even when it is fixed and pagelocked, you must convert this handle to a pointer. This is done by calling GlobalLock, using as a parameter the handle returned by GlobalAlloc. The block is freed with a call to GlobalFree, passing in the same handle returned by the original GlobalAlloc.

As explained earlier, a protected mode pointer consists of a selector and an offset. Some Windows API functions, such as GlobalPageLock, take a selector parameter, not a handle. To obtain a selector from a far pointer, use the SELECTOROF macro provided in WINDOWS.H. Better yet, the GlobalAllocPtr and GlobalAllocFree macros in WINDOWSX.H combine the allocation and lock (handle dereference). The GlobalAllocPtr macro combines a call to GlobalAlloc with a subsequent call to GlobalLock, returning a pointer. The GlobalFreePtr macro combines a call to GlobalHandle (which converts a selector to a handle), GlobalUnlock, and GlobalFree.

Note that you should not use the C library `malloc` function instead of `GlobalAlloc` to allocate an interrupt-safe buffer. The problem is not the attributes flag. As explained earlier in Chapter 13, `malloc` allocates moveable memory, using exactly the same flags you would pass to `GlobalAlloc` if you were going to fix and pagelock the memory later. The problem is that `malloc` doesn't usually allocate a segment via `GlobalAlloc`. Instead, `malloc` acts as a sub-segment allocator, usually returning an offset into an already-allocated segment. So when you call `GlobalPageLock`, it will fix and pagelock the entire segment, not just your portion of it. And the golden rule of Windows memory management is to never fix, and never ever pagelock, memory if it's not absolutely necessary.

A Safe Pagelock Function

There is nothing tricky about `SafePageLock`, shown in the following paragraph of code. It takes a single `WORD` parameter, which is the selector of the buffer you want to pagelock, and performs the three steps outlined earlier:

- Repeatedly calls `GlobalDosAlloc` until all memory below 1Mb has been allocated.
- Calls `GlobalPageLock` to fix and pagelock the caller's buffer. This call is now safe because the allocated blocks, which completely fill up the area below 1Mb, will prevent the heap manager from moving our buffer below 1Mb.
- Repeatedly calls `GlobalFree` to free all memory blocks below 1Mb that were allocated earlier.

```
UINT SafePageLock( HGLOBAL sel )
{
    WORD i, rc;
    static WORD SelArray[ 1024 ];

    memset( SelArray, 1024 * sizeof(WORD), 0 );
    for (i=0; i < 1024; i++)
    {
        SelArray[i] = LOWORD( GlobalDosAlloc( 1024 ) );
        if (!SelArray[i])
            break;
    }
```

```
    rc = GlobalPageLock( sel );

    for (i=0; i < 1024; i++)
    {
       if (!SelArray[i] )
          break;
       GlobalFree( SelArray[i] );
    }

    return rc;
}
```

The following code fragment uses SafePageLock to fix and pagelock a driver's code and data segments:

```
_asm mov myds, ds
_asm mov mycs, cs
SafePageLock( myds );
SafePageLock( mycs );
```

And the next code fragment dynamically allocates an interrupt-safe buffer the right way, by combining a call to GlobalAlloc to get moveable memory with a subsequent call to SafePageLock:

```
HGLOBAL hnd;
UINT sel, bufsize, flags;
bufsize = 8192;
char far *pBuffer;

flags = GMEM_MOVEABLE | GMEM_NODISCARD | GMEM_SHARE;
hnd = GlobalAlloc( flags, bufsize );
pBuffer = GlobalLock( hnd );
sel = SELECTOROF( pBuffer );
SafePageLock( sel );
```

Note that no SafePageUnlock function is necessary, because the Windows API function GlobalPageUnlock has no undesirable side effects.

Installing an Interrupt Handler

The proper way to install an interrupt handler from a Windows DLL driver is through the DOS `Set Vector` call (`INT 21h AH=25h`). A DOS driver written in a high-level language like C can use a library function like _dos_setvect to make this DOS call. However, the Windows-specific versions of the VC++ 1.x run-time library don't contain _dos_setvect because the library implementation of the function isn't compatible with Windows.

That leaves two alternatives: make the DOS call through the library routine `intdosx`, which is available to Windows programs, or write your own version of _dos_setvect. I've chosen the latter approach because it is trivial to code and is more efficient than using `intdosx`. (See the sidebar "Initialize Those Registers!")

If your Windows-specific C library supports a high-level call like _dos_setvect, feel free to use it. But if it doesn't, call `INT 21h` with AH=25h from assembly. Almost all compilers that generate Windows applications also support embedded assembly, which makes this trivial. Here's a C function that installs a handler using embedded assembly.

Initialize Those Registers!

If you do choose to use `intdosx`, you must carefully initialize both `SREGS.es` and `SREGS.ds`. The easiest way to do this is through the `segread` function. This step is necessary because during the `intdosx` call the DS and ES registers are loaded from the `SREGS` structure, and an invalid value in a segment register will cause a processor exception.

The following code calls DOS `Set Vector` through `intdosx`.

```
typedef void (FAR interrupt *VOIDINTPROC);
void DosSetIntVector( BYTE vector, VOIDINTPROC pHandler )
{
  struct SREGS SegRegs;
  union REGS InRegs, OutRegs;
  segread( &SegRegs );
  SegRegs.ds = SELECTOROF( pHandler );  InRegs.x.bx = OFFSETOF( pHandler );
  InRegs.h.ah = 0x25;
  intdosx( &InRegs, &OutRegs, &SegRegs );
}
```

```
void InstallHandler( void far *myHandler, int      intNumber  )
{
    asm
    {
    mov     ah, 25h
    mov     bl, intNumber
    push    ds                  ; don't lose this!
    lds     si,  myHandler
    int     21h
    pop     ds                  ; put DS back!
    }
}
```

Although you may need assembly code to install the handler, the handler itself can be written entirely in C using the interrupt keyword. This keyword instructs the compiler to generate special prolog and epilog code. The prolog pushes all registers onto the stack and loads DS with the data segment. The epilog pops all registers from the stack and returns with an IRET instruction. These entry and exit sequences are necessary for the handler to work properly.

As under DOS, the interrupt handler should not call any C library functions nor any DOS or BIOS services. In addition, the only Windows functions that can be called safely are listed in Table 15.1.

The New Driver: An Overview

To demonstrate these techniques, I've modified the example driver so that the UART's receive and transmit buffers are serviced by an interrupt handler. The driver and ISR communicate through circular buffers. The data area of each buffer is dynamically allocated by

Table 15.1	Windows functions that can safely be called from an interrupt handler.
Function Type	**Function Name**
Messaging Functions	OutputDebugStr
	PostMessage
	PostAppMessage
Multimedia Functions	timeGetSystemTime
	timeSetEvent
	timeKillEvent
	midiOutShortMsg
	midiOutLongMsg

the driver (just to show the technique). The main driver stores a pointer to each buffer in the DEVICECONTEXT structure so that the ISR will know how to find and manipulate the buffer.

The largest changes are in the DeviceOpen and DeviceClose routines. I use these routines as hooks to install and remove the ISR. The DeviceOpen routine allocates the buffers, sets up the DEVICECONTEXT structure, and installs the ISR. The DeviceClose routine reverses these steps, un-installing the ISR and freeing the buffers.

The New Driver: The Code

To convert the polled driver of the last chapter to an interrupt, one must:

- add information about the ISR and the interrupt to the DEVICECONTEXT structure,
- add code in DeviceOpen to allocate interrupt-safe buffers and install the ISR,
- change the code in DeviceRead so that it retrieves its data from the receive buffer instead of directly from the device,
- change the code in DeviceWrite so that it copies its data in the transmit buffer instead of writing directly to the device,
- create an ISR to service the interrupt, and
- add code in DeviceClose to deallocate the buffers and deactivate the ISR.

The New DEVICECONTEXT

The following code shows the C declaration for the new DEVICECONTEXT structure. As in earlier examples the driver will define a separate static instance of this structure for each supported device. This definition adds fields for the interrupt to be serviced (Irq), substructures that describe the two ring buffers (RxBuf and TxBuf), and storage for the old interrupt vector so that the driver can properly restore the system state when it removes the ISR.

```
typedef struct
{
   char far *Start;
   WORD     Size;
   WORD     In;
   WORD     Out;
} BUFINFO;
```

```
typedef struct
{
        WORD            usDevNumber;
        WORD            usIoBase;
        BYTE            bIrq;
        BYTE            bFlags;
        HWND            hwnd;
        VOIDINTPROC     pfOldHandler;
        BUFINFO         RxBuf;
        BUFINFO         TxBuf;
        DRIVERPARAMS    params;
} DEVICECONTEXT, FAR *HDEVICE;
```

The buffer structures include storage for a pointer to the buffer, the buffer's size, a next-in index (In), and a next-out index (Out). Figure 15.5 illustrates how these data structures will be used while the driver is running.

Allocating an Interrupt-safe Buffer

The DeviceOpen function allocates the buffers to be used by the interrupt handler. The size for each buffer is taken from the DRIVERPARAMS structure. The buffers are allocated with the GMEM_FIXED flag so that the region is safe for use at interrupt time, a critically important step, although the code is relatively simple:

```
hDevice->RxBuf.Size = hDevice->params.usReadBufSize;
hDevice->RxBuf.Start = GlobalAllocPtr( GMEM_SHARE |
                                       GMEM_MOVEABLE |
                                       GMEM_NODISCARD,
                                       hDevice->RxBuf.Size  );
if (!hDevice->RxBuf.Start)
{
   OutputDebugString( "ERROR GlobalAlloc Rx\n");
   return (HDEVICE)-1;
}
SafePageLock( (HGLOBAL)SELECTOROF( hDevice->RxBuf.Start ) );
```

Notice that I've used the GlobalAllocPtr macro in place of an explicit call to GlobalLock. I've also allocated the buffer as GMEM_MOVEABLE and used SafePageLock to fix and pagelock it, without moving it below 1Mb.

Finally, DeviceOpen initializes the ring pointers:

```
hDevice->RxBuf.In = hDevice->RxBuf.Out = 0;
```

Installing the ISR

Once the buffer is built, DeviceOpen turns to the task of installing the ISR. To handle the general case, DeviceOpen must disable interrupts, save the existing vector, install the new vector, and then enable interrupts.

The first step is to disable interrupts from the device by masking the device's interrupt level (sometimes called IRQ for Interrupt Request Level) in the Programmable Interrupt Controller (PIC). The PC uses two PICs, termed master and slave, which are daisy-chained together. The mask register of the master PIC is located at I/O port 21h and controls IRQs 0–7. Hardware IRQs 8–15 are controlled by the slave PIC at I/O

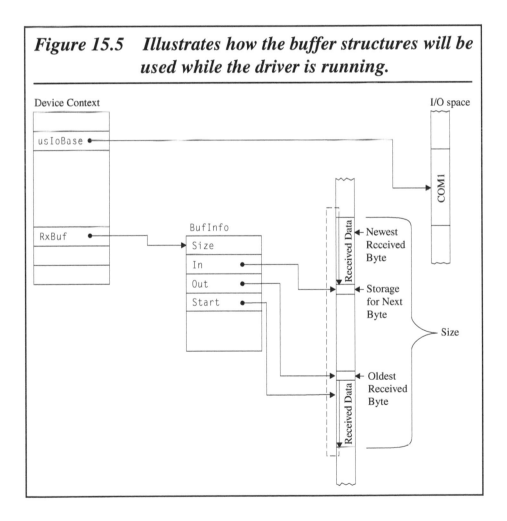

Figure 15.5   *Illustrates how the buffer structures will be used while the driver is running.*

port A1h. The mask registers are both bitmapped, where a 1 in a bit position disables the interrupt line, and a 0 enables the interrupt line. For the master, IRQs 0 to 7 correspond directly to bits 0 to 7. For the slave, IRQ 8 corresponds to bit 0, IRQ 9 corresponds to bit 1, etc. (See the sidebar "IRQ 2 versus IRQ 9".)

Now that the device can't generate an interrupt, it's safe to install the interrupt handler. Using the DOS `Set Vector` and `Get Vector` services (through my home-brew functions), `DeviceOpen` saves a copy of the current vector and then installs a vector that points to the new ISR:

```
hDevice->pfOldHandler = DosGetIntVector( bVector );
DosSetIntVector( bVector, DeviceIsr );
```

With the ISR properly installed, it's safe to enable interrupts. That means programming the device to generate interrupts and also unmasking the interrupt in the interrupt controller.

IRQ 2 Versus IRQ 9

Developers of Windows device drivers often handle IRQ 9 incorrectly. Here's the problem: The original IBM PC bus supports only IRQs 0 through 7, using a single interrupt controller. The two-controller design used by the AT bus successor to the PC is used by all of today's systems. The AT design doesn't support IRQ 2 because the designers used the IRQ 2 input of the master controller to connect it to the slave controller. IBM wanted old PC cards that used IRQ 2 to work in an AT, so the IRQ 2 bus signal was re-routed to the IRQ 9 input on the slave controller. The AT BIOS was also updated so that the default IRQ 9 interrupt handler did nothing but call the IRQ 2 handler.

With this backward-compatible design, the same hardware device and software using IRQ 2 in an older PC bus system automatically uses IRQ 9 on today's AT bus system. On the PC bus, the device asserts IRQ 2 on the bus, the signal goes to IRQ 2 on the interrupt controller, and the processor vectors to the IRQ 2 handler. On an AT bus, that same device asserts IRQ 2 on the bus, but that signal goes to IRQ 9 on the controller, so the processor vectors to the IRQ 9 handler. Then the default BIOS handler for IRQ 9 calls the IRQ 2 handler.

So IBM's improved AT bus design didn't require hardware manufacturers to change their cards or software developers to rewrite their drivers. That was a good idea when PC bus and AT bus systems were both in use, but almost a decade later there are no PC bus systems. And although many hardware vendors still refer to their cards as using IRQ 2, the card really uses IRQ 9.

It is very important to make this distinction when developing a Windows driver because hooking IRQ 2 when the device really uses IRQ 9 simply doesn't work under Windows. This worked under DOS because the BIOS handler for IRQ 9 called the IRQ 2 handler. Under Windows, the real mode BIOS IRQ 9 handler doesn't even see the interrupt if a Windows driver has hooked IRQ 9. If you're writing a Windows driver for an IRQ 9 device, hook IRQ 9 during installation and unmask the interrupt level on the slave controller.

Programming the COM port to generate interrupts is a two-step process:

* enable the interrupt internally in the UART, and

* enable the interrupt externally (using a spare UART output to gate the signal on the serial card).

The code looks like this:

```
outp( hDevice->usIoBase + UART_REG_IER, UART_IER_RXRDY );
outp( HDevice->usIoBase + UART_REG_MCR, UART_MCR_OUT2 );
```

Finally, `DeviceOpen` clears the appropriate mask bit in the interrupt controller, again paying attention to the chained controllers.

Processing Interrupts

If you are familiar with ring buffers, the ISR in Listing 15.5 (see page 359) will be straightforward. The handler first determines the exact cause of the interrupt and then branches to service either a receive ready or a transmit complete. The receive ready case reads a byte from the UART, copies the byte to the receive ring buffer and updates the buffer indices. In addition, if the receive buffer is full, the ISR uses `PostMessage` to post a message to the window whose handle was provided by the caller in `DeviceOpen`. The transmit complete case either pulls the next byte out of the transmit ring buffer and writes it to the UART or, if the transmit buffer is empty, disables the UART's transmit interrupt.

Although it's invisible here, the ISR isn't really talking to the physical PIC. The End Of Interrupt (EOI) write will actually be intercepted by Windows (using some of the 386 protection hardware) and redirected to a Virtual PIC, the VPICD. See Chapter 7 for more information about the VPICD.

Both `DeviceGetReadStatus` and `DeviceRead` in `INTBASIC.C` are slightly modified. `DeviceGetReadStatus` compares buffer indices to decide if characters are available. The `DeviceRead` routine just copies from the ring buffer to the calling program's buffer. `DeviceGetWriteStatus` and `DeviceWrite` have similar, transmit-oriented modifications.

Cleaning Up

The `DeviceClose` routine handles all the clean-up activities. Of course there's a natural symmetry between `DeviceClose` and `DeviceOpen`. Inverting the install sequence, this function should disable interrupts, install the original vector, re-enable interrupts, and then dispose of the buffer memory.

Summary

If you are familiar with interrupt-driven drivers under DOS, you should find the basic driver of this chapter quite accessible. (In fact, the ISR could easily be a DOS ISR.) If you test the performance of this driver and a comparable DOS driver, you may be surprised at the difference. The Windows driver will be significantly slower than its DOS cousin.

Although this chapter's ISR looks like it is written directly on the hardware, it really isn't. Windows is using the 386 protection hardware to insert a non-trivial layer of virtualizing software between your code and the hardware. This layer introduces some very significant service delays. A VxD is your best alternative if you need better response time from the driver.

Listing 15.1   INTBASIC.H

```
#define FLAGS_ON_SLAVE_PIC    0x01
#define FLAGS_OPEN            0x02
#define FLAGS_RXQOVER         0x04

#define MASTER_PIC_CTRL       0x20
#define MASTER_PIC_MASK       0x21
#define SLAVE_PIC_CTRL        0xA0
#define SLAVE_PIC_MASK        0xA1
#define EOI                   0x20

#define SET( value, mask )    value |= mask
#define CLR( value, mask )    value &= (~mask)

typedef struct
{
    WORD    usReadBufSize;
} DRIVERPARAMS, FAR * PDRIVERPARAMS;

typedef struct
{
    WORD    version;
} DRIVERCAPS, FAR * PDRIVERCAPS;
typedef PDRIVERCAPS FAR * PPDRIVERCAPS;

typedef void (FAR interrupt *VOIDINTPROC)();

typedef struct
{
    char far *Start;
    WORD    Size;
    WORD    In;
    WORD    Out;
} BUFINFO;

typedef struct
{
    WORD        usDevNumber;
    WORD        usIoBase;
    BYTE        bIrq;
    BYTE        bFlags;
    HWND        hwnd;
    VOIDINTPROC pfOldHandler;
    BUFINFO     RxBuf;
    BUFINFO     TxBuf;
    DRIVERPARAMS params;
} DEVICECONTEXT, FAR *HDEVICE;
```

Listing 15.1 (continued)   `INTBASIC.H`

```
HDEVICE FAR PASCAL DeviceOpen( HWND hwnd );
int FAR PASCAL DeviceClose( HDEVICE );
int FAR PASCAL DeviceGetWriteStatus( HDEVICE, LPWORD pusStatus );
int FAR PASCAL DeviceGetReadStatus( HDEVICE, LPWORD pusStatus );
int FAR PASCAL DeviceWrite( HDEVICE, LPBYTE lpData, LPWORD pcBytes );
int FAR PASCAL DeviceRead( HDEVICE, LPBYTE lpData, LPWORD pcBytes );
int FAR PASCAL DeviceSetDriverParams( HDEVICE, PDRIVERPARAMS pParms );
int FAR PASCAL DeviceGetDriverParams( HDEVICE, PDRIVERPARAMS pParms );
int FAR PASCAL DeviceGetDriverCapabilities( HDEVICE, PPDRIVERCAPS ppDriverCaps );

extern DEVICECONTEXT Device1;
```

Listing 15.2   `UART.H`

```
#define UART_REG_THR        0x00
#define UART_REG_RDR        0x00
#define UART_REG_IER        0x01
#define UART_REG_IIR        0x02
#define UART_REG_LCR        0x03
#define UART_REG_MCR        0x04
#define UART_REG_LSR        0x05
#define UART_REG_BAUDLO     0x00
#define UART_REG_BAUDHI     0x01

#define UART_IIR_NONE       0x01
#define UART_IIR_THRE       0x02
#define UART_IIR_RXRDY      0x04
#define UART_IER_THRE       0x02
#define UART_IER_RXRDY      0x01
#define UART_MCR_OUT2       0x08
#define UART_MCR_LOOP       0x10
#define UART_LSR_THRE       0x20
#define UART_LCR_DLAB       0x80
#define UART_LCR_8N1        0x03
#define UART_LSR_RXRDY      0x01
#define BAUD_1200           0x60
#define BAUD_110            0x0417L
```

Listing 15.3   `ISR.H`

```
void interrupt FAR DeviceIsr( void );
```

Listing 15.4   `INTBASIC.C`

```c
#include <dos.h>
#include <conio.h>
#include <windows.h>
#include <windowsx.h>
#include "intbasic.h"
#include "uart.h"
#include "isr.h"
#include "malloc.h"

#define DOS_GET_INT_VECTOR          0x35
#define DOS_SET_INT_VECTOR          0x25

DEVICECONTEXT Device1 = { 0, 0x3F8, 4, 0, NULL };
DRIVERPARAMS DefaultParams = { 1024 };
DRIVERCAPS DriverCaps = { 0x0101 };

BOOL ValidHandle( HDEVICE hDevice );
VOIDINTPROC DosGetIntVector( BYTE Irq );
void DosSetIntVector( BYTE Irq, VOIDINTPROC pHandler );
void interrupt FAR DeviceIsr( void );
UINT SafePageLock( HGLOBAL sel );

HDEVICE FAR PASCAL _export DeviceOpen( HWND hwnd )
{
    HDEVICE    hDevice;
    BYTE       bVector, mask;
    WORD       mycs, myds;

    OutputDebugString( "DeviceOpen\n");

    hDevice = &Device1;

    if (hDevice->bFlags & FLAGS_OPEN)
    {
        OutputDebugString( "ERROR already open\n");
        return (HDEVICE)-1;
    }
```

Listing 15.4 (continued)   INTBASIC.C

```c
hDevice->params = DefaultParams;
hDevice->hwnd = hwnd;

hDevice->RxBuf.Size = hDevice->params.usReadBufSize;
hDevice->RxBuf.Start = GlobalAllocPtr( GMEM_SHARE |
                                       GMEM_MOVEABLE |
                                       GMEM_NODISCARD,
                                       hDevice->RxBuf.Size  );
if (!hDevice->RxBuf.Start)
{
    OutputDebugString( "ERROR GlobalAlloc Rx\n");
    return (HDEVICE)-1;
}
SafePageLock( (HGLOBAL)SELECTOROF( hDevice->RxBuf.Start ));
hDevice->RxBuf.In = hDevice->RxBuf.Out = 0;

hDevice->TxBuf.Size = hDevice->params.usReadBufSize;
hDevice->TxBuf.Start = GlobalAllocPtr( GMEM_SHARE |
                                       GMEM_MOVEABLE |
                                       GMEM_NODISCARD,
                                       hDevice->TxBuf.Size  );
if (!hDevice->TxBuf.Start)
{
    OutputDebugString( "ERROR GlobalAlloc Tx\n");
    return (HDEVICE)-1;
}
SafePageLock( (HGLOBAL)SELECTOROF( hDevice->TxBuf.Start ));
hDevice->TxBuf.In = hDevice->TxBuf.Out = 0;

_asm mov myds, ds
_asm mov mycs, cs
SafePageLock( myds );
SafePageLock( mycs );

if (hDevice->bIrq < 8)
{
    mask = _inp( MASTER_PIC_MASK );
    SET( mask, 1 < hDevice->bIrq );
    _outp( MASTER_PIC_MASK, mask );
}
else
{
    SET( hDevice->bFlags, FLAGS_ON_SLAVE_PIC );
    mask = _inp( SLAVE_PIC_MASK );
    SET( mask, 1 < (hDevice->bIrq-8) );
    _outp( SLAVE_PIC_MASK, mask );
}
```

Listing 15.4 (continued) `INTBASIC.C`

```
    if (hDevice->bIrq < 8)
        bVector = hDevice->bIrq + 0x08;
    else
        bVector = hDevice->bIrq - 8 + 0x70;
    hDevice->pfOldHandler = DosGetIntVector( bVector );
    DosSetIntVector( bVector, DeviceIsr );

    // Configure UART.
    _outp( hDevice->usIoBase+UART_REG_IER, 0 );
    _outp( hDevice->usIoBase+UART_REG_LCR, UART_LCR_DLAB );
    _outp( hDevice->usIoBase+UART_REG_BAUDLO, BAUD_1200 );
    _outp( hDevice->usIoBase+UART_REG_BAUDHI, 0 );
    _outp( hDevice->usIoBase+UART_REG_LCR, UART_LCR_8N1 );
    _outp( hDevice->usIoBase+UART_REG_IER, UART_IER_RXRDY );
    _outp( hDevice->usIoBase+UART_REG_MCR, UART_MCR_OUT2 );

    // Unmask interrupt at PIC.
    if (hDevice->bIrq < 8)
    {
        mask = _inp( MASTER_PIC_MASK );
        CLR( mask, (1 < hDevice->bIrq) );
        _outp( MASTER_PIC_MASK, mask );
    }
    else
    {
        mask = _inp( SLAVE_PIC_MASK );
        CLR( mask, (1 < (hDevice->bIrq-8)) );
        _outp( SLAVE_PIC_MASK, mask );
    }

    SET( hDevice->bFlags, FLAGS_OPEN );

        return hDevice;
}
```

Listing 15.4 (continued)   INTBASIC.C

```c
int FAR PASCAL _export DeviceClose( HDEVICE hDevice )
{
    BYTE    mask, bVector;

    OutputDebugString( "DeviceClose\n" );

    if (!ValidHandle( hDevice ))
        return -1;

    if ((hDevice->bFlags & FLAGS_OPEN) == 0)
        return FALSE;

    CLR( hDevice->bFlags, FLAGS_OPEN );

    // Disable UART interrupts.
    _outp( hDevice->usIoBase + UART_REG_IER, 0 );
    _outp( hDevice->usIoBase + UART_REG_MCR, 0 );

    if (hDevice->bIrq < 8)
    {
        mask = _inp( MASTER_PIC_MASK );
        SET( mask, 1 < hDevice->bIrq );
        _outp( MASTER_PIC_MASK, mask );
    }
    else
    {
        SET( hDevice->bFlags, FLAGS_ON_SLAVE_PIC );
        mask = _inp( SLAVE_PIC_MASK );
        SET( mask, 1 < (hDevice->bIrq-8) );
        _outp( SLAVE_PIC_MASK, mask );
    }

    if (hDevice->bIrq < 8)
        bVector = hDevice->bIrq + 0x08;
    else
        bVector = hDevice->bIrq - 8 + 0x70;
    DosSetIntVector( bVector, hDevice->pfOldHandler );

    GlobalFreePtr( hDevice->RxBuf.Start  );
    GlobalFreePtr( hDevice->TxBuf.Start  );

    return 0;
}
```

Listing 15.4 *(continued)*   *INTBASIC.C*

```c
int FAR PASCAL _export DeviceGetWriteStatus( HDEVICE hDevice,
                                             LPWORD pusStatus )
{
    OutputDebugString( "DeviceGetWriteStatus\n" );

    if (!ValidHandle( hDevice ))
        return -1;

    if ((hDevice->bFlags & FLAGS_OPEN) == 0)
        return -1;

    if (_inp( hDevice->usIoBase + UART_REG_LSR ) & UART_LSR_THRE)
    {
        *pusStatus = 1;        // ready to transmit
    }

    else
    {
        *pusStatus = 0;        // not ready to transmit
    }

    return 0;
}

int FAR PASCAL _export DeviceGetReadStatus( HDEVICE hDevice, LPWORD pusStatus )
{
    OutputDebugString( "DeviceGetReadStatus\n" );

    if (!ValidHandle( hDevice ))
        return -1;

    if ((hDevice->bFlags & FLAGS_OPEN) == 0)
        return -1;

    if (hDevice->RxBuf.In != hDevice->RxBuf.Out)
    {
        *pusStatus = 1;        // data ready
    }
    else
    {
        *pusStatus = 0;        // no data ready
    }

    return 0;
}
```

Listing 15.4 (continued)   INTBASIC.C

```c
int FAR PASCAL _export DeviceWrite( HDEVICE hDevice, LPBYTE lpData,
                                    LPWORD pcBytes )
{
    WORD   i;
    char  ier;

    OutputDebugString( "DeviceWrite\n" );

    if (!lpData)
    {
        OutputDebugString( "ERROR\n" );
        return -1;
    }

    if (!ValidHandle( hDevice ))
    {
        OutputDebugString( "ERROR\n" );
        return -1;
    }

    if ((hDevice->bFlags & FLAGS_OPEN) == 0)
    {
        OutputDebugString( "ERROR\n" );
        return -1;
    }

    for (i=0; i < *pcBytes; i++)
    {
    hDevice->TxBuf.Start[ hDevice->TxBuf.In++ ] = lpData[ i ];
    if (hDevice->TxBuf.In >= hDevice->TxBuf.Size)
       hDevice->TxBuf.In = 0;
    }

    if (UART_LSR_THRE & _inp( hDevice->usIoBase+UART_REG_LSR ))
    {
        ier = _inp( hDevice->usIoBase+UART_REG_IER );
        if ((UART_IER_THRE & ier) == 0)
        {
            _outp( hDevice->usIoBase+UART_REG_IER, ier | UART_IER_THRE );
        }
    }

    return 0;
}
```

Listing 15.4 (continued)   `INTBASIC.C`

```c
int FAR PASCAL _export DeviceRead( HDEVICE hDevice, LPBYTE lpData,
                                   LPWORD pcBytes )
{
    WORD    cBytes, i;

    OutputDebugString( "DeviceRead\n" );

    if (!lpData)
    {
        OutputDebugString( "ERROR\n" );
        return -1;
    }

    if (!ValidHandle( hDevice ))
    {
        OutputDebugString( "ERROR\n" );
        return -1;
    }

    if ((hDevice->bFlags & FLAGS_OPEN) == 0)
    {
        OutputDebugString( "ERROR\n" );
        return -1;
    }

    cBytes = *pcBytes;

    for (i=0; i < cBytes; i++)
    {
        if (hDevice->RxBuf.In == hDevice->RxBuf.Out)
            break;
        lpData[i] = hDevice->RxBuf.Start[ hDevice->RxBuf.Out++ ];
        if (hDevice->RxBuf.Out >= hDevice->RxBuf.Size)
            hDevice->RxBuf.Out = 0;
        *pcBytes--;
    }

    return 0;
}
```

Listing 15.4 (continued) *INTBASIC.C*

```c
int FAR PASCAL _export DeviceSetDriverParams( HDEVICE hDevice,
                                              PDRIVERPARAMS pParams )
{
    OutputDebugString( "DeviceSetDriverParams\n");

    if (!pParams)
        return -1;

    if (!ValidHandle( hDevice ))
        return -1;

    if ((hDevice->bFlags & FLAGS_OPEN) == 0)
        return -1;

    hDevice->params = *pParams;

    return 0;
}

int FAR PASCAL _export DeviceGetDriverParams( HDEVICE hDevice,
                                              PDRIVERPARAMS pParams )
{
    OutputDebugString( "DeviceGetDriverParams\n");

    if (!pParams)
        return -1;

    if (!ValidHandle( hDevice ))
        return -1;

    if ((hDevice->bFlags & FLAGS_OPEN) == 0)
        return -1;

    *pParams = hDevice->params;

    return 0;
}
```

Listing 15.4 (continued)   `INTBASIC.C`

```
int FAR PASCAL _export DeviceGetDriverCapabilities( HDEVICE hDevice,
                                                    PPDRIVERCAPS ppDriverCaps )
{
    OutputDebugString( "DeviceGetDriverCapabilities\n");

    if (!ppDriverCaps)
        return -1;

    if (!ValidHandle( hDevice ))
        return -1;

    if ((hDevice->bFlags & FLAGS_OPEN) == 0)
        return -1;

    *ppDriverCaps = &DriverCaps;

    return 0;
}

BOOL ValidHandle( HDEVICE hDevice )
{
    return (hDevice == &Device1);
}

VOIDINTPROC DosGetIntVector( BYTE vector )
{
    WORD  selHandler, offHandler;

    _asm
    {
        mov  al, vector
        mov  ah, DOS_GET_INT_VECTOR
        push es
        int  21h
        mov  offHandler,bx
        mov  selHandler,es
        pop  es
    }
    return( MAKELP( selHandler, offHandler ) );
}
```

Listing 15.4 (continued)   *INTBASIC.C*

```c
void DosSetIntVector( BYTE vector, VOIDINTPROC pHandler )
{
    WORD  offHandler, selHandler;

    selHandler = SELECTOROF( pHandler );
    offHandler = OFFSETOF( pHandler );

    _asm
    {
        mov  al, vector
        mov  ah, DOS_SET_INT_VECTOR
        mov  dx, offHandler
        mov  bx, selHandler
        push ds
        mov  ds, bx
        int  21h
        pop  ds
    }
}

UINT SafePageLock( HGLOBAL sel )
{
    WORD i, rc;
    static WORD SelArray[ 1024 ];

    memset( SelArray, 1024 * sizeof(WORD), 0 );
    for (i=0; i < 1024; i++)
    {
        SelArray[i] = LOWORD( GlobalDosAlloc( 1024 ) );
        if (!SelArray[i])
            break;
    }

    rc = GlobalPageLock( sel );

    for (i=0; i < 1024; i++)
    {
        if (!SelArray[i])
            break;
        GlobalFree( SelArray[i] );
    }

    return rc;
}
```

Listing 15.5 ISR.C

```
#include <conio.h>
#include <windows.h>
#include "intbasic.h"
#include "uart.h"

void interrupt FAR DeviceIsr( void )
{
    BYTE  ier, intid;
    LPBYTE buf;
    DEVICECONTEXT *hDevice;

    hDevice = &Device1;

    while( TRUE )
    {
        intid = _inp( hDevice->usIoBase + UART_REG_IIR );
        if (intid == UART_IIR_NONE)
            break;

        if (intid == UART_IIR_RXRDY)
        {
            if ((hDevice->RxBuf.In+1==hDevice->RxBuf.Out)
                || ((hDevice >RxBuf.Out == 0)
                && (hDevice->RxBuf.In == hDevice->RxBuf.Size 1)))
            {
                PostMessage( hDevice->hwnd, WM_USER, 0, NULL );
            }
            buf = hDevice->RxBuf.Start;
            buf[ hDevice->RxBuf.In++ ] = _inp( hDevice->usIoBase +
                                               UART_REG_RDR );
            if (hDevice->RxBuf.In >= hDevice->RxBuf.Size)
                hDevice->RxBuf.In = 0;
        }

        else if (intid == UART_IIR_THRE)
        {
            if (hDevice->TxBuf.In==hDevice->TxBuf.Out)
            {
                ier = _inp( hDevice->usIoBase + UART_REG_IER );
                _outp( hDevice->usIoBase + UART_REG_IER,
                    ier & (~UART_IER_THRE) );
            }
```

Listing 15.5 (continued)   *ISR.C*

```
        else
        {
            buf = hDevice->TxBuf.Start;
            _outp( hDevice->usIoBase+UART_REG_THR,
                buf[ hDevice->TxBuf.Out++ ] );
            if (hDevice->TxBuf.Out >= hDevice->TxBuf.Size)
                hDevice->TxBuf.Out = 0;
        }
    }
}

    if (hDevice->bFlags & FLAGS_ON_SLAVE_PIC)
    _outp( SLAVE_PIC_CTRL, EOI );
    _outp( MASTER_PIC_CTRL, EOI );

}
```

Listing 15.6   *INTBASIC.MAK*

```
all: intbasic.dll

# DRIVER DLL

intbasic.obj: intbasic.c intbasic.h
    cl -c -W3 -ASw -GD2s -Zi -Oi $*.c

isr.obj: isr.c intbasic.h
    cl -c -W3 -ASw -GD2s -Zi -Oi $*.c

intbasic.dll: intbasic.def intbasic.obj isr.obj
    link intbasic+isr,intbasic.dll,intbasic.map /CO /MAP, \
        sdllcew libw /nod/noe,intbasic.def
    implib intbasic.lib intbasic.dll intbasic.def
    copy intbasic.dll \windows\driver.dll
```

Listing 15.7   *INTBASIC.DEF*

```
LIBRARY       DRIVER
DESCRIPTION   "Basic Interrupt-Driven Driver"
EXETYPE       WINDOWS
DATA          PRELOAD MOVEABLE SINGLE
CODE          PRELOAD MOVEABLE NONDISCARDABLE
```

Chapter 16

Driver DLLs: Using DMA

All of the drivers in earlier chapters have relied on the processor to transfer data to and from the device, either with IN/OUT operations on a port address, or with read/write operations on a memory address. Devices that manipulate large blocks of data, such as disk controllers, are often capable of transferring data directly to memory using Direct Memory Access (DMA), thereby reducing the load on the data bus.

Windows driver DLLs that use DMA are somewhat uncommon because of the difficulties implicit in assuring that the DMA controller device, which always writes to a physical address, is writing into the right logical address. This chapter explains the requirements for a DMA buffer and shows how to write a driver that uses DMA to transfer data.

DMA Buffer Requirements

A buffer used for a DMA transfer, either driver-to-device or device-to-driver, must meet several strict requirements. The DMA buffer must be:

- physically contiguous,
- fixed and pagelocked, and
- aligned on a 64Kb boundary.

These requirements are necessary because the DMA controller has no knowledge of selectors or pages and performs no address translation. The controller is programmed with a starting physical address and simply increments (or decrements) that address with each byte transferred in order to generate the next physical address. For more details on the exact reason for each of the above requirements, refer to the section "System DMA Buffer Requirements" in Chapter 6.

How to Allocate a DMA Buffer

Chapter 15 showed how to allocate fixed and pagelocked memory. A search of all Windows API or DPMI calls reveals no way to specify 64Kb alignment, but there are several usable work-arounds. A small buffer is less likely to cross a 64Kb boundary, so in this case a good strategy is to keep allocating (fixed and pagelocked) buffers until you get a suitable one, then deallocate the unused ones.

The larger the buffer, however, the greater the chance the buffer will span a 64Kb boundary. To get a large buffer, allocate a buffer twice as big as you need and then use the half that doesn't span the 64Kb boundary.

That leaves the last requirement: physically contiguous pages. There is absolutely no Windows API or DPMI call to allocate memory with this attribute. (One API can do this, discussed in Chapter 6, but it's available only to VxDs.) There is another problem as well. Even if such a buffer could be allocated, the driver must obtain its physical address to program the DMA controller's base address register, and there is no Windows API or DPMI call to obtain a physical address. The closest you can get is a linear address, using `GetSelectorBase`.

DMA DOS Applications Under Windows

It's interesting to note that a DOS application that does DMA transfers works fine under Windows, whereas a Windows application must overcome the contiguous pages and physical address obstacles in order to do the same task. How is this possible? The secret is the Virtual DMA Device (or VDMAD), a VxD that operates behind the scenes. VDMAD's main reason for existence is to make sure DOS applications can do DMA transparently just as they did under DOS, even though V86 mode memory translation is radically different.

VDMAD does this by trapping all accesses to the DMA controller and caching the data internally instead of letting it go through to the controller. VDMAD is particularly interested in the controller's base address register. VDMAD knows that because a DOS application is running in V86 mode, the address a DOS application programs into this register is really a linear address, not a physical address. So VDMAD translates this linear address into a physical address and writes that to the controller's address register. In addition, VDMAD pagelocks the entire buffer and verifies that it is physically contiguous.

If the buffer is not contiguous (and it rarely is), VDMAD substitutes the physical address of its own buffer, which meets all DMA requirements. At this point VDMAD tells the controller to start with the transfer. When the transfer is over — and if the VDMAD-owned buffer was used — VDMAD copies data to the DOS application's original address. Despite all this interaction, the DOS application sees nothing but a DMA transfer as usual — except that the transfer is much slower because of the double buffering.

DMA Windows Applications Can Use this Knowledge

Because VDMAD traps accesses by Windows applications as well as DOS applications, Windows applications can use this knowledge of VDMAD interaction to overcome the contiguous pages and physical address problems explained above. Basically, a Windows application:

- allocates a buffer,
- gets its linear address,
- programs the DMA controller with that linear address, and
- relies on VDMAD to make everything work out right.

The only trick here is that a Windows application can't use just any buffer. The DMA controller's base address register is 24 bits. Both `GlobalAlloc`'ed buffers and those buffers statically allocated in a driver's data segment are generally located above 2Gb in linear address space where addresses can't be represented in 24 bits. The proper method is to use a `GlobalDosAlloc`'ed buffer, which is guaranteed to have a linear address below 1Mb that fits into 24 bits.

Using Virtual DMA Services Is Better

The solution described above is easy to implement, but there is a price: `GlobalDosAlloc` takes up precious linear memory under 1Mb. Another solution that avoids this problem is to use VDMAD's own DMA buffer, just for the duration of the transfer, then copy the data to the driver buffer. Borrowing the VDMAD buffer is possible because the VDMAD exports services for Win16 applications. This API, available through `INT 4Bh`, is known as Virtual DMA Services (VDS).

VDS services are not available to Win32 applications, only to Win16 applications.

The VDS interface is listed in Table 16.1. In summary, VDS includes functions to pagelock a buffer, request use of the VDS buffer, and copy data between the VDS buffer and another buffer.

Table 16.1 *Virtual DMA Services (VDS).*

All Services		
`INT 4B, AH=81h, AL=Function, DX=flags, ES:DI->DMA Descriptor`		
DMA Descriptor		
`DWORD`	Size	
`DWORD`	Offset	
`DWORD`	Segment/Selector	
`DWORD`	Physical Address	
Name	**Function**	**Description**
`Get Version`	02h	returns size of DMA buffer and whether or not memory is physically contiguous
`Lock DMA Region`	03h	pagelocks buffer; if buffer not ok for DMA, borrows VDS buffer, returns its physical address
`Unlock DMA Region`	04h	pageunlocks buffer
`Scatter/Gather Lock`	05h	pagelocks multiple regions
`Scatter/Gather Unlock`	06h	pageunlocks multiple regions
`Request DMA Buffer`	07h	borrow VDS buffer for DMA use
`Release DMA Buffer`	08h	return VDS buffer to VDS
`Copy Into DMA Buffer`	09h	copy data into VDS buffer
`Copy Out Of DMA Buffer`	0Ah	copy data from VDS buffer
`Disable DMA Translation`	0Bh	tells VDS that address programmed into controller is physical not linear
`Enable DMA Translation`	0Ch	tells VDS that address programmed into controller is linear not physical

To borrow the VDS buffer for a DMA transfer, a driver calls `Request DMA Buffer`, which returns the physical address of the VDS buffer (which meets all DMA requirements). Before programming this address into the controller's base address register, the driver calls `Disable DMA Translation`. This tells VDS that the address is already a physical address and needs no translation. Note that a Windows driver that uses VDS for buffer services interacts with the DMA controller — programming address, count, mode, etc. — in the exact same manner that a DOS DMA application would.

When borrowing the VDS buffer, if the transfer is from memory to the device, then the driver must call `Copy Into DMA Buffer` before starting the transfer. This call copies data from the driver buffer to the VDS buffer. If the transfer is in the opposite direction, then the driver calls `Copy Out Of DMA Buffer` after the transfer completes. In both cases, the driver re-enables translation with `Enable DMA Translation` and relinquishes the borrowed buffer with `Release DMA Buffer`.

It is possible that a buffer that a driver allocates as fixed and pagelocked also happens to be physically contiguous. If so, then a driver that always requests the VDS buffer is incurring the performance penalty of an extra copy (before or after the transfer) unnecessarily.

A better alternative, which is no more difficult to code, is to use `Lock DMA Region` instead of `Request DMA Buffer`. `Lock DMA Region` combines several useful functions. The driver passes in a buffer pointer, and VDS first checks to see if the buffer is 64Kb aligned and is physically contiguous. If both conditions are met, VDS then locks all the pages in the buffer and returns with the buffer's physical address. The buffer now meets all DMA requirements. If the buffer doesn't meet DMA requirements, VDS returns with the physical address of its own DMA buffer.

When using `Lock DMA Region`, the driver calls `Disable DMA Translation`, programs the physical address into the controller, and starts the transfer. When the transfer is complete, the driver calls `Enable DMA Translation` (and `Release DMA Buffer`, as a result of the `Lock`, if the VDS buffer was used as a result of the `Lock`). A call to `Copy Into DMA Buffer` or `Copy Out Of DMA Buffer` is unnecessary, because VDS does this copy automatically if the VDS buffer was used. If the driver's buffer was used for the transfer, then there is no copy, which is more efficient.

Using the VDS buffer for a DMA transfer is far from an ideal solution. The single VDS/VDMAD buffer must be shared among all Windows and DOS applications that are using DMA, so a driver may have to wait for the buffer to become available. More importantly, using this intermediate buffer results in an extra data copy operation. This is the case whether the buffer is used implicitly with the `GlobalDosAlloc` and invisible VDMAD interaction or used explicitly via VDS.

The best solution is to write a VxD to allocate the DMA buffer and have your driver DLL use the services of the VxD to obtain the buffer's physical address. Chapter 11 presented such a VxD and Win16 application.

Summary

While DMA transfers offer significant performance advantages under other operating systems, unless you are willing to write a helper VxD or do all of the DMA inside a VxD, you will probably not see the same advantages under Windows. Unless you can force a contiguous, fixed, and pagelocked buffer, DMA transfers will incur an extra copy operation after the DMA transfer, a cost that more than offsets the normal advantages of DMA transfers.

## Chapter 17

Driver DLLs:
Using Real Mode Services

Windows applications run in protected mode, but they can and do use real mode DOS and BIOS services. Whether a DOS application running under Windows calls the C library `read` function, or a Windows application calls the Windows `lread` function, the read eventually boils down to a simple call to DOS through `INT 21h`, just as it does in a program running under DOS. In addition to using DOS and BIOS services, Windows applications may also use other real mode services such as TSRs or DOS device drivers.

To properly execute real mode code from a protected mode application the programmer must overcome a number of obstacles. First and most obvious is that the processor must be switched from protected mode to V86 mode and then back again. Addressing and other differences create more subtle obstacles. Protected mode data may live above 1Mb where it is inaccessible to real mode code, complicating parameter passing. Also, any parameters returned by the real mode code in segment registers will cause an exception when the processor switches back into protected mode, since they aren't valid selectors.

Windows application programmers rarely need to worry about any of the above issues. By intercepting `INT 21h` calls and doing the work necessary to take care of all of these issues, the Windows kernel makes it very easy for Windows applications to use DOS and BIOS services. Windows driver DLLs don't lead such a protected life.

Although many Windows applications never use any real mode services other than DOS, a Windows driver DLL may use an existing real mode TSR or DOS device driver that provides support for a hardware device. In this case, the driver developer needs a good understanding of the translation issues mentioned above, because the Windows kernel cannot provide this same level of transparent support for TSRs and device drivers it knows nothing about. In addition, there are a few INT 21h services that Windows doesn't support (Table 17.1), so the driver developer will also need to provide translation when using one of these unsupported services.

Windows does help with some of this work; Windows will automatically switch processor modes as necessary whenever an application issues a software interrupt. However, Windows can't provide automatic buffer translation for unknown services because it doesn't know which registers contain pointers. If a Windows driver DLL needs to exchange pointers with a real mode service, it must do some of the translation work itself.

The following sections explore several alternate techniques for calling real mode services and passing parameters to real mode services.

Talking to a DOS Device Driver

By definition, a DOS device driver presents a specific interface: an application accesses the device driver as if it were a file, using DOS Open, Read, Write, and Close calls. An application can either make these DOS calls directly through INT 21h or use the C run-time low-level file functions (_open, _read, etc.). In addition to these standard calls, many DOS device drivers support an extended interface through IOCTL (I/O Control) commands. The C run-time offers no support for issuing IOCTL commands to device drivers, so if you need to issue IOCTLs you'll have to issue the DOS IOCTL command through embedded assembly.

Windows applications access DOS device drivers through this same file interface. The Windows kernel traps INT 21h and provides any necessary translation for DOS device driver access. For example, Windows looks at the buffer address passed to the DOS device driver in both Read and Write calls and checks to see if this buffer lives above 1Mb. If so, this buffer is unacceptable — DOS device drivers execute in V86 mode and the process can't access anything above 1Mb. So Windows substitutes the address of its own buffer, below 1Mb, before calling the DOS device driver.

In addition, even if the Windows application's buffer happened to be located below 1Mb, the buffer address supplied by the Windows application would be a protected mode address, not a real mode address usable by the DOS device driver. So Windows must also convert protected mode pointers to real mode pointers.

Table 17.1 *DOS functions not supported or partially supported by Windows.*

Service	Description
INT 20h	Terminate program
INT 25h	Absolute disk read
INT 26h	Absolute disk write
INT 27h	Terminate and stay resident
INT 21h	Func 00h Terminate process
INT 21h	Func 0Fh Open file with FCB
INT 21h	Func 10h Close file with FCB
INT 21h	Func 14h Sequential read
INT 21h	Func 15h Sequential write
INT 21h	Func 16h Create file with FCB
INT 21h	Func 21h Random read
INT 21h	Func 22h Random write
INT 21h	Func 23h Get file size
INT 21h	Func 24h Get relative record
INT 21h	Func 25h Get interrupt vector (supported, but gets protected mode interrupt vector)
INT 21h	Func 27h Random block read
INT 21h	Func 28h Random block write
INT 21h	Func 35h Set interrupt vector (supported, but sets protected mode interrupt vector)
INT 21h	Func 38h Get country page (supported, but returns real mode call address)
INT 21h	Func 44h Subfunc 02 (fails if buffer address > 1Mb and buffer size > 4Kb)
INT 21h	Func 44h Subfunc 03 (fails if buffer address > 1Mb and buffer size > 4Kb)
INT 21h	Func 44h Subfunc 04 (fails if buffer address > 1Mb and buffer size > 4 Kb)
INT 21h	Func 44h Subfunc 05 (fails if buffer address > 1Mb and buffer size > 4 Kb)
INT 21h	Func 65h Get extended country info (supported, but returns real mode call address)
INT 21h	Func 67h Set handle count

In addition to file functions like Open, Read, Write, and Close, many DOS device drivers also support device-specific functionality through the DOS IOCTL function call (INT 21h, AH=44h). A specific subfunction is specified in the AL register. Examples of IOCTL subfunctions include Receive Control Data (AL=2) and Write Control Data (AL=3). Both of these subfunctions, and some others as well, use a single buffer parameter passed in DS:DX. Because the registers used by each subfunction for the buffer parameter are defined by the IOCTL interface, Windows is able to perform automatic translation for all IOCTL buffers. However, the device driver is free to interpret the buffer contents in any way it chooses. So a device driver may view the Receive Control Data buffer as a structure, and that structure could contain pointers. In this case, Windows does not know that the buffer contains pointers that need translation.

That means a Windows application that calls, for example, the driver's Read Control Data through the IOCTL must handle translation of these embedded pointers itself.

Special Handling for IOCTLs

This section will explain what a Windows driver DLL needs to do to issue the Read Control Data IOCTL call to a DOS device driver that does use embedded pointers in this buffer. In this example, the buffer passed to the device driver via DS:DX is not just a character buffer but a CONTROL_DATA structure, shown in the following code. The CONTROL_DATA structure contains a pointer to an array of ints. The device driver will fill in the array, allocated by the DLL, with a list of supported speeds.

```
typedef struct
{
    void far *speeds;
    int numspeeds;
} CONTROL_DATA;
```

It's not strictly necessary to allocate the CONTROL_DATA structure itself below 1Mb because the automatic buffer translation provided by Windows knows about the pointer in DS:DX. But Windows does not know that CONTROL_DATA contains a pointer, so the Windows driver DLL must ensure that the speeds array lives below 1Mb and also that the speeds pointer is a real mode pointer.

The Windows API function GlobalDosAlloc exists specifically to provide real mode buffers and will always allocate below (linear) 1Mb. The DWORD return value from GlobalDosAlloc provides both a selector for addressing the buffer in protected mode and a segment for addressing the buffer in real mode. The Windows driver DLL uses both portions because it needs a protected mode pointer for normal buffer access and a real mode pointer to give to the device driver. The offset portion of the pointer in both cases is always zero.

The LOWORD macro extracts the selector, and MAKELP turns the selector into a far pointer. The Windows driver DLL uses this pointer to initialize the buffer and to access the buffer after the device driver fills it in. The real mode segment is extracted with the HIWORD macro and formed into a real mode pointer using MAKELP. Just be careful not to dereference this real mode pointer — it's not valid in protected mode.

The following code uses the methods described previously to retrieve device information from a DOS device driver. After the speeds array is filled in by the device driver, the code scans the array to determine if the device supports high speeds.

```
BOOL SupportsHighSpeed( void )
{
    CONTROL_DATA cdata;
    BOOL highspeed = 0;
    WORD far *speeds;
    WORD numspeeds = 8;
    DWORD dw;
    WORD i;

#define IOCTL_READ_CONTROL_DATA    2
    dw = GlobalDosAlloc( numspeeds * sizeof( WORD ) );
    cdata.numspeeds = numspeeds;
    cdata.speeds = MAKELP( HIWORD( dw ), 0 );
    speeds = MAKELP( LOWORD( dw ), 0 );
    DosIoctl( IOCTL_READ_CONTROL_DATA, &cdata, sizeof( cdata ));
    for (i=0; i < numspeeds; i++)
    {
        if (*speeds++ > 9600)
            return( TRUE );
    }
}
```

Windows is only able to provide automatic translation for DOS device drivers because it knows that these drivers are accessed through INT 21h. Even when the driver is accessed through INT 21h, Windows can only provide perfect translation when it knows exactly which parameters are expected. Windows knows everything it needs to know about read and write, but not always IOCTL.

TSRs are a different matter, as each TSR has its own interface with its own method of parameter passing. For this reason, interfacing Windows code to a TSR usually requires more work than interfacing to a DOS device driver. The next section will explore this topic.

Talking to TSRs

TSRs are nearly always accessed via a software interrupt, with parameters passed in registers. In some cases, parameters are instead passed on the stack. In this alternate approach, an initial call through a software interrupt returns one or more function addresses. Applications later call these addresses directly, passing parameters on the stack.

The first method, in which an interrupt is used and parameters are passed in registers, is the simpler of the two, so I'll explain it first. To issue a software interrupt, a Windows driver DLL uses the same method a DOS program does: either a run-time library function like int86 or embedded assembly.

I prefer to use embedded assembly because the int86 function uses pointer parameters, and passing pointers to any library functions in a DLL means worrying about SS != DS issues. (Refer to Chapter 13 for more on this issue.)

Windows traps all software interrupt instructions and first determines if the software interrupt is a supported interrupt, like DOS or BIOS, that requires special translation handling. (Refer to Table 17.1 for a list of unsupported interrupts.) If no special handling is required, Windows does nothing but switch the processor into V86 mode and call the software interrupt handler in the V86 mode IVT. When the real mode software interrupt handler issues an IRET, Windows switches the processor back to protected mode and the Windows application continues executing.

If your Windows driver DLL doesn't need to pass any pointers to the TSR and the TSR doesn't return any pointers to your driver, then your DLL need only initialize processor registers and issue the software interrupt. If pointers are exchanged, your DLL must do some extra work.

Passing Data via Buffers

A Windows driver DLL must take special precautions when calling a DOS TSR and exchanging data via a buffer. If the Windows driver DLL supplies the buffer and the TSR fills it in, the precautions involve both allocating real-mode-addressable memory and passing the buffer address to the TSR. If the transfer is the other way around (where the TSR owns the buffer and gives the Windows driver DLL the buffer's address), the precautions involve correctly forming a pointer to access the real mode buffer.

The TSR Owns the Buffer

A pointer returned by a TSR through some pair of registers is, by definition, a real mode pointer. A Windows driver DLL can't use this pointer directly — doing so results in a protection violation. As a protected mode application, a Windows driver DLL must use a protected mode pointer. The trick is to turn the TSR's real mode pointer into a protected mode pointer which the driver can use.

A real mode pointer consists of a segment and an offset; a protected mode pointer consists of a selector and an offset. A segment and a selector differ in how the processor transforms each into a linear address. In real mode, the processor computes a linear address by performing a simple arithmetic calculation: `physical address = (segment<<4)+offset`. In protected mode, there is no arithmetic relationship between a selector and a linear address. Instead, each protected mode selector has an associated base address (stored in a descriptor table maintained by the operating system), and the processor adds the offset portion of the pointer to this base address to get the linear address.

The basic idea behind transforming a real mode pointer to a protected mode pointer is this: the driver creates a protected mode selector with a base address equal to the same linear address generated by the real mode pointer. Windows provides a set of selector API functions to perform this conversion: `AllocSelector`, `SetSelectorBase`, and `SetSelectorLimit`.

As its name suggests, `AllocSelector` allocates a protected mode selector. The single parameter is a template selector. To create a selector to address data, pass in the value in the `DS` register. To create a code selector (this is much less common), use the value in `CS`. Failure to pass the right selector as a parameter usually results in a selector of the wrong type, followed by a protection violation when using the selector.

`SetSelectorBase` performs the actual conversion. Given a linear address that corresponds to the real mode pointer, `SetSelectorBase` will update the allocated selector so that its base is at that linear address.

There is one more important step in setting up the protected mode selector: setting its limit with `SetSelectorLimit`. Unlike a real mode segment, a selector has an associated length, or limit. A memory access past this limit results in a protection violation. For maximum protection, the Windows driver DLL should set this limit to the size of the allocated buffer.

The following code encapsulates this series of Windows API calls in a single function, which converts a real mode segment and offset to a protected mode pointer.

```
void far *RealPtrToProtPtr( WORD seg, WORD off )
{
    char far *ptrProt;
    WORD myDs,sel;

    _asm mov myDs, ds
    sel = AllocSelector( myDs );

    // Set selector's linear address to (seg < 4)+offset
    base = (seg << 4) + off;
    SetSelectorBase( sel, base );

    // Set selector limit to 64K.
    SetSelectorLimit( sel, 64*1024 );

    return( MAKELP( sel, 0 ) );
}
```

The Windows Application Owns the Buffer

Transferring data in the other direction, from Windows driver DLL to TSR, raises exactly the same issues as those described in the earlier section on IOCTL handling. A buffer passed from a Windows driver DLL to a TSR must be located below 1Mb, because when executing the TSR in V86 mode, the processor can address only 1Mb of memory. As described earlier (in the section on IOCTL handling), GlobalDosAlloc should be used to allocate such a buffer, and the DWORD return value is used to build both a protected mode pointer and a real mode pointer to address the buffer.

Once the buffer is allocated, the driver must pass the buffer's address to the TSR. By convention, TSRs expect parameters in registers, not on the stack. This means the Windows application must load the real mode segment and a zero offset into whichever pair of registers the TSR expects. A DOS application calling a TSR would either use assembly language to load the processor registers directly and issue the software interrupt, or use the C library _intdosx function. Whenever a Windows driver DLL passes a buffer to a TSR, calling the TSR is not that simple and requires using DPMI.

Calling the TSR using DPMI

The Windows designers realized it was important to allow Windows applications and drivers to communicate with TSRs, so Windows provides a set of services to facilitate this. These services aren't part of the normal Windows API, but are instead part of the DPMI interface supported by Windows through INT 31h. DPMI selector services were introduced in Chapter 14's discussion of memory-mapped devices. This section will introduce another DPMI service, Simulate Real Mode Interrupt, which will let our driver call the TSR and pass it a buffer pointer. The following pseudocode shows the calling parameters for the DPMI Simulate Real Mode Interrupt function.

```
AX = 0300h
BL = interrupt number
BH = flags
        Bit 0 = 1 to reset interrupt controller and A20 line
CX = number of words copied from prot. mode stack to real mode stack
ES:DI = far pointer to real mode call structure
```

Simulate Real Mode Interrupt passes register information through the real mode call structure. This structure contains a field for every processor register and is similar to the REGS structure used with the C library function _intdosx. The following code shows the declaration in C.

```
struct
{
  unsigned long edi;
  unsigned long esi;
  unsigned long ebp;
  unsigned long res1;
  unsigned long ebx;
  unsigned long edx;
  unsigned long ecx;
  unsigned long eax;
  unsigned short flags;
  unsigned short es;
  unsigned short ds;
  unsigned short fs;
  unsigned short gs;
  unsigned short ip;
  unsigned short cs;
  unsigned short sp;
  unsigned short ss;
} REAL_MODE_CALL_STRUC;
```

The driver fills these register fields as required by the TSR. (Note that you don't have to `GlobalDosAlloc` to make the structure itself real mode addressable, because DPMI will take care of this.) Then the driver uses embedded assembly to fill the actual processor registers — not the ones in the call structure — as required by the DPMI service, and issues an `INT 31h`. Software interrupt handlers don't use stack parameters, so `CX` will usually be zero.

When the Windows driver DLL issues the `INT 31h`, DPMI first copies the contents of the real mode call structure to an intermediate area which is addressable in V86 mode. Next, DPMI switches to V86 mode, then copies each field of the call structure to the proper processor register. Last, DPMI issues the requested software interrupt.

The TSR runs, blissfully unaware that it was invoked by a protected mode application. When the TSR returns, DPMI switches to protected mode and copies from the intermediate call structure back to the Windows driver DLL's original structure. So any information passed from the TSR to the Windows driver DLL via a register shows up afterward in the real mode call structure. To examine the buffer contents after the TSR returns, the driver uses the protected mode pointer built earlier with `MAKELP` and the protected mode selector.

The following example illustrates passing a buffer to a TSR. In this example, the TSR is called through `INT 14h` and expects a pointer to the buffer in `ES:BX`.

```
DWORD dw;
WORD seg, sel;
char far *buf;
REAL_MODE_CALL_STRUC RmCallStruc;

dw = GlobalDosAlloc( 256 );
seg = HIWORD( dw );
RmCallStruc.es = seg;
RmCallStruc.ebx = 0;
_asm
{
    mov ax, 0300h          // DPMI func Simulate Real Mode Int
    mov bl, 14h            // Software Interrupt Number
    xor bh, bh             // flags
    xor cx, cx             // num word passed on stack
    mov es, SEG RmCallStruc
    mov di, OFF RmCallStruc
    int 31h
}
buf = MAKELP( LOWORD( dw ), 0 );
// buf can now access data filled in by TSR
```

Calling a TSR via an Address

A less common method of calling a real mode TSR is through an address. Your program gets the address by calling the TSR once through a software interrupt, and the interrupt handler returns one or more function addresses. One example of this strategy is the NDIS 2.0 (Network Driver Interface Standard) interface between a protocol stack TSR and a network card device driver. Calling a service through an address is just a minor variation on the themes described above. Instead of using Simulate Software Interrupt, the Windows application uses DPMI Call Real Mode Procedure With Far Return Frame (INT 31h, AX=301h). The CS and IP fields of the call structure specify the address of the real mode procedure.

Be sure when you first retrieve this address from the TSR that you do not treat it as a pointer. Yes, it has a segment and an offset, but the two don't form a valid pointer as long as you're in protected mode.

The following code fragment obtains a real mode function address through software interrupt 50h then makes a call to the function using DPMI Call Real Mode Procedure With Far Return Frame. In addition, the call passes two parameters to the real mode function. The first parameter is a function code (to tell the TSR what to do), and the second is a pointer to a buffer. This buffer must be GlobalDosAlloc'ed, and the buffer address we pass to the TSR is, of course, the real mode segment returned by GlobalDosAlloc.

```
// TSR expects function code in AX
REAL_MODE_CALL_STRUC RmCallStruc;
DWORD dw;
WORD seg;

mov RmCallStruc.eax = 7h;                // Function: Get Entry Point
// Call DPMI Simluate Real Mode Interrupt through INT 31h
_asm
{
    mov ax, 0300h                        // DPMI function
    mov bl, 50h                          // int number
    mov bh, 0                            // flags
    xor cx, cx                           // stack words
    mov es, SEG RmCallStruc
    mov di, OFF RmCallStruc
    int 31h
}
```

```
// TSR returned the entry point address in CX:DX
// Move it to CS:IP of call structure
RmCallStruc.cs = (WORD)RmCallStruc.ecx;
RmCallStruc.ip = (WORD)RmCallStruc.edx;

// Allocate memory for buffer to give to TSR
dw = GlobalDosAlloc( sizeof( MyStruc ) );
seg = LOWORD( dw );

// Call TSR entrypoint, whose function prototype is:
// void pascal TsrEntry( WORD FunctionCode, char far *pBuffer )
_asm
{
    // Push parameters onto stack using
    // pascal (left-to-right) calling convention
    mov ax, 1
    push ax
    push seg
    mov ax, 0
    push ax

    // Use DPMI Call Real Mode Procedure With Far Return Frame
    mov ax, 0303h                   // DPMI function
    xor bh, bh                      // flags
    mov cx, 2                       // stack words
    mov es, SEG RmCallStruc
    mov di, OFF RmCallStruc
    int 31h

}
```

TSR Calls a Windows Application

Another common interaction between a Windows application and a TSR involves giving the TSR a callback address. The TSR saves the address and calls the function later during the execution of an interrupt handler. Usually the TSR passes information to the callback function through registers, like a software interrupt would. To get this job done from a Windows driver DLL, you need DPMI `Allocate Real Mode Callback Address` (INT 31h, AX=03h).

Unfortunately, it's not enough for the Windows driver DLL to give the TSR a real mode pointer. Because Windows is a multitasking environment, the TSR must some-how make certain the correct VM is running when it performs the callback.

Windows distinguishes between two types of TSRs: local and global. A local TSR is mapped into the address space of a single VM, and is created when a user creates a new DOS VM and then loads a TSR from the command line. A user can also load a TSR local to the System VM, via the WINSTART.BAT file. The other type of TSR is a global TSR, loaded before Windows begins. A global TSR is mapped into the address space of all VMs — that is, the system VM and any DOS VMs created later. There is a single copy of a global TSR (code and data) in memory, but each VM has its own linear address that maps to this single copy in physical memory.

TSRs do their magic by hooking interrupt vectors, both hardware and software. When a global TSR hooks a hardware interrupt, there is no particular VM associated with the handler. So Windows calls the TSR's interrupt handler immediately, in the context of whatever VM was interrupted. If a global TSR's interrupt handler simply services a hardware interrupt and has no interaction with an application, then it doesn't matter which VM was interrupted. For example, the BIOS keyboard interrupt handler (IRQ 1) reads a key from the keyboard controller and stores it in the BIOS keyboard buffer — it does not call an application to give it the key.

But if a global TSR also calls to an application, through a callback, then it matters very much indeed which VM was interrupted. Remember, each VM has its own address space. Suppose the TSR interrupt handler is using VM1's address space (because VM1 happened to be the current VM at the time of the interrupt), and the callback is in VM2's address space. Then the callback won't work because the callback address is valid only when VM2 is executing, not when VM1 is executing. The TSR will call an address that points to garbage in VM1, and the system will probably crash.

Solution to Callbacks

Making the TSR Windows-aware will solve this problem. A Windows-aware TSR won't use the callback directly, but instead will wait until the System VM (where Windows applications live) is the current VM. The TSR can force the System VM to be scheduled through an INT 2Fh Switch VMs and Callback interface offered by the VMM, and the VMM will call back into the TSR when the System VM is current. With the System VM running, the TSR can safely use the DPMI callback, which will in turn trigger a mode switch from V86 mode to protected mode. Finally, with the processor in protected mode and the right VM active, the callback in the Windows application can execute.

If you are unable to modify the TSR you're using in order to make it Windows-aware, you must write a second helper TSR which *is* Windows-aware and does the VM switch on behalf of the original TSR.

Thus, when a Windows driver DLL communicates with a TSR through a callback, the communication involves this sequence of steps:

The Windows driver DLL:

1. Uses DPMI `Allocate Real Mode Callback Address` to allocate a callback.

2. Gives the DPMI callback address to the TSR.

   TSR at interrupt time:

3. If current VM is System VM, go to step 5.

4. If current VM is not System VM, use `INT 2Fh Switch VMs and Callback` to force System VM to be scheduled.

   System VM is now current:

5. Calls DPMI callback address, DPMI switches to protected mode and calls Windows driver DLL.

   Windows driver DLL:

6. Callback executes in protected mode, in the System VM, and accesses TSR data through the real mode call structure.

7. Callback adjusts the real mode call structure `CS:IP` and returns.

8. DPMI switches back to V86 mode and returns to TSR.

Callback Coding Details

The code for the Windows driver DLL will be affected in two places: when registering with the TSR for a callback (usually in an initialization or open function) and in the driver callback itself. In addition, the TSR needs additional code to handle the VM switch.

The following code illustrates how a Windows driver DLL would use the DPMI `Allocate Real Mode Callback Address` service. In order to register with the TSR for a callback, the Windows driver DLL first needs to obtain a real mode callback address, through DPMI `Allocate Real Mode Callback Address`. The input parameters to this service are a protected mode pointer to the function to be called and a protected mode pointer to a real mode call structure. DPMI returns the real mode segment and offset of a stub function which, when called from V86 mode, will switch into protected mode and then call your driver. The driver then gives the real mode segment and offset returned by DPMI to the TSR.

```
void far *pfFoo;
WORD CallbackSeg;
WORD CallbackOff;

pFoo = &Foo;
_asm
{
    mov ax, 0303h          // DPMI Allocate RM Callback
    mov es, SEG RmCallStruc
    mov di, OFFSET RmCallStruc
    mov si, pFoo
    push ds
    mov ds, pFoo+2
    int 31h
    pop ds
    mov CallbackSeg, cx
    mov CallbackOff, dx
}
// Give CallbackSeg and Callback Off to TSR as callback
```

Before using the callback address from the Windows driver DLL (which is really a DPMI allocated callback, as shown above), the TSR must check the current VM. If the System VM is current, the TSR calls through the callback address and returns. If another VM is current, the TSR uses the INT 2Fh service Switch VMs and Callback. (The following code lists the parameters for this service.) This INT 2Fh service is just another way of calling the VMM's Call_Priority_VM_Event service — a fact that gives the parameters more meaning.

The Calling Interface for Switch VMs and Callback

```
INT 2Fh
AX=1685h (function code Switch VMs and Callback)
BX=switch to this VM (id)
ES:DI=address of function to call when VM is current
CX=flags
DX:SI=priority boost
Flags: PEF_Wait_For_STI (0001h) to wait until interrupts are enabled
       PEF_Not_Crit (0002h) to wait until critical section is unowned
Priority boost: Cur_Run_VM_Boost (00000004h) to run the VM for its full
                   time slice
                Low_Pri_Device_Boost (00000010h) to give the VM
                   moderate priority over other VMs
                High_Pri_Device_Boost (00001000h) to give the VM
                   significant priority over other VMs
                Critical_Section_Boost (00100000h) to give the VM
                   same priority as if in critical section
                Time_Critical_Boost (00400000h) to give the VM
         higher priority than a critical section
```

To use this service, a TSR tells the VMM (through INT 2Fh) which VM to schedule, what kind of priority boost to give the VM (to make it get scheduled faster), and an address to call when the VM switch has occurred. Because the code that will execute is a Windows driver DLL, the TSR needs to schedule VM1, which is the System VM. A TSR doesn't usually need to use a priority boost, so this parameter would usually be zero. Last, the callback address given to VMM is the same one given to the TSR by the Windows driver DLL.

After making the INT 2Fh call, the TSR returns. The VMM will schedule the System VM, and the System VM will eventually become the current VM. At that time, VMM will call the callback address registered with Switch VMs and Callback. When that happens, the callback in the Windows driver DLL will finally execute in protected mode and in the right VM. The following code fragment illustrates the use of Switch VMs and Callback in a TSR.

```
; TSR uses Switch VMs and Callback
mov  ax, 1683h      ; function Get Current VM
int  2fh
cmp  CallbackVM, 1 ; is SYSVM current?
jz   same_VM       ; yes, no need to switch

mov  ax, 1685h      ; function Switch VM and Callback
mov  bx, 1          ; switch to SYSVM
mov  cx, 0          ; flags
mov  dx, 0
mov  si, 0          ; priority boost
; callback address registered by Windows driver goes in ES:DI
mov  di, WORD PTR CallbackAddr+2
mov  es, di
mov  di, WORD PTR CallbackAddr
int  2fh
jmp  xit

same_VM:
pushf               ; SYSVM is current, use callback directly
call DWORD PTR:CallbackAddr
xit:
```

Once the System VM is current, the callback used by the TSR can execute — in the right VM. This callback is actually a DPMI stub function with several important duties. The stub immediately preserves all V86 mode register values (by copying them into a call structure) and then switches into protected mode. Next, the stub loads ES:DI with a pointer to the call structure and loads DS:SI with a protected mode pointer that addresses the real mode stack. Finally, the stub calls the Windows driver DLL, at the address originally registered by DPMI Allocate Real Mode Callback Address.

The driver now executes. Because the driver is running in interrupt context, the usual prohibitions apply: no DOS or BIOS functions and only Windows API functions from the interrupt-safe list in Chapter 15. Also, any pointers in the real mode call structure (pointed to by ES:DI) are real mode pointers and thus can't be used directly but must be translated into protected mode pointers. Earlier in this chapter, I've explained how to perform this conversion using AllocSelector, SetSelectorBase, and SetSelectorLimit.

The driver can access register values passed from the TSR by examining the appropriate field of the call structure (pointed to by ES:DI). The driver can also modify the call structure to return register values to the TSR. Moreover, if the TSR communicates with the driver through stack parameters instead of register parameters, the driver can even access the real mode stack, using the protected mode pointer in DS:SI. For example, if the TSR pushes a single word value onto the stack and does a far call to the callback address, the (real mode) stack looks like:

```
SS:SP+4 - parameter
SS:SP+2 - return address of TSR (segment)
SS:SP   - return address of TSR (offset)
```

Thanks to the DPMI stub function, when the Windows driver DLL executes, these same values can be accessed relative to DS:SI:

```
DS:SI+4 - parameter
DS:SI+2 - return address of TSR (segment)
DS:SI   - return address of TSR (offset)
```

When the driver finishes, it can't just exit with a simple iret instruction. When it returns and DPMI switches back to V86 mode, DPMI restores all V86 mode registers from the call structure, including CS and IP. That means the real mode code will resume execution at the CS:IP value in the call structure. Normally you want the TSR to resume execution at the instruction *following* the far call into the callback. Notice these desired CS and IP values are on the real mode stack pointed to by DS:SI, at locations SI+2 and SI. So the Windows driver DLL retrieves the desired CS and IP from the real mode stack and places them in the real mode call structure before doing the iret.

The following code illustrates how to fix up the `CS:IP` in the call structure. This fix up should be completed immediately before leaving the callback in the Windows driver DLL.

```
Callback:
; do your own thing
; access real mode call structure via ES:DI
; if parameters needed from real mode stack,
; use DS:SI
call DoYourOwnThing

; Extract proper real mode CS and IP from
; top of real mode stack, pointed to by DS:SI.
; Put CS and IP values into real mode call structure
cld
lodsw
mov    WORD PTR es:[di.RM_IP], ax ; real mode IP
lodsw
mov    WORD PTR es:[di.RM_CS], ax ; real mode CS
add    WORD PTR es:[di.RM_SP], 4  ; toss old CS:IP from stack
iret
```

Summary

The DPMI services make it possible for a Windows driver DLL to communicate with DOS TSRs and device drivers. If you already have a DOS driver, then modifying it to be Windows-aware may be your shortest development path.

If, however, you are creating a DOS-based driver from scratch, the information in this chapter should make it obvious that a driver that is called via software interrupt and that expects all parameters in registers will be the easiest to implement and support.

Listing 17.1   DOSTSR.H

```
typedef struct
{
    WORD    usReadBufSize;
} DRIVERPARAMS, FAR * PDRIVERPARAMS;

typedef struct
{
    WORD    version;
} DRIVERCAPS, FAR * PDRIVERCAPS;
typedef PDRIVERCAPS FAR * PPDRIVERCAPS;

typedef struct
{
    WORD        usDevNumber;
    BOOL        bFlags;
    DRIVERPARAMS params;
} DEVICECONTEXT, FAR *HDEVICE;

HDEVICE FAR PASCAL DeviceOpen( void );
WORD FAR PASCAL DeviceClose( HDEVICE );
WORD FAR PASCAL DeviceGetWriteStatus( HDEVICE, LPWORD pusStatus );
WORD FAR PASCAL DeviceGetReadStatus( HDEVICE, LPWORD pusStatus );
WORD FAR PASCAL DeviceWrite( HDEVICE, LPBYTE lpData, LPWORD pcBytes );
WORD FAR PASCAL DeviceRead( HDEVICE, LPBYTE lpData, LPWORD pcBytes );
WORD FAR PASCAL DeviceSetDriverParams( HDEVICE, PDRIVERPARAMS pParms );
WORD FAR PASCAL DeviceGetDriverParams( HDEVICE, PDRIVERPARAMS pParms );
WORD FAR PASCAL DeviceGetDriverCapabilities( HDEVICE, PPDRIVERCAPS ppDriverCaps );
```

Listing 17.2   UART.H

```
#define UART_REG_THR        0x00
#define UART_REG_RDR        0x00
#define UART_REG_IER        0x01
#define UART_REG_IIR        0x02
#define UART_REG_LCR        0x03
#define UART_REG_MCR        0x04
#define UART_REG_LSR        0x05
#define UART_REG_BAUDLO     0x00
#define UART_REG_BAUDHI     0x01

#define UART_IIR_NONE       0x01
#define UART_IIR_THRE       0x02
#define UART_IIR_RXRDY      0x04
#define UART_IER_THRE       0x02
#define UART_IER_RXRDY      0x01
#define UART_MCR_OUT2       0x08
#define UART_MCR_LOOP       0x10
#define UART_LSR_THRE       0x20
#define UART_LCR_DLAB       0x80
#define UART_LCR_8N1        0x03
#define UART_LSR_RXRDY      0x01
#define BAUD_1200           0x60
```

Listing 17.3   `DOSTSR.C`

```c
#include <io.h>
#include <fcntl.h>
#include <sys\types.h>
#include <sys\stat.h>
#include <errno.h>
#include <stdlib.h>

#include <windows.h>
#include <conio.h>
#include "dostsr.h"

#define FLAGS_OPEN              0x04

#define TSR_FUNC_OPEN           0x00
#define TSR_FUNC_READSTATUS     0x00
#define TSR_FUNC_WRITESTATUS    0x00
#define TSR_FUNC_READ           0x00
#define TSR_FUNC_WRITE          0x00
#define TSR_FUNC_GETPARAMS      0x00
#define TSR_FUNC_GETCAPS        0x00


#define SET( value, mask )    value |= mask
#define CLR( value, mask )    value &= (~mask)

DEVICECONTEXT Device1 = { 0 };
DRIVERPARAMS DefaultParams = { 1024 };
DRIVERCAPS DriverCaps = { 0x0101 };

BOOL ValidHandle( HDEVICE hDevice );
WORD DosGetStatus( WORD hnd, WORD InOut, BOOL *pReady );
WORD DosReadOrWrite( WORD hnd, WORD ReadOrWrite, LPBYTE lpBuf, LPWORD pcbBytes );
WORD DosGetDeviceData( WORD hnd, WORD *pData );

HDEVICE FAR PASCAL _export DeviceOpen(   )
{
    HDEVICE    hDevice;
    WORD       usData;

    OutputDebugString( "DeviceOpen\n" );

    hDevice = &Device1;

    if (hDevice->bFlags & FLAGS_OPEN)
        return -1;

    hDevice->usDosHandle = open( "com1" , O_BINARY | O_RDWR );
    if (hDevice->usDosHandle == -1)
        return -1;

    hDevice->params = DefaultParams;

    SET( hDevice->bFlags, FLAGS_OPEN);

    return hDevice;
}
```

Listing 17.3 (continued)   DOSTSR.C

```
WORD FAR PASCAL _export DeviceClose( HDEVICE hDevice )
{
    OutputDebugString( "DeviceClose\n" );

    if (!ValidHandle( hDevice ))
        return -1;

    if ((hDevice->bFlags & FLAGS_OPEN) == 0)
        return -1;

    CLR( hDevice->bFlags, FLAGS_OPEN );

    close( hDevice->usDosHandle );

    return 0;
}

WORD FAR PASCAL _export DeviceGetWriteStatus( HDEVICE hDevice, LPWORD pusStatus )
{
    BOOL    bReady;

    OutputDebugString( "DeviceGetWriteStatus\n" );

    if (!ValidHandle( hDevice ))
        return -1;

    if ((hDevice->bFlags & FLAGS_OPEN) == 0)
        return -1;

    DosGetStatus( hDevice->usDosHandle, DOS_STATUS_OUT, &bReady );
    if (bReady)
    {
        *pusStatus = 1;         // ready to transmit
    }
    else
    {
        *pusStatus = 0;         // not ready to transmit
    }

    return 0;
}
```

Listing 17.3 (continued)   *DOSTSR.C*

```c
WORD FAR PASCAL _export DeviceGetReadStatus( HDEVICE hDevice, LPWORD pusStatus )
{
    BOOL    bReady;

    OutputDebugString( "DeviceGetReadStatus\n");

    if (!ValidHandle( hDevice ))
        return -1;

    if ((hDevice->bFlags & FLAGS_OPEN) == 0)
        return -1;

    DosGetStatus( hDevice->usDosHandle, DOS_STATUS_IN, &bReady );
    if (bReady)
    {
        *pusStatus = 1;         // data ready
    }
    else
    {
        *pusStatus = 0;         // no data ready
    }

    return 0;
}

WORD FAR PASCAL _export DeviceWrite( HDEVICE hDevice, LPBYTE lpData, LPWORD pcBytes )
{
    OutputDebugString( "DeviceWrite\n");

    if (!lpData)
        return -1;

    if (!ValidHandle( hDevice ))
        return -1;

    if ((hDevice->bFlags & FLAGS_OPEN) == 0)
        return -1;

    DosReadOrWrite( hDevice->usDosHandle, DOS_WRITE, lpData, pcBytes );

    return 0;
}
```

Listing 17.3 (continued)   DOSTSR.C

```c
WORD FAR PASCAL _export DeviceRead( HDEVICE hDevice, LPBYTE lpData, LPWORD pcBytes )
{
    WORD        i;

    OutputDebugString( "DeviceRead\n");

    if (!lpData)
        return -1;

    if (!ValidHandle( hDevice ))
        return -1;

    if ((hDevice->bFlags & FLAGS_OPEN) == 0)
        return -1;

    DosReadOrWrite( hDevice->usDosHandle, DOS_READ, lpData, pcBytes );

    return 0;
}

WORD FAR PASCAL _export DeviceSetDriverParams( HDEVICE hDevice,
                                               PDRIVERPARAMS pParams )
{
    OutputDebugString( "DeviceSetDriverParams\n");

    if (!pParams)
        return -1;

    if (!ValidHandle( hDevice ))
        return -1;

    if ((hDevice->bFlags & FLAGS_OPEN) == 0)
        return -1;

    hDevice->params = *pParams;

    return 0;
}

WORD FAR PASCAL _export DeviceGetDriverParams( HDEVICE hDevice,
                                               PDRIVERPARAMS pParams )
{
    OutputDebugString( "DeviceGetDriverParams\n");

    if (!pParams)
        return -1;

    if (!ValidHandle( hDevice ))
        return -1;

    if ((hDevice->bFlags & FLAGS_OPEN) == 0)
        return -1;

    *pParams = hDevice->params;

    return 0;
}
```

Listing 17.3 (continued) *DOSTSR.C*

```c
WORD FAR PASCAL _export DeviceGetDriverCapabilities( HDEVICE hDevice,
                                                     PPDRIVERCAPS ppDriverCaps )
{
    OutputDebugString( "DeviceGetDriverCapabilities\n");

    if (!ppDriverCaps)
        return -1;

    if (!ValidHandle( hDevice ))
        return -1;

    if ((hDevice->bFlags & FLAGS_OPEN) == 0)
        return -1;

    *ppDriverCaps = &DriverCaps;

    return 0;
}

BOOL ValidHandle( HDEVICE hDevice )
{
    return (hDevice == &Device1);
}

WORD DosGetDeviceData( WORD hnd, WORD *pData )
{
    WORD    rc = 0;
    WORD    data;

    _asm
    {
        mov     ah, 0x44
        mov     al, 0x00
        mov     bx, hnd
        int     21h
        jnc     ok
        mov     rc, ax
        jmp     xit
ok:     mov     data, ax
        jmp     xit
xit:
    }

    *pData = data;

    return rc;
}
```

Listing 17.3 (continued)   DOSTSR.C

```
WORD DosGetStatus( WORD hnd, WORD InOut, BOOL *pReady )
{
    WORD    rc = 0;
    BYTE  stat;

    *pReady = 0;
    _asm
    {
        mov     ax, InOut
        mov     ah, 0x44
        mov     bx, hnd
        int     21h
        jnc     ok
        mov     rc, ax
        jmp     xit
ok:     mov     stat, al
        jmp     xit
xit:
    }

    *pReady = (stat == 0xFF ? TRUE : FALSE );

    return rc;
}

WORD DosReadOrWrite( WORD hnd, WORD ReadOrWrite, LPBYTE lpBuf, LPWORD pcbBytes )
{
    WORD    rc = 0;
    WORD    cBytes = *pcbBytes;

    _asm
    {
        mov     ax, ReadOrWrite
        xchg    ah, al
        mov     bx, hnd
        mov     cx, cBytes
        push    ds
        lds     di, lpBuf
        mov     dx, di
        int     21h
        pop     ds
        jnc     ok
        mov     rc, ax
        jmp     xit
ok:     mov     cBytes, ax
        jmp     xit
xit:
    }

    *pcbBytes = cBytes;

    return rc;

}
```

Listing 17.3 (continued)   DOSTSR.C

```
#ifdef DOS
main()
{
    char abOut[4], abIn[4];
    unsigned short status;
    HDEVICE hDev;
    unsigned short cb;

    hDev = DeviceOpen();
    DeviceGetWriteStatus( hDev, &status );
    cb = 3;
    abOut[0] = 'a';
    abOut[1] = 't';
    abOut[2] = '\r';
    DeviceWrite( hDev, abOut, &cb );
    DeviceGetReadStatus( hDev, &status );
    DeviceRead( hDev, abIn, &cb );
    DeviceClose( hDev );
}
#endif
```

Listing 17.4   DOSTSR.MAK

```
all: dostsr.dll

# DRIVER DLL

dostsr.obj: dostsr.c dostsr.h
  cl -c -W3 -ASw -Gsw2 -Oi $*.c

dostsr.dll: dostsr.def dostsr.obj
  link dostsr,dostsr.dll,dostsr.map /CO /MAP,sdllcew libw /nod/noe,dostsr.def
  implib driver.lib dostsr.dll
  copy dostsr.dll \windows\driver.dll
```

Listing 17.5   DOSTSR.DEF

```
LIBRARY      DRIVER
DESCRIPTION  "DLL To Interface to DOS TSR"
EXETYPE      WINDOWS
DATA         PRELOAD MOVEABLE SINGLE
CODE         PRELOAD MOVEABLE DISCARDABLE
```

## Chapter 18

Thunks:
Calling from 32-bit to 16-bit

Chapter 13 eplained that Win32 DLLs can perform only very limited types of hardware interaction. Although a Win32 DLL may issue IN and OUT instructions safely when running under Windows 95 (but not under NT), a Win32 DLL may not access a memory-mapped device, perform DMA transfers, or handle hardware interrupts. To *properly* implement these tasks under Win32 you should write a true device driver — a VxD for Windows 95 and a kernel-mode driver for Windows NT.

If you must support Windows NT, you really must write a driver. But if you're concerned only about Windows 95, there is an alternative to writing a VxD. You can put the hardware access in a 16-bit DLL (using the techniques in Chapters 14 through 17), and then write a translation layer to connect the Win32 application to the 16-bit DLL. The translation layer is called a "thunk". Note that Windows 95 uses flat thunks, not to be confused with the universal thunks supported by Win32 or the generic thunks supported by Windows NT. From now on I'll usually just say thunk, but I will always mean *flat* thunk.

The rest of this chapter will examine:

• What is a flat thunk?

• What tasks are performed by the thunk layer?

• How does the thunk layer do its "magic"?

• What are the steps for implementing a thunk layer?

• How is a thunk layer built?

What is a Flat Thunk?

Suppose you're writing a Win32 application, and you need to call some functions in a Win16 DLL. Figure 18.1 shows what you want to do.

Because of the 32-bit/16-bit boundary shown in Figure 18.1, simply calling from APP32 to DLL16 won't work. In order to successfully call from 32-bit down to 16-bit, you must address such issues as: pointer translation (flat vs segment:offset), stack addressing (SS:ESP vs SS:SP), and code segment size (16-bit or 32-bit). A thunk is a layer of code that handles these issues; that does the "magic" necessary to allow 32-bit code to call 16-bit code. Although flat thunks can be used in the other direction, 16-to-32, I'll discuss only 32-to-16 here, because hardware access functions are in 16-bit code.

You should encapsulate the thunk layer in a 32-bit DLL. Create a 32-bit DLL that contains the same set of exported functions as the 16-bit DLL you want to call. In each of the Win32 DLL's exported functions, the function in the 32-bit DLL calls the analogous function in the 16-bit DLL.

Figure 18.1   Showing why 32-bit applications can't call directly to a 16-bit DLL.

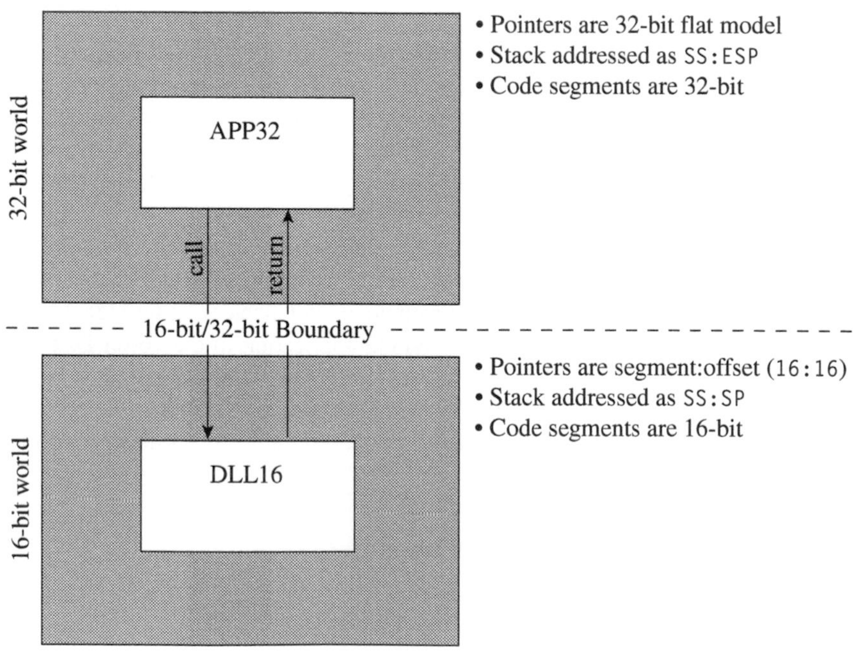

- Pointers are 32-bit flat model
- Stack addressed as SS:ESP
- Code segments are 32-bit

- Pointers are segment:offset (16:16)
- Stack addressed as SS:SP
- Code segments are 16-bit

The 32-bit DLL doesn't call the 16-bit DLL directly, but goes through a thunk layer (Figure 18.2). If the flat-to-segmented pointer conversions and stack switching mentioned above sounds too complicated, don't worry — you don't have to write the code in the thunk layer. Thunks are automatically generated by the Microsoft Thunk Compiler. You provide a "thunk script" (a file containing modified function prototypes) as input, and the thunk compiler produces code (an assembly language source file) as output. It is this code, linked into both the 32-bit DLL and the 16-bit DLL, that acts as the thunk layer.

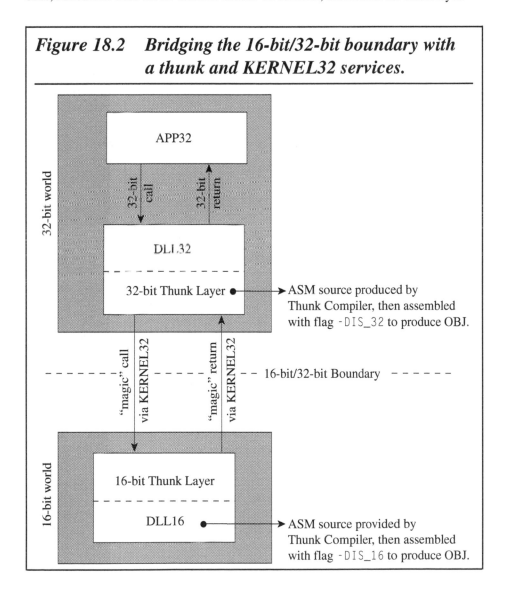

Figure 18.2   *Bridging the 16-bit/32-bit boundary with a thunk and KERNEL32 services.*

The Thunk Compiler is not provided with the Visual C++ package. It's only available in the Win32 SDK, which is itself only available with the MSDN CD Professional Subscription. You'll also need the Microsoft Assembler (MASM) to assemble the thunk compiler's output.

The assembly language source produced by the thunk compiler serves double duty; it's used on both sides of the 16-bit/32-bit boundary. In building the thunk, the assembly file is assembled first as 32-bit code (using the flag -DIS_32), producing a 32-bit OBJ which is linked into the Win32 DLL. Then the same ASM file is assembled again as 16-bit code (using the flag -DIS_16), producing a 16-bit OBJ, and linked into the 16-bit DLL.

Thunk Layer Tasks

The flat thunk layer generated by the Thunk Compiler performs these tasks:

- translates pointer parameters,
- translates integer parameters,
- switches from 32-bit to 16-bit stack and back again,
- transfers control from the 32-bit calls to the 16-bit target and back again, and
- translates return values to the appropriate 32-bit representation.

When a pointer parameter is passed from 32-bit code to 16-bit code via a thunk, the pointer must be translated from a flat (0:32) pointer to a far (16:16) pointer. Because a flat pointer is a linear address, the thunk layer's translation involves allocating a selector and setting its base address equal to the flat pointer value.

A simpler translation must be performed on integer-sized parameters, because an integer is 32 bits for 32-bit code but only 16 bits for 16-bit code. To handle an integer parameter, the 32-bit caller would push a 32-bit argument on the stack, but the called 16-bit function would pop only 16 bits off the stack. The thunk code must adjust the stack to contain the truncated (16-bit) version of the integer instead.

After converting parameters, the thunk layer prepares for the trip to 16-bit land by switching from a 32-bit stack to a 16-bit stack (i.e. from a stack addressed by SS:ESP to a stack addressed by SS:SP). This translation also involves selector allocation and manipulation. On the return trip, from 16-bit code back to 32-bit code, the thunk code reverses the process to return to the original 32-bit stack.

Once the parameters and stack have been modified, the thunk layer transfers control from 32-bit to 16-bit, but not directly from the thunk compiler code. Instead, the thunk compiler code makes a call into a KERNEL32 function; KERNEL32 completes the 32-to-16 transition, using some fancy stack manipulation to push the segment and offset of the 16-bit target onto the stack, and then uses a RETF to essentially "jump" into the 16-bit world. (The next section will explain more about KERNEL32's role in the thunk.)

Once the call has returned to 32-bit land, the thunk layer converts the 16-bit callee's return value, if it had one. A 16-bit function returns a 32-bit value in two registers, DX:AX. But its 32-bit caller expects a 32-bit return value to be found in EAX, so the thunk layer must copy the return value from DX:AX to EAX.

Thunk Layer Magic

In this section, we'll examine a simple thunk script and the code produced by the thunk compiler to see how a thunk layer performs its magic.

Below is a thunk script for a 16-bit DLL with a single exported function named DLL16Foo, which has an integer parameter and a void pointer parameter and returns an unsigned long.

```
unsigned long DLL16Foo(int nThunk, void *lpvoidThunk)
{
    lpvoidThunk = input;
}
```

The thunk script contains something that looks like a function, but acts more like a function prototype. Inside the function "body" is additional information about the function's pointer parameters, specifying each pointer parameter as an input parameter, an output parameter, or both. The input keyword directs the thunk compiler to generate code to translate a 32-bit flat pointer to a far pointer before calling the 16-bit DLL, the output keyword directs the thunk compiler to generate code that translates a far pointer "returned" by the 16-bit DLL to a flat pointer usable by 32-bit code, and the inout keyword results in code that does both. By default, the thunk compiler treats all pointers as input. The following fragment shows the assembly code generated for the DLL16Foo thunk script.

```
public DLL16Foo@32
DLL16Foo@32:
    mov   cl,0
public IIDLL16Foo@8
IIDLL16Foo@8:
    push  ebp
    move  bp,esp
    push  ecx
    sub   esp,60
    push  word ptr [ebp+8] ;nThunk: dword->word
    call  SMapLS_IP_EBP_12 ;lpvoidThunk: flat->16:16
    push  eax
    call  dword ptr [pfnQT_Thunk_X2to16]
    shl   eax,16
    shrd  eax,edx,16
    call  SUnMapLS_IP_EBP_12
    leave
    retn  8
```

Note that although the thunk script gave the 16-bit DLL's function name as DLL16Foo, the name of the function in the generated assembly code is different: DLL16Foo@32. This is an example of "name decoration" for the PASCAL naming/calling convention. (The 32 refers not to 32-bits, but to the total number of bytes used for parameters.) Because functions exported by a 16-bit DLL are always declared as PASCAL, the code in the Win32 DLL that calls DLL16Foo actually results in compiled code that calls DLL16Foo@32.

Immediately following the function name declaration is a mov instruction and another function name declaration:

```
DLL16Foo@32:
    mov cl,0
public IIDLL16Foo@8
IIDLL16Foo@8:
```

This second function, IDLL16Foo@8, is a helper function which expects the CL register to contain a "function number" parameter. If the thunk script included multiple function prototypes, the thunk compiler code for each of them would have a similar MOV CL instruction, but with a different operand, followed by a jump to IIDLL16Foo@8. So IIDLL16Foo@8 serves as a common intermediate function for all of the 16-bit DLL exported functions.

The first few instructions in IIDLL16Foo@8 are standard prologue code for setting up the stack frame and reserving storage for local stack variables. The PUSH ECX puts the "function number" parameter from its immediate caller (in this case DLL16Foo@32) on the stack in preparation for a call to another subroutine later. The next push, PUSH WORD PTR [EBP+8], is the translation of the 32-bit caller's first parameter, an integer. The 32-bit caller pushed a DWORD onto the stack, which now lives at EBP+8, and this thunk code takes only a WORD of that parameter and pushes it onto the stack as an integer parameter for the 16-bit callee.

The 32-bit caller's pointer parameter lives at EBP+12, and the next instruction, CALL SMapLS_IP_EBP_12, calls a subroutine to translate this pointer. If DLL16Foo was declared such that the pointer parameter ended up at EBP+8 instead (e.g. if there was no integer parameter), then the thunk compiler would have generated code to call SMapLS_IP_EBP_8 instead.

SMapLS_IP_EBP_12 is an undocumented function exported by KERNEL32.DLL. This function translates the 32-bit flat pointer located at EBP+12 to an equivalent 16-bit far pointer. Its sibling functions — EBP_8, EBP_10, etc. — act similarly for pointers located at EBP+8, EBP+10, etc. A flat pointer is really a linear base address, so the translation involves nothing more than allocating a 16-bit selector and setting the selector's base address equal to the value at EBP+12. SMapLS_IP_EBP_12 doesn't actually allocate selectors, but uses the next available Local Descriptor Table (LDT) selector from a pool of already-allocated selectors. This selector is returned to the pool when the thunk compiler code calls SUnMapLS_IP_EBP_12. This cleanup call happens after the thunk has returned from the call down to the 16-bit DLL.

Now that the parameters are translated, the switch from 32-bit to 16-bit happens as part of this line:

```
call dword ptr [pfnQT_Thunk_X2to16]
```

This call through a table of function pointers eventually results in a call to another undocumented KERNEL32 function called QT_Thunk. QT_Thunk is passed the 16:16 address of the real DLL16Foo function. It is QT_Thunk that performs the switch from a 32-bit stack to a 16-bit stack and then jumps to the 16:16 address of the real DLL16Foo function in the real 16-bit DLL.

The process of initializing the table of function pointers mentioned above involves quite a lot of black magic. I won't go into detail, but in short, the 32-bit DLL must call a special initialization function, called ThunkConnect32, in its DllMain. The ThunkConnect32 function is also generated by the Thunk Compiler and "connects" the 32-bit DLL to the 16-bit DLL by initializing the table with the 16:16 address of each of the 16-bit DLL's exported functions. (These addresses aren't known until run time, when the 16-bit DLL has been loaded.) ThunkConnect32 uses yet another undocumented KERNEL32 function, Connect32, to obtain these 16:16 addresses for the table.

Creating a Thunk Layer, Step by Step

To create a thunk DLL, follow the following procedure:

1. Create a thunk script by modifying the 16-bit DLL's header file to include input, output, and inout information about each exported function's parameters.

2. Create a 32-bit DLL with a set of exported functions that match the 16-bit DLL's exported functions.

3. Create a DLL entry point in the 32-bit DLL (usually called DllEntryPoint) which calls X_ThunkConnect32, where X is the name of the thunk script.

4. Add a new exported function to the 16-bit DLL, called DllEntryPoint, which calls X_ThunkConnect16, where X is the name of the thunk script.

The above procedure, as well as the build procedure in the following section, must be followed exactly. Deviation will most likely result in a thunk that doesn't build, doesn't work, or both.

I'll explain each of these steps in more detail, using the 16-bit SKELETON DLL from Chapter 13 as an example. The 32-bit thunk DLL will be called `SKEL32.DLL` and will consist of: `SKEL32.C` [the 32-bit DLL source file (Listing 18.3, page 408)]; `SKEL32.H` [the header used by Win32 applications (Listing 18.2, page 407)]; `SKEL32.DEF` [the module definition file (Listing 18.5, page 410)]; and `SKELETON.THK` [the thunk script (Listing 18.1, page 405)].

The Thunk Script

The starting point for the script file is the 16-bit DLL's header file, `SKELETON.H`. The first step in creating `SKELETON.THK` from `SKELETON.H` is to add the following line:

```
enablemapdirect3216 = true;      //creates 32 to 16 thunk
```

This tells the thunk compiler the direction of the thunk — in this case, from 32-bit to 16-bit. The next step is to take each function in `SKELETON.H`, modify its counterpart in `SKELETON.THK` to include a "function body" containing parameter information. The function definition is also modified to remove any declaration keywords (such as `export`, `far`, `pascal`). Thus, the `DeviceGetWriteStatus` definition in `SKELETON.H` (Listing 18.6, page 411), shown below:

```
int FAR PASCAL DeviceGetWriteStatus( HDEVICE hDevice, LPWORD usStatus)
```

is transformed into this in `SKELETON.THK`

```
int DeviceGetWriteStatus( HDEVICE hDevice, LPWORD pusStatus)
{
   // the hDevice pointer is used as input by the 16-bit DLL
   hDevice=input;
}
```

All of the functions except for `DeviceGetDriverCapabilities` use `input` pointer parameters. `DeviceGetDriverCapabilities` uses one `input` and one `output` parameter:

```
int DeviceGetDriverCapabilites( HDEVICE hDevice,
                                PDRIVERCAPS *ppDriverCaps )
{
   // the hDevice pointer is used as input by the 16-bit DLL
   hDevice=input;
   ppDriverCaps=output;
}
```

In addition to function prototypes, the real SKELETON.H contains typedefs (HDEVICE, PDRIVERPARAMS, etc.) and includes WINDOWS.H for additional typedefs (LPBYTE, LPWORD, etc.). Because WINDOWS.H contains a lot of other stuff that the thunk compiler wouldn't understand, SKELETON.THK doesn't actually include WINDOWS.H. Instead, SKELETON.THK directly contains all the necessary typedefs, extracted from WINDOWS.H and SKELETON.H, as shown in the following code fragment.

```
typedef unsigned char BYTE;
typedef unsigned short WORD;
typedef unsigned long DWORD;
typedef BYTE far* LPBYTE;

typedef struct
{
   WORD usDevNumber;
} DEVICECONTEXT
typedef DEVICECONTEXT FAR *HDEVICE;

typedef struct
{
   WORD  usReadBufSize;
} DRIVERPARAMS;
typedef DRIVERPARAMS FAR *PDRIVERPARAMS;

typedef struct
{
   WORD version;
} DRIVERCAPS;
typedef DRIVERCAPS FAR *PDRIVERCAPS;
typedef PDRIVERCAPS FAR *PPDRIVERCAPS;
```

SKEL32.C

The 16-bit SKELETON.DLL exports nine functions, so SKEL32.C will contain the same nine functions, but with the suffix "32" added to the function name. Here's an example of one of those nine functions in SKEL32.C:

```
#define DLLEXPORT __declspec( dllexport )

DLLEXPORT int APIENTRY DeviceGetWriteStatus32( HDEVICE hDevice,
                                               LPWORD pusStatus)
{
   return DeviceGetWriteStatus( hDevice, pusStatus );
}
```

The DLLEXPORT technique used above to declare an exported function in a 32-bit DLL is the method recommended by VC++ 4.x. See your compiler documentation for details on declaring an exported function.

SKEL32.C also contains a DLL entry point, DllEntryPoint which does nothing but call the function SKELETON_ThunkConnect32 (which will be provided by the assembly language thunk module). Note that the SKELETON_ prefix comes from t parameter on the thunk compiler command line. When processing the script file, the thunk compiler automatically adds this prefix to the name of each function it creates in the assembly language module.

SKELETON_ThunkConnect32 takes four parameters: the name of the 16-bit DLL ("SKELETON.DLL"), the name of the 32-bit DLL ("SKEL32.DLL"), and the hInst and dwReason parameters provided by DllEntryPoint's caller. The following fragment shows the code for DllEntryPoint.

```
// function prototype for function provided by assembly thunk module
BOOL WINAPI SKELETON_ThunkConnect32(LPSTR pszDll16, LPSTR pszDll32, DWORD hIinst,
                                    DWORD dwReason);

BOOL WINAPI DllEntryPoint(DWORD hInst, DWORD dwReason, DWORD wReserved)
{
    if (!(SKELETON_ThunkConnect32("SKELETON.DLL", "SKEL32.DLL", hInst, dwReason)))
    {
        return FALSE;
    }
    return TRUE;
}
```

SKELETON.C

The source for the 16-bit SKELETON.DLL must be modified slightly also, to add a new export function to SKELETON.C. This function must be named DllEntryPoint. It acts as the mirror image of its counterpart in the 32-bit DLL, calling SKELETON_Thunk-Connect16 instead of SKELETON_ThunkConnect32. It passes exactly the same parameters in exactly the same order. SKELETON_ThunkConnect16 is also provided by the assembly language thunk module. The following fragment shows the code for DllEntryPoint.

```
// function prototype for function provided by assembly thunk module
BOOL FAR PASCAL __export SKELETON_ThunkConnect16(LPSTR pszDll16, LPSTR pszDll32,
                                                 WORD hInst, DWORD dwReason);

BOOL FAR PASCAL __export DllEntryPoint(DWORD dwReason, WORD hInst, WORD wDS,
                                       WORD wHeapSize, DWORD dwReserved1,
                                       WORD wReserved2)
{
   if (!(SKELETON_ThunkConnect16("SKELETON.DLL", "SKEL32.DLL", hInst, dwReason)))
   {
      return FALSE;
   }
   return TRUE;
}
```

Building the Thunk Layer, Step by Step

Building a thunk layer consisting of a 16-bit and a 32-bit DLL is more complicated than building normal 16-bit and 32-bit DLLs. Though the two makefiles (SKEL16.MAK and SKEL16.MAK) hide the complexity, it's worth a closer look at the steps involved.

The 16-bit DLL must be built first. This is necessary because it's SKEL16.MAK that executes the thunk compiler to produce the assembly source file SKELETON.ASM, which is required by both makefiles. The thunk compiler command line used by SKEL16.MAK is

```
thunk -t SKELETON -o skeleton.asm skeleton.thk
```

The -t flag specifies a "base name" which the thunk compiler prefixes to the names of the ThunkConnect16 and ThunkConnect32 functions in the assembly output file. The above command line results in functions named SKELETON_ThunkConnect16 and SKELETON_ThunkConnect32, which matches the names used in the DLLEntry-Point code in SKELETON.C and SKEL32.C.

SKEL16.MAK then assembles SKELETON.ASM, using the /DIS_16 flag and naming the object file thk16.obj. The /DIS_16 flag produces code that implements the 16-bit side of the thunk layer depicted in Figure 18.2.

```
ml /DIS_16 /c /W3 /Fo thk16.obj skeleton.asm
```

The link step used by SKEL16.MAK isn't any different than building a normal 16-bit DL — other than linking in the thunk code in THK16.OBJ — but there is one final step which is unusual. The DLL must be marked as compatible with Windows 95, by running the resource compiler and using the -40 option. Without this mark, Windows 95 will refuse to load the 32-bit DLL.

The makefile for the 32-bit DLL, SKEL32.MAK, looks almost exactly like a makefile for a normal 32-bit DLL. The only difference is assembling the source generated by the thunk compiler, which resides in the 16-bit DLL's directory. This time the /DIS_32 flag is used to produce code that implements the 32-bit side of the thunk layer.

```
ml /DIS_32 /c /W3 /Fo thk32.obj ..\16\skeleton.asm
```

I've also included a sample Win32 console application which utilizes the 32-bit DLL and, indirectly, the 16-bit DLL. The application does nothing more than call the functions DeviceOpen32 and DeviceClose32. These functions are implemented in SKEL32.DLL, which in turn calls the analogous function in the 16-bit SKELETON.DLL. Note that the application is completely unaware of the thunking: the functions it uses are all in SKEL32.DLL, and it links only with SKEL32.LIB.

Summary

If your driver must support Win32 applications but you're not ready to make the transition to writing a VxD, or if you have already created a 16-bit driver DLL, a thunk layer might be your best option. Developing a thunk DLL may not be a lot of creative fun, and you do have to be careful to get all the steps right, but if you follow carefully the procedures outlined in this chapter, you can create a thunk DLL that allows you to keep hardware access in a DLL while still supporting Win32 applications.

Listing 18.1   *SKELETON.THK*

```
enablemapdirect3216 = true;

typedef unsigned char BYTE;
typedef unsigned short WORD;
typedef unsigned long DWORD;
typedef BYTE    *LPBYTE;
typedef DWORD   *LPDWORD;
typedef WORD    *LPWORD;

typedef struct
{
    WORD      usDevNumber;
} DEVICECONTEXT;

typedef DEVICECONTEXT *HDEVICE;

typedef struct
{
    WORD      usReadBufSize;
} DRIVERPARAMS;
typedef DRIVERPARAMS * PDRIVERPARAMS;

typedef struct
{
    WORD      version;
} DRIVERCAPS;
typedef DRIVERCAPS * PDRIVERCAPS;


HDEVICE DeviceOpen( void )
{
}
int DeviceClose( HDEVICE hDevice )
{
    hDevice=input;
}

int DeviceGetWriteStatus( HDEVICE hDevice, LPWORD pusStatus )
{
    hDevice=input;
    pusStatus=input;
}

int DeviceGetReadStatus( HDEVICE hDevice, LPWORD pusStatus )
{
    hDevice=input;
    pusStatus=input;
}

int DeviceWrite( HDEVICE hDevice, LPBYTE lpData, LPWORD pcBytes )
{
    hDevice=input;
    lpData=input;
    pcBytes=input;
}
```

Listing 18.1 (continued)   *SKELETON.THK*

```
int DeviceRead( HDEVICE hDevice, LPBYTE lpData, LPWORD pcBytes )
{
    hDevice=input;
    lpData=input;
    pcBytes=input;
}

int DeviceSetDriverParams( HDEVICE hDevice, PDRIVERPARAMS pParms )
{
    hDevice=input;
    pParms=input;
}

int DeviceGetDriverParams( HDEVICE hDevice, PDRIVERPARAMS pParms )
{
    hDevice=input;
    pParms=input;
}

int DeviceGetDriverCapabilities( HDEVICE hDevice, PDRIVERCAPS *ppDriverCaps )
{
    hDevice=input;
    ppDriverCaps=output;
}
```

Listing 18.2   SKEL32.H *(32-bit DLL)*

```
#ifndef SKELETON_H
#define SKELETON_H

#include <windows.h>

typedef struct
{
    WORD      usDevNumber;
} DEVICECONTEXT, FAR *HDEVICE;

typedef struct
{
    WORD      usReadBufSize;
} DRIVERPARAMS, FAR * PDRIVERPARAMS;

typedef struct
{
    WORD      version;
} DRIVERCAPS, FAR * PDRIVERCAPS;
typedef PDRIVERCAPS FAR * PPDRIVERCAPS;

#ifndef DLL

#define DLLIMPORT __declspec( dllimport )

DLLIMPORT HDEVICE APIENTRY DeviceOpen32( void );
DLLIMPORT int APIENTRY DeviceClose32( HDEVICE );
DLLIMPORT int APIENTRY DeviceGetWriteStatus32( HDEVICE, LPWORD pusStatus );
DLLIMPORT int APIENTRY DeviceGetReadStatus32( HDEVICE, LPWORD pusStatus );
DLLIMPORT int APIENTRY DeviceWrite32( HDEVICE, LPBYTE lpData, LPWORD pcBytes );
DLLIMPORT int APIENTRY DeviceRead32( HDEVICE, LPBYTE lpData, LPWORD pcBytes );
DLLIMPORT int APIENTRY DeviceSetDriverParams32( HDEVICE, PDRIVERPARAMS pParms );
DLLIMPORT int APIENTRY DeviceGetDriverParams32( HDEVICE, PDRIVERPARAMS pParms );
DLLIMPORT int APIENTRY DeviceGetDriverCapabilities32( HDEVICE,
                                        PPDRIVERCAPS ppDriverCaps );

#endif

#endif
```

Listing 18.3   SKEL32.C *(32-bit DLL)*

```c
#include <windows.h>
#include "..\16\skeleton.h"

DEVICECONTEXT Device1 = { 0 };
DRIVERPARAMS DefaultParams = { 1024 };

// function prototype for function provided by assembly thunk module
BOOL WINAPI SKELETON_ThunkConnect32(LPSTR pszDll16, LPSTR pszDll32, DWORD hIinst,
                                    DWORD dwReason);

BOOL WINAPI DllEntryPoint(DWORD hInst, DWORD dwReason, DWORD wReserved)
{
    if (!(SKELETON_ThunkConnect32("SKELETON.DLL", "SKEL32.DLL", hInst, dwReason)))
    {
        return FALSE;
    }
    return TRUE;
}

#define DLLEXPORT __declspec( dllexport )

DLLEXPORT void APIENTRY DeviceOpen32( void )
{
    DeviceOpen();
}

DLLEXPORT int APIENTRY DeviceClose32( HDEVICE hDevice )
{
    return DeviceClose( hDevice );
}

DLLEXPORT int APIENTRY DeviceGetWriteStatus32( HDEVICE hDevice, LPWORD pusStatus )
{
    return DeviceGetWriteStatus( hDevice, pusStatus );
}

DLLEXPORT int APIENTRY DeviceGetReadStatus32( HDEVICE hDevice, LPWORD pusStatus )
{
    return DeviceGetReadStatus( hDevice, pusStatus );
}

DLLEXPORT int APIENTRY DeviceWrite32( HDEVICE hDevice, LPBYTE lpData, LPWORD pcBytes )
{
    return DeviceWrite( hDevice, lpData, pcBytes );
}

DLLEXPORT int APIENTRY DeviceRead32( HDEVICE hDevice, LPBYTE lpData, LPWORD pcBytes )
{
    return DeviceRead( hDevice, lpData, pcBytes );
}
```

Listing 18.3 (continued)   *SKEL32.C (32-bit DLL)*

```
DLLEXPORT int APIENTRY DeviceSetDriverParams32( HDEVICE hDevice,
                                                PDRIVERPARAMS pParms )
{
    return DeviceSetDriverParams( hDevice, pParms );
}

DLLEXPORT int APIENTRY DeviceGetDriverParams32( HDEVICE hDevice,
                                                PDRIVERPARAMS pParms )
{
    return DeviceGetDriverParams( hDevice, pParms );
}

DLLEXPORT int APIENTRY DeviceGetDriverCapabilities32( HDEVICE hDevice,
                                                      PPDRIVERCAPS ppDriverCaps )
{
    return DeviceGetDriverCapabilities( hDevice, ppDriverCaps );
}
```

Listing 18.4   *SKEL32.MAK (32-bit DLL)*

```
all: skel32.dll

!message
!message +++++++++++++++++++++++++++++++++++++++++++++++++++++++++++++++
!message + To make the file dll32.dll, you will need to have the        +
!message + Microsoft Thunk compiler and the Microsoft Macro Assembler   +
!message + (ML) on the path.                                            +
!message +++++++++++++++++++++++++++++++++++++++++++++++++++++++++++++++
!message

skel32.obj: skel32.c skel32.h
    cl -c -W3 -Z7 -Od -DWIN32 -D_WIN32 -D_MT -D_DLL $*.c

thk32.obj:  ..\16\skeleton.asm
  ml /DIS_32 /c /W3 /Fo thk32.obj ..\16\skeleton.asm
```

Listing 18.4 (continued)   *SKEL32.MAK (32-bit DLL)*

```
# Build rule for the DLL
skel32.dll: skel32.def skel32.obj thk32.obj
    link /NODEFAULTLIB /INCREMENTAL:NO /PDB:NONE /RELEASE \
    -debug:full -debugtype:cv -align:0x1000 -dll \
    -base:0x1C000000           \
    -entry:_DllMainCRTStartup@12 \
    -out:skel32.dll         \
    -implib:skel32.lib   \
    skel32.obj thk32.obj thunk32.lib libc.lib oldnames.lib kernel32.lib


# Build rule for EXE
$(PROJ).EXE: $(BASE_OBJS) $(PROJ_OBJS) $(DLLNAME).dll
    $(link) $(linkdebug) $(guilflags4) \
    $(BASE_OBJS) $(PROJ_OBJS) $(guilibsdll) $(EXTRA_LIBS) \
    $(DLLNAME).lib \
    -out:$(PROJ).exe $(MAPFILE)


# Rules for cleaning out those old files
clean:
    del *.bak *.pdb *.obj *.res *.exp *.map *.sbr *.bsc
```

Listing 18.5   *SKEL32.DEF (32-bit DLL)*

```
LIBRARY    SKEL32

DATA       READ WRITE

EXPORTS
    SKELETON_ThunkData32
```

Listing 18.6   SKELETON.H (16-bit DLL)

```c
#ifndef SKELETON_H
#define SKELETON_H

#include <windows.h>

typedef struct
{
    WORD        usDevNumber;
} DEVICECONTEXT, FAR *HDEVICE;

typedef struct
{
    WORD        usReadBufSize;
} DRIVERPARAMS, FAR * PDRIVERPARAMS;

typedef struct
{
    WORD        version;
} DRIVERCAPS, FAR * PDRIVERCAPS;
typedef PDRIVERCAPS FAR * PPDRIVERCAPS;

HDEVICE FAR PASCAL DeviceOpen( void );
int FAR PASCAL DeviceClose( HDEVICE );
int FAR PASCAL DeviceGetWriteStatus( HDEVICE, LPWORD pusStatus );
int FAR PASCAL DeviceGetReadStatus( HDEVICE, LPWORD pusStatus );
int FAR PASCAL DeviceWrite( HDEVICE, LPBYTE lpData, LPWORD pcBytes );
int FAR PASCAL DeviceRead( HDEVICE, LPBYTE lpData, LPWORD pcBytes );
int FAR PASCAL DeviceSetDriverParams( HDEVICE, PDRIVERPARAMS pParms );
int FAR PASCAL DeviceGetDriverParams( HDEVICE, PDRIVERPARAMS pParms );
int FAR PASCAL DeviceGetDriverCapabilities( HDEVICE, PPDRIVERCAPS ppDriverCaps );

#endif
```

Listing 18.7   SKELETON.C *(16-bit DLL)*

```c
#include <windows.h>
#include "skeleton.h"

DEVICECONTEXT Device1 = { 0 };
DRIVERPARAMS DefaultParams = { 1024 };

BOOL FAR PASCAL __export SKELETON_ThunkConnect16(LPSTR pszDll16, LPSTR pszDll32,
                                          WORD hInst, DWORD dwReason);

BOOL FAR PASCAL __export DllEntryPoint(DWORD dwReason, WORD hInst, WORD wDS,
                                    WORD wHeapSize, DWORD dwReserved1,
                                    WORD wReserved2)
{
    if (!(SKELETON_ThunkConnect16("SKELETON.DLL",    // name of 16-bit DLL
                             "SKEL32.DLL",        // name of 32-bit DLL
                             hInst, dwReason)))
    {
        return FALSE;
    }
    return TRUE;
}


HDEVICE FAR PASCAL _export DeviceOpen( void )
{
    OutputDebugString( "DeviceOpen\n");

    return &Device1;
}

int FAR PASCAL _export DeviceClose( HDEVICE hDevice )
{
    OutputDebugString( "DeviceClose\n");

    return 0;
}

int FAR PASCAL _export DeviceGetWriteStatus( HDEVICE hDevice, LPWORD pusStatus )
{
    OutputDebugString( "DeviceGetWriteStatus\n");

    return 0;
}

int FAR PASCAL _export DeviceGetReadStatus( HDEVICE hDevice, LPWORD pusStatus )
{
    OutputDebugString( "DeviceGetReadStatus\n");

    return 0;
}

int FAR PASCAL _export DeviceWrite( HDEVICE hDevice, LPBYTE lpData, LPWORD pcBytes )
{
    OutputDebugString( "DeviceWrite\n");

    return 0;
}
```

Listing 18.7 (continued)   SKELETON.C *(16-bit DLL)*

```
int FAR PASCAL _export DeviceRead( HDEVICE hDevice, LPBYTE lpData, LPWORD pcBytes )
{
    OutputDebugString( "DeviceRead\n" );

    return 0;
}

int FAR PASCAL _export DeviceSetDriverParams( HDEVICE hDevice, PDRIVERPARAMS pParms )
{
    OutputDebugString( "DeviceSetDriverParams\n" );

    return 0;
}

int FAR PASCAL _export DeviceGetDriverParams( HDEVICE hDevice, PDRIVERPARAMS pParms )
{
    OutputDebugString( "DeviceGetDriverParams\n" );

    return 0;
}

int FAR PASCAL _export DeviceGetDriverCapabilities( HDEVICE hDevice,
                                            PPDRIVERCAPS ppDriverCaps )
{
    OutputDebugString( "DeviceGetDriverCapabilities\n" );

    return 0;
}
```

Listing 18.8   SKEL16.MAK *(16-bit DLL)*

```
WIN32SDK_BINW16 = \win32sdk\binw16

all: skeleton.dll

!message
!message +++++++++++++++++++++++++++++++++++++++++++++++++++++++++++++++++++++++
!message + To make the 16-bit skeleton.dll, you will need to have the   +
!message + Microsoft Thunk compiler and the Microsoft Macro Assembler   +
!message + (ML) on the path.                                           +
!message +++++++++++++++++++++++++++++++++++++++++++++++++++++++++++++++++++++++
!message
```

Listing 18.8 (continued)   SKEL16.MAK *(16-bit DLL)*

```
skeleton.obj: skeleton.c skeleton.h
    cl -c -W3 -ASw -GD2s -Oi $*.c

skeleton.asm: skeleton.thk
    thunk -t SKELETON -o skeleton.asm skeleton.thk

thk16.obj: skeleton.asm
    ml /DIS_16 /c /W3 /Fo thk16.obj skeleton.asm

skeleton.dll: skeleton.def skeleton.obj thk16.obj
    link skeleton+thk16,skeleton.dll,skeleton.map /MAP,sdllcew libw
                                          /nod/noe,skeleton.def

    $(WIN32SDK_BINW16)\rc -40 skeleton.dll
    mapsym skeleton
    implib skeleton.lib skeleton.dll
    copy skeleton.dll \windows\driver.dll
```

Listing 18.9   SKELETON.DEF *(16-bit DLL)*

```
LIBRARY      Skeleton
DESCRIPTION "Skeleton Driver"
EXETYPE      WINDOWS
DATA         PRELOAD MOVEABLE SINGLE
CODE         PRELOAD MOVEABLE DISCARDABLE

EXPORTS
DllEntryPoint            @1 RESIDENTNAME
SKELETON_ThunkData16     @2

IMPORTS
C16ThkSL01     = KERNEL.631
ThunkConnect16 = KERNEL.651
```

Chapter 19

Driver DLLs: Using Timers

Drivers often need to use some sort of timer service, either to gain control of the processor on a periodic basis, or to measure elapsed time. The timer services available under DOS were well understood. DOS drivers hooked the timer interrupt for periodic notification and used the C run-time, DOS, or BIOS services for measuring elapsed time. Windows driver DLLs — both 16-bit and 32-bit — also have timer services available. This chapter will examine the periodic timer and elapsed time mechanisms available to Windows driver DLLs.

Timers for Periodic Notification

Drivers use timers to gain control of the processor on a periodic basis in order to poll a device, to update some variables, or even to refresh the screen. Under DOS, the only way to get a periodic notification is to hook the timer interrupt, which normally occurs every 55 ms — 18.2 times per second. A DOS application also has the option of reprogramming the PC timer hardware so that the interrupt rate is faster.

A Windows driver that needs periodic control has several different options, from using Windows API functions that hook timer interrupts to using a VxD. The following sections explore each of these options and explain the limitations of each. You shouldn't be surprised to learn that achieving precise timing control under Windows is more difficult than under DOS.

Using *SetTimer*

The familiar Windows API timer function SetTimer is available to both Win32 and Win16 DLLs. The timer created by this call can either post a WM_TIMER message or invoke a callback function when the timer expires.

Unfortunately, SetTimer is not practical for applications that require immediate notification, because SetTimer communicates with the timer handler via the messaging system — not an interrupt. An indeterminate amount of time can elapse between the timer's expiration and the processing of the WM_TIMER message or the invocation of the callback function. Windows hooks the hardware timer interrupt to implement these timers, but all that interrupt handler does is set a flag to indicate that a timer event has occurred. Later, the application enters its message loop and calls GetMessage. At this point, lots of time may have elapsed already since the interrupt. This message delay can be surprisingly long because even if the timer event flag is set, GetMessage only returns a WM_TIMER message if no other messages are in the application's message queue. Windows considers WM_TIMER messages low priority.

The same delay occurs even when you use SetTimer with the callback function option (instead of the WM_TIMER option) because Windows still treats the timer as a low priority event. The callback function is not called directly by the Windows timer interrupt handler. Again, the handler sets the timer event flag and GetMessage later checks this flag. But instead of returning a WM_TIMER message, in this case, GetMessage directly calls the callback function.

Hooking *INT 1Ch* and *INT 8h*

A 16-bit Windows driver DLL can choose to avoid the delay described above by hooking the timer interrupt directly. Many DOS applications hook the software timer interrupt (INT 1Ch) instead of the hardware timer interrupt (INT 8h). This works under DOS because the INT 8h handler in the BIOS issues an INT 1Ch after processing the timer interrupt. Hooking INT 1Ch won't work under Windows, even for a 16-bit DLL. The Windows handler for the hardware timer interrupt does pass the interrupt on to the BIOS INT 8h handler, but the BIOS runs in V86 mode; when the INT 8h handler calls the INT 1Ch handler, the processor is still in V86 mode. So a Windows driver that has hooked INT 1Ch won't see this interrupt because the Windows driver runs in protected mode.

This initialization means a 16-bit driver DLL should hook INT 8h, the hardware timer interrupt handler. Windows calls all protected mode INT 8h handlers first before switching to V86 mode and calling the BIOS INT 8h handler. Unfortunately, this solution has the same limitation that any hardware interrupt does: the only useful Windows API function available at interrupt time is PostMessage. (See Chapter 15 for a complete discussion of the restrictions imposed at interrupt time.)

An INT 8h handler should perform only truly time-critical actions and defer other actions (like updating the client window) by calling PostMessage with a user-defined message. The window procedure then finishes the processing when it retrieves the message.

Don't Depend on 18.2 Ticks per Second

The INT 8h handler solution is far from perfect. Not only is it available only to 16-bit DLLs, but the handler isn't guaranteed to be called every 55 ms. The actual hardware timer interrupt is serviced by a VxD, the Virtual Timer Device (VTD). The VTD then simulates timer interrupts for VMs. Because VMs are seeing simulated interrupts and not the real thing, the frequency of timer interrupts will vary.

VTD gives the foreground VM (the VM with the display and keyboard focus) 18.2 timer ticks per second — that is, a normal rate. But each background VM gets many fewer than 18.2 ticks per second, usually around three or four. In other words, INT 8h handlers running under Windows, whether in a DOS application or a Win16 driver DLL, cannot depend on receiving an interrupt every 55 ms.

Using timeSetEvent: Pros and Cons

The most accurate periodic notification available to a Windows driver DLL is provided by timeSetEvent. This is one of the Windows multimedia functions, available to both Win16 and Win32 code. Before using timeSetEvent, your code should call timeGetDevCaps determine the timer's minimum period, and then timeBeginPeriod to program the timer resolution.

According to timeGetDevCaps, the minimum timer period is 1 ms. While not as good as the minimum period achievable under DOS (see the sidebar "Reprogramming the 8254 Timer"), it's good enough for many drivers. Note, however, that this resolution isn't guaranteed — it is possible for a callback to be delayed. In fact, actual performance of timeSetEvent varies considerably between Windows 3.x and Windows 95, even though timeGetDevCaps returns the same information under both versions.

Reprogrammining the 8254 Timer

To receive more frequent interrupts, a DOS application may reprogram the 8254 timer chip — up to a maximum interrupt frequency of about 1 million times per second. To avoid "breaking" the standard 55 ms time-base, the application's INT 8h handler must track the number of interrupts and call the original INT 8h handler every 55 ms, not every interrupt. While Win16 applications may use this same trick, the VTD does provide an API to increase the timer rate to a frequency of 1 ms. A Win16 application can use VTDAPI_Begin_Min_Int_Period to set the timer rate, then VTDAPI_Start_User_Timer to register a callback.

Under Windows 3.x, the timer latency doesn't vary much, and only occasionally are callbacks delayed — up to roughly 10 ms. The worst-case latency under Windows 95 is much worse — it can be on the order of a hundred milliseconds! This variation is created by the Windows 95 preemptive thread scheduling mechanism. Windows 95 queues all threads of the same priority together and runs each thread from that queue for its entire time slice before starting the next thread. If the time slice is 30 ms, and there are three threads ahead of the timer callback thread, then the timer callback thread will be delayed by 90 ms.

Although `timeSetEvent` is now a standard part of Windows (beginning with Windows 3.1), it is not packaged as part of the "normal" Windows DLLs (USER, KERNEL, and GDI) that all Windows applications link with. When using `timeSetEvent`, be sure to: include `MMSYSTEM.H` in your source (to get the function prototype); add `MMSYSTEM.LIB` to the import libraries listed in your link command. Both files should be provided by your Windows compiler vendor.

If All Else Fails ... Use a VxD

If reprogramming the 8254 is out of the question under Windows, and `timeSetEvent` isn't really accurate at 1 ms, then how can a driver DLL get an accurate high-frequency timer? Unfortunately, it can't. Thus, if you need accurate high-resolution timing, write a VxD. The timing services available to a VxD provide 1 ms resolution and aren't subject to the whims of the thread scheduler.

A VxD can use the VMM `Set_Global_Time_Out` service to force a callback function to be executed after a certain number of milliseconds. This creates a one-shot timer. The VxD can call `Set_Global_Time_Out` again in the callback to start another timer, thus providing a continuously running timer.

Normally the resolution of this timeout is 20 ms, but a VxD can get a better resolution, up to 1 ms, by calling `VTD_Begin_Min_Int_Period`. This service will return with an error if the requested resolution is not supported. Be aware that increasing the timeout interval can seriously degrade system performance. When the VxD is finished with its timing job, it should call `VTD_End_Min_Int_Period` to return the timer frequency to its original value.

Measuring Elapsed Time

Timing services are also used to measure the duration of an event. Because PC system hardware doesn't include anything as nifty as a stopwatch, applications must derive elapsed times by capturing an event's start and end times and calculating the difference.

Under DOS, there are several different ways to query the current system time. The highest level service, DOS Get Time (INT 21h Func 2Ch), returns time in hours, minutes, seconds, and hundredths of seconds — inconvenient for calculating time differences. The BIOS Get Tick Count service (INT 15h Func 1Ah) returns time in a more convenient form: ticks (55 ms) since power up. Programs can also directly read the current BIOS tick count in the BIOS data area. All of these methods boil down to accessing the same information: the timer tick count, updated every 55 ms by the BIOS INT 1Ch handler (called by the BIOS INT 8h handler).

Choices: GetTickCount, timeGetTime, and QueryPerformanceCounter

A 16-bit Windows driver DLL can query the BIOS tick count with a call to the Windows API function GetTickCount. If a 16 bit only solution with a resolution of 55 ms is enough for your application, this method will suffice.

The multimedia timeGetTime service, however, offers significant advantages. It's available to both Win16 and Win32 DLLs and has a much better resolution — 1 ms. Plus, its easier to call a Windows API function than to issue a software interrupt — even in 16-bit code.

If you're not supporting Windows 3.x at all, you can use the QueryPerformanceCounter function offered by the Win32 API. This function (which doesn't rely on counting timer interrupts but instead reads the free-running timer hardware) has an incredible resolution of 0.8 microseconds! This is one of the few areas where a Win32 driver DLL gets performance as good as a DOS application.

Summary

Windows wasn't designed to be a real time operating system, and the behavior of the various timing functions clearly reflects that. If your application only needs periodic notifications and can live with occasional latencies, your best and easiest alternative is to use the multimedia functions (which makes sense — after all, sound and video need to be near real time). If occasional latencies aren't acceptable, you'll have to write a VxD. For measuring elapsed time, use timeGetTime if you must support both Windows 3.x and Windows 95. Use the more accurate QueryPerformanceCounter if you're supporting only Windows 95.

Appendix A

Intel Architecture

8086/8088 and Real Mode

The Intel 8088, the first processor to be used in a PC, has 16-bit registers. A direct addressing scheme using 16 bits allows access to only 2^{16} or 64Kb of memory. Yet the 8088 can address up to 2^{20} or 1Mb of memory, because the processor uses a memory architecture called *segmentation*. All memory references involve both a 16-bit segment and a 16-bit offset. The segment specifies the base of a 64Kb region, and the offset specifies the byte within the region. Each of the possible regions is 16 bytes apart, which means the last region starts at 64Kb × 16 or 1Mb. This combined segment and offset address is known as a *logical address*.

Internally, the processor forms a physical address by shifting the segment left by four bits and then adding the offset, resulting in a 20-bit physical address. It is this physical address that the processor outputs onto the bus. Memory devices don't have knowledge of segments or offsets and understand only physical addresses. This address translation process is illustrated in Figure A.1.

421

This original addressing scheme used by the 8088 is now referred to as *real mode*. Real mode has several limitations that make it unsuitable for a sophisticated operating system.

- An address space limited to 1Mb is no longer adequate.

- The fixed relationship between a logical address and a physical address makes it difficult to implement moveable memory.

- There are no hardware protection mechanisms, allowing a buggy program to crash the entire system.

80286 and Protected Mode

The next generation Intel processor, the 80286, addresses the deficiencies of the 8088/8086. Like the 8086, the 80286 has 16-bit registers, uses a segmented architecture, and supports real mode. The improvement is a new operating mode known as *protected mode*. Protected mode offers advanced features such as access to 16Mb of memory, more flexible address translation, and various protection mechanisms.

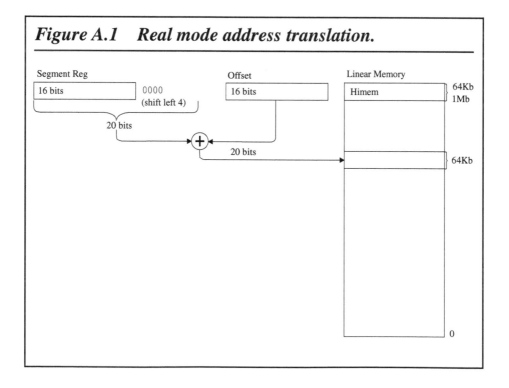

Figure A.1   Real mode address translation.

The segmented architecture, where memory references consist of a segment and an offset, is still used in protected mode. However, the address translation mechanism, which translates a logical (segment and offset) address into a physical address, is more sophisticated than in real mode. A segment is now called a selector. Instead of shifting a selector by a fixed amount to form a physical address, the processor uses a selector as an index into a descriptor table. The descriptor stored in the table, not the selector itself, determines the selector's base address. This layer of indirection between a selector and a physical address facilitates the implementation of moveable memory, which is a necessity for multitasking operating systems.

Because the 80286 uses 16-bit registers, an offset can only address 64Kb, which means a segment is still limited to a maximum size of 64Kb. But the segment's base address, stored in the segment descriptor, is a 24-bit value. The processor generates a 24-bit physical address by adding together the 24-bit base address and the 16-bit offset, so the processor can address 16Mb (2^{24}) of memory.

Protected mode also offers several types of protection mechanisms which prevent one program from interfering with another program or from interfering with the operating system. The three mechanisms are:

- the ability to isolate the operating system from applications,

- the ability to isolate user applications from each other, and

- the ability to use protected mode to enforce the proper use of segments, so that an errant program can't execute from a data segment or address a location beyond the limit of the segment.

Selectors and Descriptors

A value stored in a segment register, known simply as a *segment* in real mode, is more precisely called a *selector* in protected mode. In protected mode, a selector specifies a descriptor, and a descriptor in turn specifies a segment's base address and length. A selector is made up of several fields, as illustrated in Figure A.2. The Table field specifies where the descriptor is located: 0 for the Global Descriptor Table (GDT), or 1 for the Local Descriptor Table (LDT). The 13-bit Index field specifies one of the 2^{13} (8192) descriptors in that table.

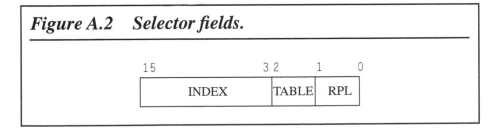

Figure A.2   Selector fields.

These two descriptor tables aren't physically located on the processor itself, as registers are, but are located in main memory. Special processor registers, the GDTR and the LDTR, hold the physical addresses of these two tables. These two registers are implicitly referenced by the LGDT/LLDT (load) and SGDT/SLDT (store) instructions.

There is only one GDT, designed to be used for selectors that are either used by all applications, or shared by applications. The LDT, on the other hand, is designed to be used for selectors that are "local" to an application. Multiple LDTs are allowed, which allows a multitasking operating system to easily isolate applications from each other by allocating a different LDT for each application. The LDTR register always holds the address of the current LDT. When the operating system switches from one application to another, it also loads the LDTR with the address of the LDT of the new application. This way an Application X can't possibly access code or data belonging to Application Y, because all memory references by Application X are resolved using X's own LDT. (Note that in all of the Windows versions, all Windows applications share the same LDT.)

A segment descriptor stored in one of these tables consists of 8 bytes, as depicted in Figure A.3. Two bytes aren't used but are needed to be compatible with the 32-bit 80386. The segment's base address takes up 3 bytes, so the highest base address is 2^{24} or 16Mb. Two bytes hold the segment's limit, or size, resulting in a maximum size of 64Kb. The remaining byte is an access byte made up of various flag bits. Some flag bits specify a segment's type: either a code segment (executable, not writable) or a data segment (read/write, or read-only). Other flag bits include the Present bit, which indicates whether the segment is present in main memory, and the Accessed bit, which is set by the processor every time a segment is loaded into a segment register.

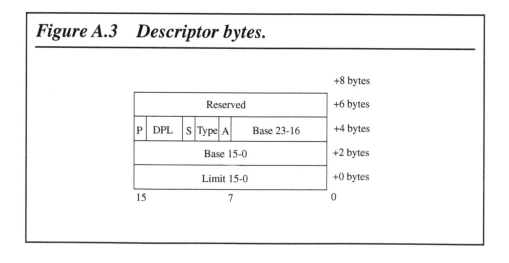

Figure A.3   Descriptor bytes.

When a segment register (CS, DS, SS, ES) is loaded, the processor reads the associated segment descriptor from the descriptor table in main memory and stores its contents in a hidden descriptor cache register. There is one descriptor cache register for each segment register, and like all registers, they are located on the processor. Future memory references involving the same segment register use the descriptor information in the cache register, instead of fetching that information from the descriptor table in main memory. These cache registers are crucial for good performance in protected mode.

Interrupts and Exceptions

The sequential execution of a program can be altered by an unexpected event, either an interrupt or an exception. Interrupts may be external, caused by a device asserting the processor's INTR pin, or internal, caused by a program executing an INT instruction. Exceptions result when an instruction completes abnormally. Both interrupts and exceptions cause control to be transferred automatically to a handler routine specified by the Interrupt Descriptor Table (IDT).

Each interrupt and exception has an identifying number N, and when interrupt/exception N occurs, the processor looks up slot N of the IDT to get the address of the handler. The IDT contains a special type of descriptor called a gate, which doesn't describe a segment (no base address or limit fields). Instead, a gate contains an address: a selector and an offset. The IDT may contain both interrupt gates and trap gates. The only difference is that an interrupt gate causes the processor to clear the interrupt flag before calling the handler, preventing further interrupts. Interrupt gates are commonly used for hardware interrupts, and trap gates for software interrupts and exceptions.

An exception that can be corrected by the handler is called a fault. When a fault occurs, the processor executes the handler specified in the IDT, then automatically modifies CS and IP so that the faulting instruction is executed again. If the fault handler has corrected the condition that caused the exception, the instruction executes correctly the second time. Operating systems often use the Segment Fault (number 11) and the Page Fault (number 14) to implement virtual memory. However, some faults, particularly the General Protection Fault (number 13), usually indicate a program bug. In this case, the fault handler may terminate the program instead of letting the instruction execute again.

Protection Mechanisms

The processor performs one set of consistency checks when executing instructions to load a segment register and a different set when executing instructions that reference memory (which always involve a segment). If one of these consistency checks fails, the processor generates an exception.

When a segment register is loaded with a descriptor, the first check made is to see if the selector indexes to a valid entry in the descriptor table. Loading a segment register with an invalid selector causes a General Protection Fault (number 13). An operating system typically fills all unused table entries with invalid descriptors, which are simply descriptors where the Access byte equals zero. Thus, a program is prevented from generating its own selector to access data or code, and must use a valid selector in the GDT or LDT given to it by the operating system. Also, if an executing program transfers control to an invalid location, usually due to a stack pointer mismatch, this will result in the CS register being loaded with an invalid selector and will immediately cause this fault.

Next, the processor ensures that the type of segment defined by the desciptor matches the segment register. This is done by examining the Type field of the descriptor. For example, the CS register must be loaded with a code segment which has type executable and not writable. The DS, ES, and SS registers must be loaded with a data segment, which has either type read/write or read-only. If these conditions are not met, a General Protection Fault is generated.

Last, the processor checks the descriptor's Present bit. If not set, the processor generates a Segment Fault. This fault may be used by an operating system to implement virtual memory. Under low memory conditions, the operating system writes a segment to disk and clears the selector's Present bit. Once the segment is written to disk, the memory formerly occupied by the segment is added to the pool of free memory. When a program later accesses a location in the original segment, a fault occurs and the operating system reloads the original segment into memory from disk. (Note that Windows 3.x and Windows 95 use *paging* to implement virtual memory, not *segmentation*.)

If none of these checks results in an exception, the selector has been loaded and the processor continues with the next instruction. Now you can appreciate why the Intel manuals list the clock cycles for the MOV DS instruction in protected mode as 17, where the same instruction takes just two clocks in real mode.

A different set of checks occurs when the processor executes an instruction that uses a memory reference. All memory references involve a segment, whether specified explicitly or implicitly. For example, PUSH and POP implicitly reference the stack segment in SS. Note that the segment cache register already contains selector information (loaded when the segment register was loaded), so the processor doesn't have to access the LDT or GDT in memory. Again, if any of these checks fail, the processor generates a General Protection Fault.

The first check for a memory reference is the limit. The processor compares the offset specified in the instruction against the segment's limit and generates an exception if the offset is greater than the limit. This protection mechanism prevents an incorrect pointer from writing past the end of a segment. Next, the type information in the segment cache register is compared to the type of memory access (read or write). For example, a write to a location in the CS segment usually results in an exception because CS usually contains an execute-only segment. The final check involves verifying the segment's privilege level.

Privilege Levels

Before executing an instruction, the processor also checks the application's privilege level. Proper use of privilege levels allows an operating system to isolate itself from applications in three different ways. Privilege levels can prevent an application from accessing specific data segments and from executing certain code segments. Privilege levels can prevent an application from executing specific instructions that affect operating system data structures like the descriptor tables. Privilege levels can also prevent an application from executing instructions that control I/O devices or disable/enable hardware interrupts.

Every segment (code or data), has a DPL or Descriptor Privilege Level. The DPL bits are stored in the segment's descriptor. Privilege levels range from 0 to 3, with 0 being the most privileged or trusted, and 3 the least privileged. System designers can use all four levels to fully isolate system components, perhaps running the operating system kernel at DPL 0, device drivers at DPL 1, the file system at DPL 2, and applications at DPL 3. However, many operating systems (like Windows) use only two levels, distinguishing only between the operating system (DPL 0) and applications (DPL 1, 2, or 3).

The basic idea of privilege levels is this: code isn't allowed to access more privileged data (data segment with a numerically lower DPL) or to transfer control to more privileged code (code segment with a numerically lower DPL). If either of these access rules is violated, the processor generates a General Protection Fault. By setting operating system data segments to DPL=0 and application code segments to DPL>0, the operating system can prevent applications from accessing data owned by the operating system. By setting operating system code segments to DPL=0 and application code segments to DPL>0, the operating system can prevent an application from calling operating system functions with a normal call instruction.

Although it's a good idea to restrict an application's access to functions in the operating system's code segments, disallowing access completely is unacceptable — applications do need to call operating system services. The solution is not to move operating system code to DPL 3 with the applications, but to use a call gate. A call gate allows less privileged code to call more privileged code, but in a way that is managed by the operating sytem. (Note that Windows 3.x and Windows 95 don't use call gates.)

A call gate works much like a software interrupt. When a software interrupt (INT n) is executed, the processor loads CS and IP from the interrupt vector located in slot N of the IDT, and resumes execution at the new location. Similarly, when a FAR CALL instruction is executed and the destination segment is really a call gate instead of a normal code selector, the processor loads CS and IP from the selector and offset fields of the call gate, and execution resumes at the new location.

An operating system can use call gates to provide controlled access to system entry points. To call an operating system service, the application makes a FAR CALL to a call gate selector, which the operating system has set up to include the address (selector and offset) of the service. From the application's point of view, the call gate selector, provided by the operating system, is the entry point. The actual entry point address is hidden from the application.

Privilege level also controls an application's ability to issue any of the privileged instructions: LGDT, LLDT, LIDT, LMSW, HALT. These instructions are only allowed for DPL=0, and cause a General Protection Fault at any other level. The first three are restricted because they manipulate the descriptor tables. The LMSW instruction is used to switch to protected mode. HALT stops processor execution.

A completely separate privilege level called I/O Privilege Level (IOPL) controls an application's ability to execute trusted instructions. Trusted instructions are those which enable or disable interrupts (STI/CLI) and access I/O devices (all forms of IN and OUT including byte, word, string). An application is allowed to issue these instructions only if DPL>IOPL, otherwise a General Protection Fault results. It is also possible for an operating system to allow IN/OUT instructions while preventing STI/CLI by having the fault handler determine the faulting instruction and reissuing it if IN or OUT, or ignoring it if CLI/STI.

80386 and Virtual-8086 Mode

The 80386 is the first 32-bit processor in the 80x86 family. Both registers and memory references are 32-bits. The 80386 supports protected mode, utilizing selectors, descriptors, and descriptor tables with several important additions, such as a 4Gb segment size, paging, and a new mode called Virtual-8086 mode.

The 64Kb segment size restriction of real mode and the 80286 protected mode has been lifted with the introduction of 32-bit offsets on the 80386. This translates to a segment size of 2^{32} or 4Gb. Although the 80386 addressing mechanism always uses segmentation, segments become invisible to the programmer if an operating system implements a flat memory model.

In flat model, the operating system creates a single code segment and a single data segment, each with a limit of 4Gb. The CS register is then loaded with the code segment selector, and DS, SS, and ES are loaded with the data segment selector. After this initial loading, the segment registers never change again. Both operating system and programs can access the entire address space with this single set of selectors. Programmers using flat model thus never need to be aware of segments at all. If paging is used, flat model doesn't mean giving up protected mode's protection mechanisms.

The 80386 paging feature adds another level of address translation, in addition to segmentation (Figure A.4). With paging disabled, the 80386 acts like a 80286 with 32-bit registers: a physical address is formed by adding an offset to a selector's base address, found in a descriptor table. With paging enabled, the 80386 adds an offset to a selector's base address, found in a descriptor table, to form a linear address. This linear address is then transformed into a physical address using page tables.

The 32-bit linear address is decomposed into three fields: a 10-bit Directory Table Entry index (DTE), a 10-bit Page Table Entry index (PTE), and a page offset. The process of combining these indexes into a physical address is best described by examining Figure A.4. The CR3 register holds the address of the page directory table. The processor adds the address of this table to the 10-bit DTE. The result is the address of a page table. A particular entry in this page table is selected by adding the 10-bit PTE to the page table's address. Each entry in the page table contains the address of a 4Kb page in physical memory. The page offset is added to the the page address, and the result is a physical address, which is output to the bus.

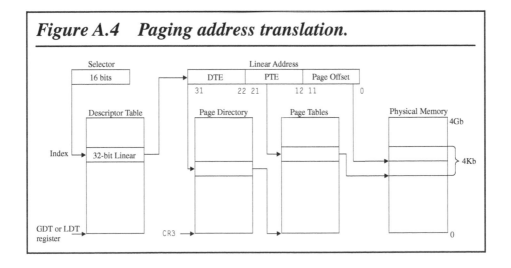

Figure A.4   Paging address translation.

An operating system can use paging to implement virtual memory. Paged virtual memory usually offers better performance than segmented virtual memory. Pages are both fixed in size and small (4Kb), whereas segments are variable in size and large (up to 64Kb). When it comes to swapping a memory block in from disk to memory, it is easier to find a free space for a small fixed-sized page than it is for a large variable-sized segment.

Demand-paged memory is implemented using the Page Fault generated when a memory reference maps to a page marked "not present" in the page tables. Under low memory conditions, the operating system writes a 4Kb page of address space to disk and clears the page's Present bit in the page table. Once the page is written to disk, the memory occupied by the page is added to the pool of free memory. When a program later accesses a location in the original page, a fault occurs and the operating system reloads the original page into memory from disk.

Paging also offers some protection mechanisms, although not as many as are available for segments. All pages are represented by a PTE in a page table (although the page table itself may be stored on disk). The PTE is illustrated in Figure A.5. A PTE contains the Present bit mentioned above, a U/S (User/Supervisor) bit, and a R/W (Read/Write) bit. A page marked as "User" can be accessed by a code segment running at any privilege level. A page marked as "Supervisor" will cause a page fault if accessed by a code segment running at Ring 3. This allows an operating system to separate operating system code from application code by running applications at Ring 3 and marking all pages used by the operating system as Supervisor. All pages are readable and executable, but the operating system can prevent Ring 3 code from writing to a page by clearing the R/W bit in the PTE.

Figure A.5   Page Table Entry (PTE)

32	12	11,10,9	8,7	6	5	4,3	2	1	0
Physical Address		Available for U/S	Intel reserved	Dirty	Accessed	Unused	U/S	Writable	Present

Virtual-8086 Mode

Virtual-8086 (V86) mode was created specifically to allow an operating system to run multiple real mode applications, while preventing each real mode application from interfering with either the operating system or other applications. On the surface, V86 mode looks like real mode. All real mode instructions, including 32-bit extensions (MOV, ADD, etc.) behave as in real mode. Protected mode instructions (LGDT, LLDT, etc.) cause an Invalid Opcode fault, as they do in real mode. However, several important differences prevent V86 mode applications from interfering with each other:

- paging is allowed,
- access to individual I/O ports can be controlled, and
- instructions that affect the interrupt flag can be trapped.

The V86 addressing scheme is basically real mode addressing plus paging. The processor shifts a segment register left by 4 bits to form a linear address, and then uses the page tables to translate this to a physical address. By maintaining different page tables for each V86 program, the operating system can map the address space of each program to a different region of physical memory. The operating system can even use virtual memory to map some or all of a V86 program's address space to disk. An exception or interrupt in V86 mode causes an automatic switch to protected mode at Ring 0, so the same page fault handler that provides virtual memory for protected mode programs works for V86 programs as well.

In real mode, access to I/O ports via IN/OUT instructions is never prohibited. In protected mode, either all IN/OUT instructions cause an exception, or all IN/OUT instructions are executed normally, depending on a program's IOPL. In V86 mode, the processor may execute some IN/OUT instructions normally, whereas others may cause an exception. The processor examines the I/O Permission Map (IOPM) to make this determination. The IOPM is a bitmapped table, 1 bit for each port address, where a 0-bit permits access to the port and a 1-bit causes a general protection fault. The operating system maintains this table, and may have multiple tables, one for each executing V86 program.

V86 programs also have an IOPL, which does *not* affect the ability to issue I/O instructions but does control execution of instructions that affect the processor's Interrupt flag. If the operating system runs a V86 program with IOPL < DPL (DPL is always 3), those instructions cause an exception. The exception handler then simulates the instruction, maintaining a "virtual" Interrupt flag for each executing V86 program. Some operating systems may choose to run with IOPL = DPL, so that these instructions are not trapped, which results in better performance.

Appendix B

Using Assembly Language with Your VxD Written in C

If you choose to write your VxDs in C using only the DDK (and not VToolsD), you'll need to write a small piece of your VxD in assembly language in order to declare your VxD's DDB (Device Descriptor Block) and Device Control Procedure. If you also use VMM or VxD services that aren't provided by either the DDK wrapper library (VXDWRAPS.CLB) or the wrapper provided with this book (WRAPPERS.CLB), you'll need to add new wrappers to the WRAPPER.ASM file discussed in this appendix.

This appendix will tell you what you need to know to write the assembly language modules you require. First I'll show you exactly how to declare a DDB and how to write a Device Control Procedure in assembly language. Next, I'll discuss adding other entry points to the assembly language module to support registered callbacks of all kinds: interrupt handlers, event handlers, port trap handler, page fault handlers, etc. Finally, I'll explain the inner workings of the WRAPPERS library so that you can easily add other services to it.

Declaring the DDB

Although written in C, your VxD will always have at least one assembly language source file which includes the VxD's DDB and Device Control Procedure. You can start with the assembly language module from any of this book's sample VxDs and modify it to suit your needs.

A VxD usually declares its Device Descriptor Block (DDB) at the top of its assembly language module, using the macro DECLARE_VIRTUAL_DEVICE. Here's the macro definition, taken from VMM.INC:

```
Declare_Virtual_Device MACRO Name, Major_Ver, Minor_Ver,\
                       Ctrl_Proc, Device_Num, Init_Order,\
                       V86_Proc, PM_Proc, Reference_Data
```

These parameters correspond one for one to the DDB fields described in the section "The Device Descriptor Block" in Chapter 4.

As an example, SKELCTRL.ASM from the SKELETON VxD in Chapter 5 uses the macro like this:

```
DECLARE_VIRTUAL_DEVICE    SKELETON, 1, 0, ControlProc,
                          UNDEFINED_DEVICE_ID,
                          UNDEFINED_INIT_ORDER
```

All VxDs are required to supply parameter values for Name, Major_Ver, Minor_Ver, and Ctrl_Proc. However, a VxD may use UNDEFINED_DEVICE_ID if it doesn't need a unique ID, and UNDEFINED_INIT_ORDER if it doesn't require a particular order in the initialization sequence. SKELETON.VXD doesn't have a V86 or PM API procedure, so those parameters are omitted. The final Reference_Data parameter is used only by IOS layered block device drivers (a kind of VxD). An IOS driver would initialize this field with a pointer to it's Driver Registration Packet. IOS drivers are not discussed in this book; see the DDK.

Coding the Device Control Procedure

A VxD's Device Control Procedure, its main entry point, usually follows the DDB declaration. Because this procedure must reside in the VxD's locked code segment, its declaration is preceded by the macro VXD_LOCKED_CODE_SEG. In SKELCTRL.ASM, the Device Control Procedure is called simply ControlProc.

```
BeginProc ControlProc
    Control_Dispatch SYS_CRITICAL_INIT, _OnSysCriticalInit, cCall, <ebx>
    Control_Dispatch SYS_VM_INIT, _OnSysVmInit, cCall, <ebx>
    Control_Dispatch SYS_VM_TERMINATE, _OnSysVmTerminate, cCall, <ebx>
    Control_Dispatch CREATE_VM, _OnCreateVm, cCall, <ebx>
    Control_Dispatch DESTROY_VM, _OnDestroyVm, cCall, <ebx>
    Control_Dispatch CREATE_THREAD, _OnCreateThread, cCall, <edi>
    Control_Dispatch DESTROY_THREAD, _OnDestroyThread, cCall, <edi>

    clc
    ret
EndProc ControlProc
```

Begin_Proc and End_Proc macros bracket the ControlProc declaration. These two macros are very complex, but you don't have to understand the implementation to use them correctly. You can think of them as being the VxD equivalent of the proc near and endp assembler directives. Just as you would bracket an assembly language procedure with proc near and endp, you bracket an assembly language control procedure with BeginProc and EndProc.

Within ControlProc, the Control_Dispatch macro generates code for a switch statement, where the control variable is the message code in EAX. The declaration of Control_Dispatch, also in VMM.INC, is:

```
Control Dispatch MACRO Service, Procedure, callc, arglst
```

The first parameter specifies the message, the second specifies the message handler, the third is the calling convention of the handler, and the last is the list of arguments to be passed to the handler. SKELCTRL.ASM uses the macro like this:

```
Control_Dispatch INIT_COMPLETE, _OnInitComplete, cCall, <ebx>
```

The first parameter is obvious, and I'll explain the leading underscore in the second parameter in a moment. The last parameter, arglst, represents the VM handle, which the VMM always places in EBX before calling a VxD with the INIT_COMPLETE message.

The calling convention parameter, cCall, is necessary because this assembly module will be linked with modules written in a different language — SKELCTRL.ASM is calling the message handler OnInitComplete, which is located in a separate C file. Calling conventions are important when mixing languages. A calling convention defines two behaviors: the order in which parameters are pushed on the stack (right to left or left to right); and who is responsible for removing parameters from the stack after the call (callee or caller). Each calling convention implies a naming convention, which defines how the compiler "mangles" the procedure name before storing it in the OBJ file.

Each high-level language compiler has a "native" calling convention, and some also support alternative conventions through keywords. The Microsoft C compiler uses the cdecl convention as its native convention: a function declared without a calling convention keyword automatically uses the cdecl convention. Microsoft C also supports the _stdcall convention through the _stdcall keyword.

Assemblers don't directly support calling conventions, which is why VMM.INC defines its own macros for calling a function in a high-level language: cCall and sCall, which match the C compiler's cdecl and _stdcall keywords. In general, it doesn't matter which calling convention is used, as long as both caller and callee use the same convention. SKELCTRL.ASM uses cCall because it's calling a C module, and cdecl is the "native" convention for C code. But the assembly module could use sCall instead — if the C module used the _stdcall keyword in the declaration of the called function.

Note that in each of the Control_Dispatch statements, the name of the message handler includes a leading underscore. The cdecl naming convention adds a leading underscore to all exported cdecl functions, which means that an assembly language module calling a C module must include this leading underscore in the name of the called fuction. The cCall macro in VMM.INC takes care of the *calling* convention, pushing parameters in the right order and removing parameters from the stack, but doesn't take care of the *naming* convention. So SKELCTRL.ASM must explicitly include the leading underscore in the cCall macro's Procedure argument. (In contrast, the sCall macro does take care of the naming convention, automatically adding a type decoration to the end of the Procedure name.)

A sure sign of a calling convention mixup is a linker error message like:

UNRESOLVED EXTERNAL: OnInitComplete@4.

The @n is a type decoration, specifying how many bytes the function uses for parameters. The name OnInitComplete@4 means OnInitComplete uses 4 bytes of parameters. This UNRESOLVED EXTERNAL error would occur if SKELCTRL.ASM used sCall when calling OnInitComplete, and OnInitComplete was declared as cdecl. sCall adds type decoration but cdecl doesn't, so at link time, the names won't match.

Adding "Thunks" to Support Callbacks from VMM/VxDs

Many VMM and VxD services require the calling VxD to register a callback function, which the VMM/VxD invokes later to notify the calling VxD that something interesting happened. For example, a VxD might call VPICD_Virtualize_IRQ to register a hardware interrupt handler; the VPICD would then call the registered handler when a hardware interrupt occurs. Or, a VxD might call the VMM service Install_IO_Handler to register a port trap handler; the VMM would then call this handler when a Ring 3 application accesses a specific I/O port.

In most cases, the parameters provided to the callback function are passed in registers, not on the stack. (A notable exception to this rule is the Configuration Manager VxD, which uses the stack to pass parameters to callback functions.) For this reason, a registered callback is usually located in an assembly language module, so that the callback can access the register parameters.

The example VxDs in this book all follow this convention. All registered callbacks are located in the VxD's assembly module (the same one containing DDB and Device Control Procedure), but the callback does minimal processing before calling a function in the VxD's C module to perform real processing. In my example VxDs, the name of the assembly callback function always ends in "Thunk", and the name of the C function it calls has the same base but ends in "Handler". Thus, the PORTTRAP example from Chapter 8 contains the function PortTrapThunk in the VxD's assembly language module, and PortTrapThunk calls PortTrapHandler which is located in the VxD's C module. (Note that this usage of the term "thunk" is *not* related to the flat thunks discussed in Chapter 18.)

Each VMM/VxD service that requires a callback uses a different set of registers to pass parameters to the callback. Therefore, when using a callback you must refer to the service's documentation to discover the register parameters and then write an appropriate assembly language "thunk". The VMM service Install_IO_Handler says the port trap handler will be called back with

```
Input:
    EAX=data (if OUT instruction)
    EBX=current VM handle
    ECX=IO type //BYTE_INPUT, BYTE_OUTPUT, WORD_INPUT, WORD_OUTPUT,
                //DWORD_INPUT, DWORD_OUTPUT, STRING_IO, REP_IO,
                //ADDR_32_IO, REVERSE_IO
    EDX=port number
    EBP=address of Client Register Structure
Output:
    EAX=data (if IN instruction)
```

So the `PortTrapThunk` function in PORTTRAP's assembly file pushes those register parameters onto the stack and calls `PortTrapHandler`, like this:

```
BeginProc PortTrapThunk

Emulate_Non_Byte_IO
cCall _PortTrapHandler, <ebx, ecx, edx, ebp, eax>
ret

EndProc PortTrapThunk
```

Notice the order of parameters in the `cCall` macro exactly matches the `PortTrapHandler` declaration in PORTTRAP's C file:

```
DWORD _stdcall PortTrapHandler(VMHANDLE hVM, DWORD IOType, DWORD Port,
                        PCLIENT_STRUCT pcrs, DWORD Data)
```

The macro takes care of pushing the parameters in the reverse order. The sample VxDs demonstrate several types of thunks and handlers:

- VXDISR uses a `VPICD_Virtualize_IRQ` callback (for `Hw_Int_Proc`) and a `Global_Event` callback
- PORTTRAP uses an `Install_IO_Handler` port trap handler
- PAGETRAP uses a `Hook_V86_Page` page fault handler

If your VxD requires a different kind of callback, see the DDK documentation for specific parameter information, then use one of these samples as a starting point.

Introducing the Wrapper Library

Although the DDK provides a library of C-callable VMM and VxD services, `VXDWRAPS.CLB`, this library only contains a small percentage of the total number of available VMM/VxD services. Many of the VxDs in this book use services that aren't included in `VXDWRAPS.CLB`, so I've developed another library, `WRAPPERS.CLB`. This library contains all other services needed by the VxDs in this book: a few more VMM services, most VPICD and VDMAD services, a few SHELL and VWIN32 services, and three IFSMgr services. A complete list of the services in `WRAPPERS.CLB` is in Table B.1.

The source file for this library is `WRAPPERS.ASM`. `WRAPPERS.H` is the header file for modules using `WRAPPERS.ASM`. (Both are found in a subdirectory of the code disk.) The Windows 95 DDK provides an example of both the assembly language module and the C header file in the `\BASE\VXDWRAPS` directory. My `WRAPPERS.ASM` and `WRAPPERS.H` are based on the DDK example.

If you need to call other VMM/VxD services that aren't in either the DDK VMMWRAPS.CLB or in my WRAPPERS.CLB, you'll need to modify WRAPPERS.ASM and WRAPPERS.H to add support for the services you need.

Table B.1   Services provided by WRAPPERS.CLB.

VMM Services	Get_Initial_Thread_Handle
	_PageReserve
	_PageCommitPhys
	_PageDecommit
	Install_IO_Handler
	Remove_IO_Handler
	Enable_Local_Trapping
	Disable_Local_Trapping
	_Assign_Device_V86_Pages
	_DeAssign_Device_V86_Pages
	_ModifyPageBits
	Hook_V86_Page
	Unhook_V86_Page
	_MapIntoV86
	_PhysIntoV86
	Map_Flat
	Call_Priority_VM_Event
	Save_Client_State
	Restore_Client_State
	Begin_Nest_Exec
	End_Nest_Exec
	Simulate_Push
	Simulate_Far_Call
IFSMgr Services	IFSMgr_Ring0_OpenCreateFile
	IFSMgr_Ring0_WriteFile
	IFSMgr_Ring0_CloseFile
VPICD Services	VPICD_Virtualize_IRQ
	VPICD_Physically_Unmask
	VPICD_Physically_Mask
	VPICD_Phys_EOI
	VPICD_Force_Default_Behavior
	VPICD_Set_Int_Request
	VPICD_Clear_Int_Request

WRAPPERS.H

WRAPPERS.H (Listing B.1, page page 448) contains a few constants and typedefs, but the two important sections are the MAKE_HEADER section and the section of macro definitions following it.

The MAKE_HEADER macro is defined in VXDWRAPS.H, (in the DDK INC32 directory). The parameters to this macro are basically the individual components of a function prototype. The macro uses preprocessor tokenizing features to generate multiple function prototypes for a single VxD service. The multiple protoypes are necessary because there are actually six different wrappers for every VxD service, one for each of the possible code segments the wrapper could be called from. For example, this call in WRAPPERS.H:

```
MAKE_HEADER(PTHCB,_stdcall,Get_Initial_Thread_Handle, (HVM hVM))
```

will expand to:

```
extern PTHCB _stdcall LCODE_Get_Initial_Thread_Handle(HVM hVM);
extern PTHCB _stdcall ICODE_Get_Initial_Thread_Handle(HVM hVM);
extern PTHCB _stdcall PCODE_Get_Initial_Thread_Handle(HVM hVM);
extern PTHCB _stdcall SCODE_Get_Initial_Thread_Handle(HVM hVM);
extern PTHCB _stdcall DCODE_Get_Initial_Thread_Handle(HVM hVM);
extern PTHCB _stdcall CCODE_Get_Initial_Thread_Handle(HVM hVM);
```

Immediately following the MAKE_HEADER section in WRAPPERS.H is another section of macro definitions. Here each service name is redefined as a macro:

```
#define Get_Initial_Thread_Handle PREPEND(Get_Initial_Thread_Handle)
```

Table B.1 (continued)	*Services provided by* WRAPPERS.CLB
VDMAD Services	VDMAD_Virtualize_Channel
	VDMAD_Set_Region_Info
	VDMAD_Set_Phys_State
	VDMAD_Phys_Mask_Channel
	VDMAD_Phys_Unmask_Channel
	VDMAD_Scatter_Lock
SHELL Services	SHELL_Resolve_Contention
	SHELL_PostMessage
VWIN32 Services	_VWIN32_QueueUserApc
	_VWIN32_SetWin32Event

The PREPEND macro (also in VXDWRAPS.H) prepends the name of the current segment to the service name: LCODE for locked code, ICODE for init code, etc. So when a C module calls Get_Initial_Thread_Handle from the locked code segment, the preprocessor actually produces a call to LCODE_Get_Initial_Thread_Handle.

The MAKE_HEADER section must precede the service name macro definitions in the header file. Because of the way the macros are implemented, reversing the order will result in incorrect function prototypes being generated by the preprocessor.

These macros in WRAPPERS.H make it easy for a C module to call a service wrapper. Defining each service name as a macro ensures that code calling the service actually calls the right wrapper, without making the calling code aware of the current segment. The MAKE_HEADER macro makes it easy for WRAPPERS.H to generate function prototypes for each of the six service wrappers, one for each segment. The next section explains how these service wrappers are implemented.

Overview of WRAPPERS.ASM

This section will discuss the details of writing a wrapper module in assembly, using WRAPPERS.ASM (Listing B.2, page page 452) as an example. Although WRAPPERS.ASM contains dozens of individual wrapper functions, this section will discuss only two of them: Get_Initial_Thread_Handle and IFSMgr_Ring0_OpenCreateFile. Get_-Initial_Thread_Handle is an example of a simple wrapper which pops parameters off the stack into registers and calls the VMM Get_Initial_Thread_Handle service. IFSMgr_Ring0_OpenCreateFile is an example of a more complicated wrapper which manipulates its caller's parameters instead of just removing them from the stack.

Like any other assembly file, WRAPPERS.ASM starts by including header files. A wrapper implementation module should always include LOCAL.INC from the DDK base\vxdwraps directory. LOCAL.INC acts as a sort of master header file and includes many other include files such as VMM.INC, DEBUG.INC, etc. A wrapper module should also include the header file of each VxD whose services are being wrapped. In the case of WRAPPERS.ASM, they are VDMAD.INC, VPICD.INC, SHELL.INC, VWIN32.INC, and IFSMGR.INC.

Next, WRAPPERS.ASM defines a few macros. WRAPPERS.ASM uses the macros StartStdCall and StartCDecl to declare _stdcall and cdecl functions, respectively. If a function prototype in WRAPPERS.H (remember that prototypes are buried in MAKE_HEADER calls) declares a wrapper function as cdecl, that function's implementation in WRAPPERS.ASM uses StartCDecl. Conversely, _stdcall in WRAPPERS.H means StartStdCall will be used in WRAPPERS.ASM.

The Start macros are similar to the BeginProc macro which SKELCTRL.ASM used to declare procedures, but the Start macros also take naming conventions into account (leading underscore for both and a trailing @ followed by the parameter size for _stdcall.) The StartCdecl macro is defined in the DDK file LOCAL.INC. However, LOCAL.INC doesn't provide StartStdCall, so I wrote my own and defined the macro at the top of WRAPPERS.ASM.

```
StartStdCall MACRO Name, Param
StartCDecl   Name&@&Param
ENDM
```

My StartStdCall takes two parameters, Name (function name) and Param (total bytes of parameters.) StartStdCall performs the type-decoration name mangling by concatenating Name and Param, and passes the concatentated name to StartCDecl. StartCDecl in LOCAL.INC does the rest of the function declaration, adding the underscore and actually generating the procedure declaration (PROC NEAR).

The code for the individual VxD wrapper functions follows the macro defiintions.

The next two sections will examine a simple wrapper, Get_Initial_Thread_Handle, and a more complicated wrapper, IFSMgr_Ring0_OpenCreateFile.

WRAPPER.ASM: Get_Initial_Thread_Handle Details

Many wrapper functions can be as simple Get_Initial_Thread_Handle. This wrapper is implemented using the _stdcall convention because that convention is slightly more efficient than cdecl and just as easy to code. Because the basic purpose of a wrapper is to move parameters from the stack into those registers expected by the real service, Get_Initial_Thread_Handle efficiently moves the parameters into registers and removes them from the stack at the same time. Having the callee remove parameters from the stack (_stdcall convention) results in slightly smaller code than having the caller remove them (cdecl convention), because the code for removing parameters appears only once in the callee, instead of appearing in every instance of the caller.

```
StartStdCall Get_Initial_Thread_Handle, 4

    pop edx          ; Get return address
    pop ebx          ; Get VMHandle
    VxDCall Get_Initial_Thread_Handle
    mov eax, edi     ; move thread hnd into return
    jmp edx          ; return addr still in edx

EndStdCall  Get_Initial_Thread_Handle, 4
```

Get_Initial_Thread_Handle is declared using the StartStdCall macro. The second macro parameter, 4, is the total size of the function arguments, in bytes. This size must match the argument sizes in the C function prototype in WRAPPERS.H. In this case, Get_Initial_Thread_Handle takes a single DWORD (4-byte) parameter.

```
MAKE_HEADER(PTHCB,_stdcall,Get_Initial_Thread_Handle, (HVM hVM))
```

The Get_Initial_Thread_Handle wrapper first pops the caller's return address from the stack and then pops the caller's single argument. A function that pops the caller's arguments off the stack must pop the return address first, because the return address was pushed on the stack last as a result of the call into the function. Get_ Initial_Thread_Handle pops this return address into the EDX register, a register which will not be used by the VMM Get_Initial_Thread_Handle service (according to the service's documentation). The caller's parameter, the VM handle, is then popped into the EBX register, as expected by the service.

With the registers set up as expected by the service, the Get_Initial_Thread_Handle wrapper uses the VXDcall macro to call the real service. This particular service returns with the thread handle in EDI, but a C caller expects the thread handle as a return value, so the wrapper moves the handle into the C return register, EAX. Last, the wrapper returns to the caller by doing a JMP to the caller's return address, still stored in EDX. Normally a function returns with a RET instruction, but in this case the wrapper has already popped the return address off the stack, and so it must use a JMP and not a RET.

The macro VMMcall and its counterpart VxDCall deserve a closer look. Both are defined in VMM.INC. The assembler will expand this call:

```
VxDCall Get_Initial_Thread_Handle
```

into these instructions

```
CD 20  int Dyna_Link_Int
0001010D dd  @@Get_Initial_Thread_Handle+0int Dyna_Link_Int
```

This instruction sequence works as a dynamic link. The first time the sequence is executed, the INT 20h handler inside the VMM expects the 2 bytes immediately following the INT instruction to hold the Device ID of the VxD being called. The handler expects the 2 bytes following the Device ID to contain the Service Number being called. In the example above, 0001 is the VxD ID of the VMM, and 010D is the service number for Get_Initial_Thread_Handle.

The INT 20h handler determines the address of the service being called by traversing the VMM's linked list of DDBs (built by the VMM as VxDs are loaded). The handler traverses the list to find a VxD with a matching Device ID. Inside the DDB is the VxD's Service Table. The handler uses the Service Number (following the INT 20h) to find the address of the specific service. The handler then dynamically replaces the 2-byte INT 20h plus the 4-byte "opcode" with a 6-byte near call to the service address. Finally, the handler restarts the instruction by backing up EIP. This time the wrapper code directly calls the VMM Get_Initial_Thread_Handle service.

WRAPPER.ASM: IFSMGR_Ring0_OpenCreateFile *Details*

Get_Initial_Thread_Handle is a simple wrapper because the underlying service returns only one piece of information (a thread handle) which is easily communicated back to the C caller through a return value. The VMM service returns with the handle in EDI, the wrapper moves it to EAX, and the C caller sees this as a return value.

IFSMgr_Ring0_OpenCreateFile is a more complicated wrapper precisely because it must return two pieces of information back to its C caller (a handle and an error code). The underlying service uses two different registers to return this information, but the wrapper can't return two registers, because it's called by C code. The wrapper can use one of the two pieces as an actual return value, but the other must be communicated back through a pointer variable. It's this extra pointer parameter that complicates the wrapper implementation.

The IFSMgr_Ring0_OpenCreateFile prototype (found in WRAPPPERS.H) looks like

```
HANDLE cdecl IFSMgr_Ring0_OpenCreateFile( BOOL bInContext, PCHAR filename,
                                 WORD mode, WORD attrib, BYTE action,
                                 BYTE flags, WORD *pError, BYTE *pAction);
```

The file handle is provided as a return value. The caller must provide a pointer to a WORD variable which the wrapper will fill in with an error code. Notice the actual IFSMgr service doesn't do anything with this pointer. The service returns the error code in a register (EAX), and it's the wrapper's job to move this register value into the location targeted by the caller's pointer parameter. Here is the wrapper implementation:

```
StartCdecl   IFSMgr_Ring0_OpenCreateFile

bInContext EQU [ebp+8]
filename EQU [ebp+12]
mode EQU [ebp+16]
attrib EQU [ebp+20]
action EQU[ebp+24]
flags EQU [ebp+28]
pError EQU [ebp+32]
pAction EQU [ebp+36]
    push ebp
    mov  ebp, esp
    mov  dl,  action
    mov  dh,  flags
    mov  cx,  attrib
    mov  bx,  mode
    mov  esi, filename
    mov  eax, RO_OPENCREATFILE
    cmp  WORD PTR bInContext, 0
    je   @f
    mov eax, RO_OPENCREAT_IN_CONTEXT
@@:
    VxDCall  IFSMgr_Ring0_FileIO
    mov esi, pError
    jnc @f
    mov WORD PTR [esi], ax        ;give caller error code
    xor eax, eax                  ;return failure to caller
@@:
    mov esi, pAction
    mov DWORD PTR [esi], ecx      ; action performed
    ;returning with handle in eax
    pop ebp
    ret

EndCdecl     IFSMgr_Ring0_OpenCreateFile
```

IFSMgr_Ring0_OpenCreateFile uses StartCdecl to declare the function as cdecl. This means the wrapper will leave the parameters on the stack. To enhance readability, the IFSMgr_RO_OpenCreateFile defines several equates (using EQU, the assembly equivalent of #define) to refer to parameters on the stack.

On entry, IFSMgr_Ring0_OpenCreateFile copies parameters from the stack to the appropriate register, as expected by the IFSMgr. After the VxDCall to the service, the wrapper checks the Carry flag. The IFSMgr sets this flag to denote that an error occurred and that EAX contains an error code. If the flag is clear (no error), the wrapper writes a zero to the location pointed to by the error code pointer and returns to the caller. The C caller's file handle return value is already initialized, because the IFSMgr puts the file handle in EAX. If the flag is set (error), the wrapper takes the IFSMgr error code in AX, copies it to the location pointed to by the error code pointer, and returns with zero in EAX. This tells the C caller that the function failed, and that a meaningful value is available in the error code parameter.

Building the Wrapper Library

Once you've modified WRAPPERS.ASM to add your own services, you'll need to rebuild the WRAPPERS.CLB library. The makefile, WRAPPERS.MAK (Listing B.3, page 464), can be found in the subdirectory on the code disk. To build WRAPPERS.CLB, type nmake -fwrappers.mak.

The only unusual thing about the makefile is that the WRAPPERS.ASM source is assembled six different times, using a different value for the SEGNUM define, to produce six different OBJs. All six of the OBJs are added to the library.

SEGNUM isn't used by WRAPPERS.ASM directly. The LOCAL.INC include file from the DDK uses the value of SEGNUM to place a wrapper function in a particular code segment, and to generate a segment-specific function name. Here is an extract from LOCAL.INC.

```
IFE SEGNUM-1
 SEGB TEXTEQU <VXD_LOCKED_CODE_SEG>
 SEGE TEXTEQU <VXD_LOCKED_CODE_ENDS>
ELSEIFE SEGNUM-2
 SEGB TEXTEQU <VXD_ICODE_SEG>
 SEGE TEXTEQU <VXD_ICODE_ENDS>
ELSEIFE SEGNUM-3
 SEGB TEXTEQU <VXD_PAGEABLE_CODE_SEG>
 SEGE TEXTEQU <VXD_PAGEABLE_CODE_ENDS>
ELSEIFE SEGNUM-4
 SEGB TEXTEQU <VXD_STATIC_CODE_SEG>
 SEGE TEXTEQU <VXD_STATIC_CODE_ENDS>
ELSEIFE SEGNUM-5
 SEGB TEXTEQU <VXD_DEBUG_ONLY_CODE_SEG>
 SEGE TEXTEQU <VXD_DEBUG_ONLY_CODE_ENDS>
ELSEIFE SEGNUM-6
 SEGB TEXTEQU <VXD_PNP_CODE_SEG>
 SEGE TEXTEQU <VXD_PNP_CODE_ENDS>
```

You can see that SEGNUM=1 corresponds to the LOCKED segment, SEGNUM=2 corresponds to the ICODE (initialization) segment, etc. LOCAL.INC uses these SEGB and SEGE equates in the definition of the StartStdCall or StartCDecl macros. As a result, when a wrapper module declares a wrapper function using one of these macros, LOCAL.INC places the wrapper function in the appropriate segment and also prepends the function name with a segment name. For example, when WRAPPERS.ASM is assembled with DSEGNUM=1, the following source code

```
StartStdCall Get_Initial_Thread_Handle, 4
```

is translated by the preprocessor into

```
 PUBLIC    _LCODE_Get_Initial_Thread_Handle@4
_LTEXT    SEGMENT
_LCODE_Get_Initial_Thread_Handle@4  PROC NEAR
```

As a result of these macros, the WRAPPERS.CLB library contains six different versions of every wrapper function, with six different names. These six names correspond to the names generated by the MAKE_HEADER macro in WRAPPERS.H. (Refer to the section "WRAPPERS.H" for an explanation of the MAKE_HEADER wrapper and the function prototypes it generates.)

Summary

The techniques described in this book allow you to write most of your VxD in C using only the DDK, without purchasing VToolsD. However, you will still need to write small portions of your VxD in assembly. This appendix demonstrates exactly how to write the assembly pieces, as well as how to extend the WRAPPERS.CLB C-callable wrapper library included with this book.

Listing B.1   WRAPPERS.H

```
/*****************************************************************************
*                                                                           *
* THIS CODE AND INFORMATION IS PROVIDED "AS IS" WITHOUT WARRANTY OF ANY      *
* KIND, EITHER EXPRESSED OR IMPLIED, INCLUDING BUT NOT LIMITED TO THE        *
* IMPLIED WARRANTIES OF MERCHANTABILITY AND/OR FITNESS FOR A PARTICULAR      *
* PURPOSE.                                                                   *
*                                                                           *
* Copyright 1993-95  Microsoft Corporation.  All Rights Reserved.           *
*                                                                           *
*****************************************************************************/

#ifndef _VMMXWRAP_H
#define _VMMXWRAP_H

#include <vxdwraps.h>

/*****************************************************************************
*                                                                           *
*    VMM services                                                           *
*                                                                           *
*****************************************************************************/
typedef DWORD VMHANDLE;
typedef DWORD EVENTHANDLE;
typedef DWORD MEMHANDLE;
typedef DWORD PTHCB;  // pointer to thread control block
typedef void *CALLBACK(void);

MAKE_HEADER(PTHCB,_stdcall,Get_Initial_Thread_Handle, (HVM hVM))
MAKE_HEADER(BOOL, _stdcall,Install_IO_Handler, (DWORD PortNum, CALLBACK Callback ))
MAKE_HEADER(BOOL, _stdcall,Remove_IO_Handler, (DWORD PortNum))
MAKE_HEADER(void, _stdcall,Enable_Local_Trapping, (VMHANDLE hVM, DWORD PortNum))
MAKE_HEADER(void, _stdcall,Disable_Local_Trapping, (VMHANDLE hVM, DWORD PortNum))
MAKE_HEADER(EVENTHANDLE,cdecl,Call_Priority_VM_Event, (DWORD PriorityBoost, VMHANDLE hVM, \
                                                DWORD Flags, void *Refdata, \
                                                CALLBACK EventCallback, \
                                                DWORD Timeout ))
MAKE_HEADER(void, cdecl, _Deallocate_Device_CB_Area, (DWORD Offset, DWORD Flags ))
MAKE_HEADER(void, cdecl, Save_Client_State, (CLIENT_STRUCT * pSavedRegs))
MAKE_HEADER(void, cdecl, Restore_Client_State, (CLIENT_STRUCT * pSavedRegs))
MAKE_HEADER(void, cdecl, Begin_Nest_Exec, (void))
MAKE_HEADER(void, cdecl, End_Nest_Exec, (void))
MAKE_HEADER(void, _stdcall, Simulate_Far_Call, (WORD seg, WORD off ))
MAKE_HEADER(void, _stdcall, Simulate_Push, (DWORD val))
MAKE_HEADER(BOOL,cdecl,_ModifyPageBits, (VMHANDLE hVM, DWORD VMLinPgNum, DWORD nPages, \
                                        DWORD bitAnd, DWORD bitOR, DWORD pType, \
                                        DWORD Flags))
MAKE_HEADER(BOOL,_stdcall,Hook_V86_Page, (DWORD PageNum, CALLBACK Callback))
MAKE_HEADER(BOOL,_stdcall,Unhook_V86_Page, (DWORD PageNum, CALLBACK Callback))
MAKE_HEADER(BOOL,cdecl,_Assign_Device_V86_Pages, (DWORD VMLinrPage, DWORD nPages, \
                                        VMHANDLE hVM, DWORD Flags))
MAKE_HEADER(BOOL,cdecl,_DeAssign_Device_V86_Pages, (DWORD VMLinrPage, DWORD nPages, \
                                        VMHANDLE hVM, WORD Flags))
MAKE_HEADER(BOOL,cdecl,_PhysIntoV86, (DWORD PhysPage, VMHANDLE hVM, DWORD VMLinPgNum, \
                                        DWORD nPages, DWORD Flags))
MAKE_HEADER(BOOL,cdecl,_MapIntoV86, (MEMHANDLE hMem, VMHANDLE hVM, DWORD VMLinPageNumber, \
                                        DWORD nPages, DWORD PageOff, DWORD Flags))
MAKE_HEADER(MEMHANDLE,cdecl,_GetNulPageHandle,(void))
MAKE_HEADER(ULONG,cdecl,_PageReserve, (ULONG page, ULONG npages, ULONG flags))
MAKE_HEADER(ULONG,cdecl,_PageCommitPhys, (ULONG page, ULONG npages, ULONG physpg, \
                                        ULONG flags))
```

Listing B.1 (continued)   WRAPPERS.H

```
MAKE_HEADER(ULONG,cdecl,_PageDecommit, (ULONG page, ULONG npages, ULONG flags))
MAKE_HEADER(void *,_stdcall,Map_Flat, (BYTE SegOffset, BYTE OffOffset ))

#define MAPFLAT(sgmnt,offst) Map_Flat(((DWORD)(&((CRS *)0)->sgmnt)), \
                                      (DWORD)(&((struct Client_Word_Reg_Struc *)0)->offst))
#define Map_Flat                    PREPEND(Map_Flat)
#define Call_Priority_VM_Event      PREPEND(Call_Priority_VM_Event)
#define Get_Initial_Thread_Handle   PREPEND(Get_Initial_Thread_Handle)
#define Install_IO_Handler          PREPEND(Install_IO_Handler)
#define Remove_IO_Handler           PREPEND(Remove_IO_Handler)
#define Enable_Local_Trapping       PREPEND(Enable_Local_Trapping)
#define Disable_Local_Trapping      PREPEND(Disable_Local_Trapping)
#define _Deallocate_Device_CB_Area  PREPEND(_Deallocate_Device_CB_Area)
#define Save_Client_State           PREPEND(Save_Client_State)
#define Restore_Client_State        PREPEND(Restore_Client_State)
#define Begin_Nest_Exec             PREPEND(Begin_Nest_Exec)
#define End_Nest_Exec               PREPEND(End_Nest_Exec)
#define Simulate_Far_Call           PREPEND(Simulate_Far_Call)
#define Simulate_Push               PREPEND(Simulate_Push)
#define _ModifyPageBits             PREPEND(_ModifyPageBits)
#define Hook_V86_Page               PREPEND(Hook_V86_Page)
#define Unhook_V86_Page             PREPEND(Unhook_V86_Page)
#define _Assign_Device_V86_Pages    PREPEND(_Assign_Device_V86_Pages)
#define _DeAssign_Device_V86_Pages  PREPEND(_DeAssign_Device_V86_Pages)
#define _PhysIntoV86                PREPEND(_PhysIntoV86)
#define _MapIntoV86                 PREPEND(_MapIntoV86)
#define _GetNulPageHandle           PREPEND(_GetNulPageHandle)
#define _PageReserve                PREPEND(_PageReserve)
#define _PageCommitPhys             PREPEND(_PageCommitPhys)
#define _PageDecommit               PREPEND(_PageDecommit)

// the following functions are really in VXDWRAPS.CLB, but aren't prototyped in VXDWRAPS.H
MAKE_HEADER(DWORD, cdecl, _Allocate_Device_CB_Area, (DWORD NumBytes, DWORD Flags ))

#define _Allocate_Device_CB_Area    PREPEND(_Allocate_Device_CB_Area)

/***************************************************************************
 *                                                                        *
 *    IFSMgr services                                                      *
 *                                                                        *
 ***************************************************************************/

#define RO_OPENCREATFILE         0xD500  /* Open/Create a file */
#define RO_OPENCREAT_IN_CONTEXT  0xD501  /* Open/Create file in current context */
#define RO_CLOSEFILE             0xD700  /* Close file */
#define RO_WRITEFILE             0xD601  /* WRite to a file */
#define RO_WRITEFILE_IN_CONTEXT  0xD603  /* Write to a file in current context */

MAKE_HEADER(HANDLE,cdecl,IFSMgr_Ring0_OpenCreateFile, (BOOL bInContext, PCHAR filename, \
                                                       WORD mode, WORD attrib, \
                                                       BYTE action, BYTE flags, \
                                                       WORD *pError, BYTE *pAction))
MAKE_HEADER(DWORD,cdecl,IFSMgr_Ring0_WriteFile, (BOOL bInContext, HANDLE filehandle, \
                                                 PVOID buf,  DWORD count, DWORD filepos, \
                                                 WORD  *perr))
MAKE_HEADER(BOOL,cdecl,IFSMgr_Ring0_CloseFile, (HANDLE filehandle,  WORD  *pError))

#define IFSMgr_Ring0_OpenCreateFile PREPEND(IFSMgr_Ring0_OpenCreateFile)
#define IFSMgr_Ring0_WriteFile PREPEND(IFSMgr_Ring0_WriteFile)
#define IFSMgr_Ring0_CloseFile PREPEND(IFSMgr_Ring0_CloseFile)
```

Listing B.1 (continued)   `WRAPPERS.H`

```c
/****************************************************************************
 *                                                                        *
 *   VPICD services                                                       *
 *                                                                        *
 ***************************************************************************/

typedef struct
{
    WORD VID_IRQ_Number;     // IRQ to virtualize
    WORD VID_Options;
    // VPICD_OPT_CAN_SHARE: allow other VxDs to virtualize IRQ also
    // VPICD_OPT_REF_DATA: pass VID_Hw_Int_Ref as param to Hw_Int_Handler
    DWORD VID_Hw_Int_Proc;   // callback for hardware interrupt
    DWORD VID_Virt_Int_Proc;
    DWORD VID_EOI_Proc;
    DWORD VID_Mask_Change_Proc;
    DWORD VID_IRET_Proc;
    DWORD VID_IRET_Time_Out;
    PVOID VID_Hw_Int_Ref;    // pass this data to Hw_Int_Handler
} VPICD_IRQ_DESCRIPTOR;

typedef DWORD IRQHANDLE;

MAKE_HEADER(IRQHANDLE,_stdcall,VPICD_Virtualize_IRQ, (VPICD_IRQ_DESCRIPTOR *pIrqDesc))
MAKE_HEADER(void,_stdcall,VPICD_Physically_Mask, (IRQHANDLE hndIrq))
MAKE_HEADER(void,_stdcall,VPICD_Physically_Unmask, (IRQHANDLE hndIrq))
MAKE_HEADER(void,_stdcall,VPICD_Force_Default_Behavior, (IRQHANDLE hndIrq))
MAKE_HEADER(void,_stdcall,VPICD_Phys_EOI, (IRQHANDLE hndIrq))
MAKE_HEADER(void,_stdcall,VPICD_Set_Int_Request, (IRQHANDLE hIRQ, MHANDLE hVM))
MAKE_HEADER(void,_stdcall,VPICD_Clear_Int_Request, (IRQHANDLE hIRQ, VMHANDLE hVM))

#define VPICD_Virtualize_IRQ          PREPEND(VPICD_Virtualize_IRQ)
#define VPICD_Physically_Mask         PREPEND(VPICD_Physically_Mask)
#define VPICD_Physically_Unmask       PREPEND(VPICD_Physically_Unmask)
#define VPICD_Force_Default_Behavior  PREPEND(VPICD_Force_Default_Behavior)
#define VPICD_Phys_EOI                PREPEND(VPICD_Phys_EOI)
#define VPICD_Set_Int_Request         PREPEND(VPICD_Set_Int_Request)
#define VPICD_Clear_Int_Request       PREPEND(VPICD_Clear_Int_Request)

/****************************************************************************
 *                                                                        *
 *   VDMAD services                                                       *
 *                                                                        *
 ***************************************************************************/

#define DMA_type_verify 0x00
#define DMA_type_write  0x04
#define DMA_type_read   0x08
#define DMA_AutoInit    0x10
#define DMA_AdrDec      0x20
#define DMA_demand_mode 0x00
#define DMA_single_mode 0x40
#define DMA_block_mode  0x80
#define DMA_cascade 0xc0
#define DMA_mode_mask   0xc0   // mask to isolate controller mode bits (above)
#define DMA_chan_sel    0x03
#define NONE_LOCKED 0
#define ALL_LOCKED  1
#define SOME_LOCKED 2
```

Listing B.1 (continued)   WRAPPERS.H

```c
typedef struct
{
    DWORD    PhysAddr;
    DWORD    Size;
} REGION;

typedef struct Extended_DDS_Struc
{
    DWORD    DDS_size;
    DWORD    DDS_linear;
    WORD DDS_seg;
    WORD RESERVED;
    WORD DDS_avail;
    WORD DDS_used;
} EXTENDED_DDS, *PEXTENDED_DDS;

typedef struct
{
    EXTENDED_DDS    dds;
    union
    {
        REGION          aRegionInfo[16];
        DWORD           aPte[16];
    };
} DDS, *PDDS;

typedef DWORD DMAHANDLE;

MAKE_HEADER(DWORD,cdecl,VDMAD_Scatter_Lock, ( VMHANDLE hVM,  DWORD Flags, PDDS pDDS, \
                                        PDWORD pPteOffset ))
MAKE_HEADER(DMAHANDLE, cdecl, VDMAD_Virtualize_Channel, (BYTE ch, CALLBACK pfCallback ))
MAKE_HEADER(void, _stdcall, VDMAD_Set_Region_Info, (DMAHANDLE DMAHandle, BYTE BufferId, \
                                        BOOL LockStatus, DWORD Region, \
                                        DWORD RegionSize, DWORD PhysAddr ))
MAKE_HEADER(void, _stdcall, VDMAD_Set_Phys_State, (DMAHANDLE DMAHandle, VMHANDLE hVM, \
                                        WORD Mode,  WORD ExtMode ))
MAKE_HEADER(void, _stdcall, VDMAD_Phys_Unmask_Channel, (DMAHANDLE DMAHandle, \
                                        VMHANDLE hVM ))
MAKE_HEADER(void, _stdcall, VDMAD_Phys_Mask_Channel, (DMAHANDLE DMAHandle))
#define VDMAD_Virtualize_Channel PREPEND(VDMAD_Virtualize_Channel)
#define VDMAD_Set_Region_Info PREPEND(VDMAD_Set_Region_Info)
#define VDMAD_Set_Phys_State PREPEND(VDMAD_Set_Phys_State)
#define VDMAD_Scatter_Lock  PREPEND(VDMAD_Scatter_Lock)
#define VDMAD_Phys_Unmask_Channel PREPEND(VDMAD_Phys_Unmask_Channel)
#define VDMAD_Phys_Mask_Channel PREPEND(VDMAD_Phys_Unmask_Channel)
```

Listing B.1 (continued)   `WRAPPERS.H`

```c
/****************************************************************************
*                                                                          *
*    SHELL services                                                        *
*                                                                          *
****************************************************************************/

MAKE_HEADER(VMHANDLE, _stdcall, SHELL_Resolve_Contention, (VMHANDLE hndOwner, \
                                                VMHANDLE hndContender, \
                                                char *DeviceName ))
MAKE_HEADER(BOOL,cdecl,_SHELL_PostMessage, (HANDLE hWnd, DWORD uMsg, WORD wParam, \
                                    DWORD lParam, CALLBACK pCallback, \
                                    void *dwRefData))

#define SHELL_Resolve_Contention PREPEND(SHELL_Resolve_Contention)
#define _SHELL_PostMessage        PREPEND(_SHELL_PostMessage)

/****************************************************************************
*                                                                          *
*    VWIN32 services                                                       *
*                                                                          *
****************************************************************************/

MAKE_HEADER(void,cdecl,_VWIN32_QueueUserApc, (void *pR3Proc, DWORD Param, PTHCB hThread))
MAKE_HEADER(BOOL,cdecl,_VWIN32_SetWin32Event, (EVENTHANDLE hEvent) )

#define _VWIN32_QueueUserApc      PREPEND(_VWIN32_QueueUserApc)
#define _VWIN32_SetWin32Event     PREPEND(_VWIN32_SetWin32Event)

#endif // _VMMXWRAP_H
```

Listing B.2   `WRAPPERS.ASM`

```asm
include local.inc
include ifsmgr.inc
include vdmad.inc
include vpicd.inc
include vwin32.inc
include shell.inc

RO_OPENCREATFILE          equ 0D500h  ; Open/Create a file
RO_OPENCREAT_IN_CONTEXT   equ 0D501h  ; Open/Create file in current context
RO_READFILE               equ 0D600h  ; Read a file, no context
RO_WRITEFILE              equ 0D601h  ; Write to a file, no context
RO_READFILE_IN_CONTEXT    equ 0D602h  ; Read a file, in thread context
RO_WRITEFILE_IN_CONTEXT   equ 0D603h  ; Write to a file, in thread context
RO_CLOSEFILE              equ 0D700h  ; Close a file
RO_GETFILESIZE            equ 0D800h  ; Get size of a file

StartStdCall    MACRO   Name, Param
StartCDecl      Name&@&Param
ENDM

EndStdCall      MACRO   Name, Param
EndCDecl        Name&@&Param
ENDM
```

Listing B.2 (continued)   WRAPPERS.ASM

```
MakeCDecl    _ModifyPageBits

MakeCDecl    _Assign_Device_V86_Pages

MakeCDecl    _DeAssign_Device_V86_Pages

MakeCDecl    _PhysIntoV86

MakeCDecl    _MapIntoV86

MakeCDecl    _GetNulPageHandle

MakeCDecl    _PageReserve

MakeCDecl    _PageCommitPhys

MakeCDecl    _PageDecommit

;  void cdecl _Deallocate_Device_CB_Area( DWORD NumBytes, DWORD Flags ))
;
MakeCDecl    _Deallocate_Device_CB_Area

;  EVENTHANDLE Call_Priority_VM_Event(DWORD PriorityBoost, VMHANDLE hVM, DWORD Flags,
;                                     void * Refdata, CALLBACK EventCallback,
;                                     DWORD Timeout );

StartCDecl  Call_Priority_VM_Event

PriorityBoost EQU [ebp+8]
hVM EQU [ebp+12]
Flags EQU [ebp+16]
Refdata EQU [ebp+20]
EventCallback EQU[ebp+24]
Timeout EQU [ebp+28]

     push  ebp
     mov   ebp, esp

     mov   eax, DWORD PTR PriorityBoost
     mov   ebx, DWORD PTR hVM
     mov   ecx, DWORD PTR Flags
     mov   edx, DWORD PTR Refdata
     mov   esi, DWORD PTR EventCallback
     mov   edi, DWORD PTR Timeout
     VMMCall Call_Priority_VM_Event
     mov   eax, esi    ; eax=event handle

     pop ebp
     ret

EndCDecl    Call_Priority_VM_Event
```

Listing B.2 (continued)   WRAPPERS.ASM

```
;  void * _stdcall Map_Flat( BYTE SegOffset, BYTE OffOffset )
;
;
StartStdCall    Map_Flat, 8

        pop    edx        ; save ret addr in unused reg
        pop    ebx        ; segment
        xor    bh,bh      ; BL=segment
        mov    ah,bl      ; AH=segment
        pop    ebx        ; offset
        xor    bh,bh      ; BL=offset
        mov    al,bl      ; AL=offset
        VMMcall Map_Flat  ; AH=seg AL=off
        jmp    edx        ; jump to caller's ret addr

EndStdCall      Map_Flat, 8

;  BOOL _stdcall Hook_V86_Page( DWORD PageNum, CALLBACK Callback )
;
StartStdCall    Hook_V86_Page, 8

        pop edx          ; save ret addr in unused reg
        pop eax          ; PageNum
        pop esi          ; Callback
        VMMCall Hook_V86_Page
        mov eax, 1       ; assume TRUE ret val
        jnc @f
        xor eax, eax     ; carry set, error, so FALSE ret val
@@:
        jmp edx          ; jump to caller's ret addr

EndStdCall      Hook_V86_Page, 8

;  BOOL _stdcall Unhook_V86_Page( DWORD PageNum, CALLBACK Callback )
;
StartStdCall    Unhook_V86_Page, 8

        pop edx          ; save ret addr in unused reg
        pop eax          ; PageNum
        pop esi          ; Callback
        VMMCall Unhook_V86_Page
        mov eax, 1       ; assume TRUE ret val
        jnc @f
        xor eax, eax     ; carry set, error, so FALSE ret val
@@:
        jmp edx          ; jump to caller's ret addr

EndStdCall      Unhook_V86_Page, 8

;  PTCB _stdcall Get_Initial_Thread_Handle( VMHANDLE hVM )
StartStdCall    Get_Initial_Thread_Handle, 4

        pop edx          ; Get return address
        pop ebx          ; Get VMHandle
        VxDCall Get_Initial_Thread_Handle
        mov eax, edi     ; move thread hnd into return
        jmp edx          ; return addr still in edx

EndStdCall      Get_Initial_Thread_Handle, 4
```

Listing B.2 (continued)   WRAPPERS.ASM

```
;  BOOL _stdcall Install_IO_Handler( PortNum, Callback  )
StartStdCall    Install_IO_Handler, 8

        pop ebx         ; save ret addr in unused reg
        pop edx         ; PortNum
        pop esi         ; Callback
        VMMCall Install_IO_Handler
        mov eax, 1      ; assume TRUE ret val
        jnc @f
        xor eax, eax    ; carry set, error, so FALSE ret val
    @@:
        jmp ebx         ; jump to caller's ret addr

EndStdCall      Install_IO_Handler, 8

;  BOOL _stdcall Remove_IO_Handler( PortNum )
;
StartStdCall    Remove_IO_Handler, 4

        pop ebx         ; save ret addr in unused reg
        pop edx         ; PortNum
        VMMCall Remove_IO_Handler
        mov eax, 1      ; assume TRUE ret val
        jnc @f
        xor eax, eax    ; carry set, error, so FALSE ret val
    @@:
        jmp ebx         ; jump to caller's ret addr

EndStdCall      Remove_IO_Handler, 4

; void _stdcall Enable_Local_Trapping( VMHANDLE hVM, DWORD PortNum )
;
StartStdCall    Enable_Local_Trapping, 8

        pop ecx         ; save ret addr in unused reg
        pop ebx         ; hVM
        pop edx         ; PortNum
        VMMcall Enable_Local_Trapping
        jmp ecx

EndStdCall      Enable_Local_Trapping, 8

; void _stdcall Disable_Local_Trapping( VMHANDLE hVM, DWORD PortNum )
;
StartStdCall    Disable_Local_Trapping, 8

        pop ecx         ; save ret addr in unused reg
        pop ebx         ; hVM
        pop edx         ; PortNum
        VMMcall Disable_Local_Trapping
        jmp ecx

EndStdCall      Disable_Local_Trapping, 8
```

Listing B.2 (continued)   WRAPPERS.ASM

```asm
;  void cdecl Save_Client_State( CLIENT_STRUCT * pSavedRegs
;
StartCdecl      Save_Client_State

pSavedRegs EQU [ebp+8]

        push ebp
        mov ebp, esp

        pushad   ; service doesn't claim to save any regs
        mov edi, pSavedRegs
        VMMcall Save_Client_State
        popad

        pop ebp
        ret

EndCdecl        Save_Client_State

;  void cdecl Restore_Client_State( CLIENT_STRUCT * pRestoredRegs
;
StartCdecl      Restore_Client_State

pSavedRegs EQU [ebp+8]

        push ebp
        mov ebp, esp

        pushad    ; service doesn't claim to save any regs
        mov edi, pSavedRegs
        VMMcall Restore_Client_State
        popad

        pop ebp
        ret

EndCdecl        Restore_Client_State

;  void cdecl Begin_Nest_Exec( void )
;
StartCdecl      Begin_Nest_Exec

        push ebp
        mov ebp, esp

        pushad         ; service doesn't claim to save any regs
        VMMcall Begin_Nest_Exec
        popad

        pop ebp
        ret

EndCdecl        Begin_Nest_Exec
```

Listing B.2 (continued)   `WRAPPERS.ASM`

```
; void cdecl End_Nest_Exec( void )
;
StartCdecl      End_Nest_Exec

    push ebp
    mov ebp, esp

    pushad          ; service doesn't claim to save any regs
    VMMcall End_Nest_Exec
    popad

    pop ebp
    ret

EndCdecl        End_Nest_Exec

; void _stdcall Simulate_Far_Call( WORD seg, WORD off )
;
StartStdCall    Simulate_Far_Call, 8

    pop eax         ; save ret addr in unused reg
    pop ecx         ; segment
    pop edx         ; offset
    VMMcall Simulate_Far_Call
    jmp eax

EndStdCall      Simulate_Far_Call, 8

; void _stdcall Simulate_Push( DWORD val )
;
StartStdCall    Simulate_Push, 4

    pop edx         ; save ret addr in unused reg
    pop eax         ; val
    VMMcall Simulate_Push
    jmp edx

EndStdCall      Simulate_Push, 4

; HANDLE cdecl IFSMgr_Ring0_OpenCreateFile( BOOL bInContext, PCHAR filename,
;                                           WORD mode, WORD attrib,
;                                           BYTE action, BYTE flags,
;                                           WORD *pError, BYTE *pAction)
StartCdecl      IFSMgr_Ring0_OpenCreateFile

bInContext EQU [ebp+8]
filename EQU [ebp+12]
mode EQU [ebp+16]
attrib EQU [ebp+20]
action EQU[ebp+24]
flags EQU [ebp+28]
pError EQU [ebp+32]
pAction EQU [ebp+36]
```

Listing B.2 (continued)   WRAPPERS.ASM

```
        push ebp
        mov  ebp, esp
        mov dl, action
        mov dh, flags
        mov cx, attrib
        mov bx, mode
        mov esi, filename
        mov eax, RO_OPENCREATFILE
        cmp WORD PTR bInContext, 0
        je @f
        mov eax, RO_OPENCREAT_IN_CONTEXT
@@:

        VxDCall IFSMgr_Ring0_FileIO
        mov esi, pError          ; esi->error code
        jnc @f
        mov WORD PTR [esi], ax   ;give caller error code
        xor eax, eax             ;return failure to caller
@@:
        mov esi, pAction
        mov DWORD PTR [esi], ecx    ; action performed
        ;returning with handle in eax

        pop ebp
        ret

EndCdecl      IFSMgr_Ring0_OpenCreateFile

StartCdecl  IFSMgr_Ring0_CloseFile

filehandle EQU [ebp+8]
pError EQU [ebp+12]

        push ebp
        mov  ebp, esp

        mov ebx, filehandle
        mov eax, RO_CLOSEFILE    ; func code
        VxDCall IFSMgr_Ring0_FileIO
        mov ecx, 1       ; assume returning true
        jnc @f
        mov esi, pError
        mov WORD PTR [esi], ax
        xor ecx, ecx     ; returning false
@@:
        mov eax, ecx     ;error code or zero

        pop ebp
        ret


EndCdecl      IFSMgr_Ring0_CloseFile
```

Listing B.2 (continued)   WRAPPERS.ASM

```
; BOOL cdecl IFSMgr_Ring0_CloseFile(HANDLE filehandle, WORD *pError)
; DWORD cdecl IFSMgr_Ring0_WriteFile(BOOL bInContext, HANDLE filehandle, PVOID buf,
;                                    DWORD count, DWORD filepos, WORD  *perr))
StartCdecl  IFSMgr_Ring0_WriteFile

bInContext EQU [ebp+8]
filehandle EQU [ebp+12]
buf EQU [ebp+16]
count EQU [ebp+20]
filepos EQU [ebp+24]
pError EQU [ebp+28]

    push ebp
    mov  ebp, esp

    mov ebx, filehandle
    mov esi, buf
    mov ecx, count
    mov edx, filepos
    mov eax, RO_WRITEFILE
    cmp WORD PTR bInContext, 0
    je  @f
    mov eax, RO_WRITEFILE_IN_CONTEXT
@@:
    VxDCall IFSMgr_Ring0_FileIO
    jnc @f
    mov esi, pError
    mov WORD PTR [esi], ax  ;give caller error code
    xor ecx, ecx           ;but byte count to zero on error
@@:
    ; ecx contains count
    mov eax, ecx

    pop ebp
    ret

EndCdecl    IFSMgr_Ring0_WriteFile

;
StartStdCall VPICD_Virtualize_IRQ, 4

    pop     edx      ; save ret addr in unused reg
    pop     edi      ; pIrqDesc
    VxDcall VPICD_Virtualize_IRQ
    jnc @f
    xor eax, eax     ; carry set, error, so zero return code
@@:
    jmp edx          ; jump to caller's ret addr

EndStdCall   VPICD_Virtualize_IRQ, 4

; void _stdcall VPICD_Physically_Mask(IRQHANDLE hndIrq))
;
StartStdCall VPICD_Physically_Mask, 4

    pop     edx  ; save ret addr in unused reg
    pop     eax  ; hndIrq
    VxDcall VPICD_Physically_Mask
    jmp edx      ; jump to caller's ret addr

EndStdCall   VPICD_Physically_Mask, 4
```

Listing B.2 (continued)   `WRAPPERS.ASM`

```
;  IRQHANDLE _stdcall VPICD_Virtualize_IRQ (VPICD_IRQ_DESCRIPTOR *pIrqDesc)
;  void _stdcall VPICD_Physically_Unmask(IRQHANDLE hndIrq))
;
StartStdCall VPICD_Physically_Unmask, 4

    pop     edx     ; save ret addr in unused reg
    pop     eax     ; hndIrq
    VxDcall VPICD_Physically_Unmask
    jmp edx         ; jump to caller's ret addr

EndStdCall    VPICD_Physically_Unmask, 4

; void _stdcall VPICD_Force_Default_Behavior(IRQHANDLE hndIrq));
; y
StartStdCall VPICD_Force_Default_Behavior, 4

    pop     edx     ; save ret addr in unused reg
    pop     eax     ; hndIrq
    VxDcall VPICD_Force_Default_Behavior
    jmp edx         ; jump to caller's ret addr

EndStdCall    VPICD_Force_Default_Behavior, 4

; void _stdcall VPICD_Phys_EOI(IRQHANDLE hndIrq))
;
StartStdCall VPICD_Phys_EOI, 4

    pop     edx             ; save ret addr in unused reg
    pop     eax             ; hndIrq
    VxDcall VPICD_Phys_EOI
    jmp edx                 ; jump to caller's ret addr

EndStdCall    VPICD_Phys_EOI, 4

; void _stdcall VPICD_Set_Int_Request(VMHANDLE hVM,, IRQHANDLE hndIrq
;
StartStdCall VPICD_Set_Int_Request, 8

    pop     edx     ; save ret addr in unused reg
    pop     ebx     ; hVM
    pop     eax     ; hndIrq
    VxDcall VPICD_Set_Int_Request
    jmp edx         ; jump to caller's ret addr

EndStdCall    VPICD_Set_Int_Request, 8

; void _stdcall VPICD_Clear_Int_Request(VMHANDLE hVM, IRQHANDLE hIrq
;
StartStdCall VPICD_Clear_Int_Request, 8

    pop     edx     ; save ret addr in unused reg
    pop     ebx     ; hVM
    pop     eax     ; hndIrq
    VxDcall VPICD_Clear_Int_Request
    jmp edx         ; jump to caller's ret addr

EndStdCall    VPICD_Clear_Int_Request, 8
```

Listing B.2 (continued) WRAPPERS.ASM

```asm
; DMAHANDLE cdecl VDMAD_Virtualize_Channel(BYTE ch  DMACALLBACK pfCallback )
StartCdecl  VDMAD_Virtualize_Channel

chan EQU [ebp+8]
pfCallback EQU [ebp+12]

    push ebp
    mov  ebp, esp

    movzx eax, BYTE PTR chan
    mov esi, pfCallback
    VxDCall VDMAD_Virtualize_Channel
    jnc @f
    xor eax, eax    ; carry set, error so zero return code
@@:
    pop ebp
    ret

EndCdecl VDMAD_Virtualize_Channel

; void _stdcall VDMAD_Set_Region_Info( DMAHANDLE DMAHandle, BYTE BufferId,
;                                      BOOL LockStatus, DWORD Region,
;                                      DWORD RegionSize, DWORD PhysAddr
;
StartStdCall VDMAD_Set_Region_Info, 24

    pop   edi     ; save ret addr in unused reg
    pop   eax     ; DMAHandle
    pop   ebx     ; BufferId
    xor   bh, bh  ; BL=BufferId
    pop   ecx     ; LockStatus
    shl   ecx, 4  ; CX=LockStatus
    xor   cl, cl  ; CH=LockStatus
    or    bx, cx  ; BX=LockStatus|BufferId
    pop   esi     ; Region
    pop   ecx     ; RegionSize
    pop   edx     ; PhysAddr
    VxDCall VDMAD_Set_Region_Info
    jmp   edi     ; jump to caller's ret addr
    ret

EndStdCall VDMAD_Set_Region_Info, 24

; void _stdcall VDMAD_Set_Phys_State( DMAHANDLE DMAHandle, VMHANDLE hVM,
;                             BYTE Mode, BYTE ExtMode
;
StartStdCall VDMAD_Set_Phys_State, 16

    pop   esi     ; save ret addr in unused reg
    pop   eax     ; DMAHandle
    pop   ebx     ; hVM
    pop   edx     ; Mode
    xor   dh, dh  ; DL=Mode
    pop   ecx     ; ExtMode
    shl   ecx, 4  ; CH=ExtMode
    xor   cl, cl  ; CH=ExtMode
    or    dx, cx  ; DX=ExtMode|Mode
    VxDcall VDMAD_Set_Phys_State
    jmp   esi     ; jump to caller's ret addr
    ret

EndStdCall   VDMAD_Set_Phys_State, 16
```

Listing B.2 (continued)   WRAPPERS.ASM

```
; void _stdcall VDMAD_Phys_Unmask_Channel( DMAHANDLE DMAHandle, VMHANDLE hVM)
;
StartStdCall  VDMAD_Phys_Unmask_Channel, 8

    pop    esi    ; save ret addr in unused reg
    pop    edx    ; DMAHandle
    pop    ebx    ; hVM
    VxDcall VDMAD_Phys_Unmask_Channel
    jmp    esi    ; jump to caller's ret addr
    ret

EndStdCall    VDMAD_Phys_Unmask_Channel, 8

; void _stdcall VDMAD_Phys_Mask_Channel( DMAHANDLE DMAHandle )
;
StartStdCall  VDMAD_Phys_Mask_Channel, 4

    pop    esi    ; save ret addr in unused reg
    pop    eax    ; DMAHandle
    VxDcall VDMAD_Phys_Mask_Channel
    jmp esi        ; jump to caller's ret addr
    ret

EndStdCall    VDMAD_Phys_Mask_Channel, 4

; DWORD cdecl VDMAD_Scatter_Lock( VMHANDLE hVM, DWORD Flags,
;                                 PDDW pDDS, PDWORD pPteOffset )
StartCdecl  VDMAD_Scatter_Lock

hVM EQU [ebp+8]
Flags EQU BYTE PTR [ebp+12]
pDDS EQU [ebp+16]
pPteOffset EQU[ebp+20]

RET_NO_LOCKED EQU 0
RET_ALL_LOCKED EQU 1
RET_PART_LOCKED EQU 2
    push ebp
    mov  ebp, esp

    mov al,  BYTE PTR Flags
    mov ebx, hVM
    mov edi, pDDS
    VxDCall VDMAD_Scatter_Lock
    jc no_lock
    jz all_lock
    mov eax, RET_PART_LOCKED
    jmp flags_checked
no_lock:
    mov eax, RET_NO_LOCKED
    jmp flags_checked
all_lock:
    mov eax, RET_ALL_LOCKED
flags_checked:
    mov ebx, pPteOffset
    mov DWORD PTR [ebx], esi
    pop ebp
    ret

EndCdecl     VDMAD_Scatter_Lock
```

Listing B.2 (continued)   *WRAPPERS.ASM*

```
; VMHANDLE cdecl SHELL_Resolve_Contention( VMHANDLE hndOwner, VMHANDLE hndContender,
;                                          char *DeviceName )
;
StartStdCall   SHELL_Resolve_Contention, 12

    pop    edx        ; save ret addr in unused reg
    pop    eax        ; hndOwner
    pop    ebx        ; hndContender
    pop    esi        ; DeviceName
    VxDcall SHELL_Resolve_Contention
    mov    eax, edi   ; move VM handle into return
    jnc    @f
    xor    eax, eax   ; carry set, error, so zero return code
@@:
    jmp    edx    ; jump to caller's ret addr
    ret

EndStdCall     SHELL_Resolve_Contention, 12


; BOOL _SHELL_PostMessage( DWORD hWnd, WORD uMsg, WORD wParam, DWORD lParam,
;                          CALLBACK pfnCallback, DWORD dwRefData );
;

MakeCDecl    _SHELL_PostMessage

; void VWIN32_QueueUserApc( void * pfnRing3APC, DWORD dwParam, PTCB hThread);
;

MakeCDecl    _VWIN32_QueueUserApc

;  BOOL VWIN32_SetWin32Event(HANDLE hEvent);
;

MakeCDecl    _VWIN32_SetWin32Event

END
```

Listing B.3   WRAPPERS.MAK

```
AFLAGS  = -coff -W2 -c -Cx -DBLD_COFF -DIS_32 -DMASM6 -Sg

OBJS    = wrapper1.obj wrapper2.obj wrapper3.obj wrapper4.obj wrapper5.obj
          wrapper6.obj
LIBING  = $(OBJS: =&^)
LIBING  = $(LIBING:&=)

target: wrappers.clb

wrappers.clb: always $(OBJS)
    if exist wrappers.clb lib @<<wrappers.lnk
/out:wrappers.clb
wrappers.clb
$(LIBING)
<<
    if not exist wrappers.clb lib @<<wrappers.lnk
/out:wrappers.clb
$(LIBING)
<<

wrapper1.obj: wrappers.asm
    ml $(AFLAGS) -DSEGNUM=1 -Fo$*.obj wrappers.asm

wrapper2.obj: wrappers.asm
    ml $(AFLAGS) -DSEGNUM=2 -Fo$*.obj wrappers.asm

wrapper3.obj: wrappers.asm
    ml $(AFLAGS) -DSEGNUM=3 -Fo$*.obj wrappers.asm

wrapper4.obj: wrappers.asm
    ml $(AFLAGS) -DSEGNUM=4 -Fo$*.obj wrappers.asm

wrapper5.obj: wrappers.asm
    ml $(AFLAGS) -DSEGNUM=5 -Fo$*.obj wrappers.asm

wrapper6.obj: wrappers.asm
    ml $(AFLAGS) -DSEGNUM=6 -Fo$*.obj wrappers.asm

always:
    @rem echo pseudotarget
```

Index